Ada

Ada

A Life and a Legacy

Dorothy Stein

The MIT Press
Cambridge, Massachusetts
London, England

This book was set in Baskerville by The MIT Press Computergraphics Department and printed and bound by Halliday Lithograph in the United States of America.

Library of Congress Cataloging in Publication Data

Stein, Dorothy.
  Ada: a life and a legacy.

(MIT Press series in the history of computing)
Bibliography: p.
Includes index.
  1. Lovelace, Ada King, Countess of, 1815–1852.
2. Mathematicians—Great Britain—Biography.
I. Title. II. Series.
QA29.L72S74 1985   510'.92'4 [B]   85–11367
ISBN 0-262-19242-X

To my sister

Contents

Preface

Because of her father, Augusta Ada Byron, later Countess of Lovelace, was a figure of romance and fascination within her own lifetime. Recently, for more complex reasons but partly because of her connection with a very different sort of man, she has become such a figure once again.

The first man, the poet Byron, so dominated the attention focused on her during her life that the mere consciousness of this connection was enough to shape her existence and even those of her children, who continued to be marked as "Byron's Grandchildren." Byron himself, who had no direct contact with his daughter after she was one month old, began the process of mythologizing their bond by using her to wrap around the third canto of *Childe Harold's Pilgrimage*, the autobiographical epic whose first two installments had established his celebrity. From then on, few biographical notices or memoirs of either father or daughter, published, or unpublished—and including those intended for very unpoetical audiences—were considered complete without the garnish of one or more of these sentimental but rather vacuous verses.

So the process of mystification grew. Disraeli, who knew even less of her than her father did, made her the eponymous heroine of his novel *Venetia*. Later she figured in a number of short biographies, prepared for popular consumption, of Byron's female connections (not including his mother, who had a reputation for being stout and unattractive). Ada was fortunate, nevertheless, in that her treatment as the subject of a full-length biography was reserved for last, even after those of her putative half-sister Medora and her unhallowed half-sister Allegra, who was only five years old when she died. She was even more fortunate to fall to the hands of an excellent Byron scholar, Doris Langley Moore.[1] Mrs. Moore, furthermore, had for the first time full use of the Lovelace Papers, by far the most important primary source of information about Ada's life, character, and abilities. In spite of, or perhaps because of, her access to this voluminous collection, Mrs. Moore's study was shaped very strongly by the obsession with self-justification that had led Ada's mother, Lady Byron, to create and preserve so many of the papers, and by her own countervailing interest in defending Byron's character and conduct. Because of this focus, and from her candidly owned lack of scientific training, Mrs. Moore accepted without examination the conventional and current assessment of Ada's ability and achievement as a mathematician and interpreter of the designs for the first modern computer.

Charles Babbage, the early-nineteenth-century polymath who first dreamed up the plans for such a computer, which he called an "Analytical Engine," is the man through whom she has been accorded most of her posthumous renown. No less apocryphal, that part of her legend was also initiated and promulgated in his autobiography by the man himself. There Babbage claimed that when Ada announced that she had translated from the French an article by L. F. Menabrea about his proposed machine, he had asked her "why she had not herself written an original paper on a subject with which she was so intimately acquainted." He then persuaded her to add some notes, for which she selected and worked out the mathematics of the illustrative examples, "except indeed that relating to the numbers of Bernouilli [*sic*] which I had offered to do to save Lady Lovelace the trouble. . . . The notes of the Countess of Lovelace extend to about three times the length of the original memoir. Their author has entered fully into almost all the very difficult and abstract questions connected with the subject."[2]

Succeeding generations of historians and computer specialists have built upon Babbage's generous tribute and its even more generous implications. In the manner of all mythology, elements have accreted to the story that have only tenuous connection with the original. The author of *The Computer Prophets*, for example, tells us that "Ada invented binary arithmetic in order to make Babbage's work more understandable to the public."[3] Yet it was not Ada but her granddaughter, Judith, Lady Wentworth, who independently thought up a method of binary arithmetic in order to assist her calculations of the bloodlines of the Arabian horses she bred.

Some of the more romantic flourishes can only have originated in the imagination, such as the contribution of Christopher Evans in *The Making of the Micro*:

She exerted an encouraging and stabilizing influence on him [Babbage]. . . . She set out to study his designs for the analytical engine in depth, filling in any blank spots by pulling them out of his head in conversation. She had money and time on her side . . . but even so it was a few years before she got it all together. When she did, she published everything in a long series of "Notes" entitled "Observations on Mr. Babbage's Analytical Engine" . . . and Babbage himself said she "seems to understand it better than I do, and is far, far better at explaining it." . . . The real significance of her Notes was probably the effect they had on Babbage. He . . . had rarely bumped up against anyone who approved of what he was doing, let alone understood it. What was particularly gratifying to him was that Lady Lovelace had taken the trouble to study his theoretical approach with the eye of a mathematician and had found no flaws in it.[4]

Perfectly sober scientists have written about Ada in a surprisingly fanciful vein. Philip and Emily Morrison, whose collection of Babbage papers, *Charles Babbage and His Calculating Engines*, has been a prime source book on the subject for over twenty years, write in their introduction:

The Countess thoroughly understood and appreciated Babbage's machine, and has provided us with the best contemporary account—an account which even Babbage recognized as clearer than his own. . . . She shared with her husband an interest in horse racing, and with Babbage she tried to develop a system for backing horses; Babbage and the Count [*sic*] apparently stopped in time, but the Countess lost so heavily that she had to pawn her family jewels.[5]

In the course of time Ada's two connections to fame have become intertwined. Before the publication of Moore's biography, Maboth Moseley's almost perversely inaccurate, distorted, and fabricated biography of Babbage was often used as a source of information about Ada in her Byron mode.[6] Since then Mrs. Moore has been taken as an authoritative source by Babbage historians. Anthony Hyman, for example, in *Charles Babbage, Pioneer of the Computer*, accepts and repeats Mrs. Moore's assertion that Ada's quarrel with Babbage arose because the latter believed her Notes "too important merely to be appended to a translation of someone else's work,"[7] rather than, as the correspondence of both makes clear, because he wished to append a defense of his own.

My interest in Ada originally stemmed from her computer connection. Around 1972 a friend who was a computer specialist mentioned her now-famous "Translation and Notes," and assured me that in it she had produced a sophisticated and polished computer program—the first in the world—and that no comparable, similar, or related work existed, either published or among her papers, leading up to or following this unique achievement. I was at once filled with a craving to see for myself the papers and correspondence, which few people at that time had looked at with the question of the origins of such an achievement in mind. As a psychologist with an interest in thinking and reasoning, and as a former computer programmer, I hoped that a careful examination would shed light not only on this particular mystery of creativity but also on the more general processes of the acquisition of mathematical concepts and the assimilation of technical innovation, which have so often been probed without success by both psychologists and scientists themselves.

As it happened, I was not able to examine the unpublished papers until 1980. Before that, however, a study of both the published translation and the original Menabrea memoir had raised a number of questions in my mind about the conventional assessment of Lady Lovelace's contribution, and had impressed me with the significance of her curiously ignored translation of a printer's error. My eventual perusal of the Babbage Papers in the British Library, the Lovelace Papers and the Somerville Papers, both on deposit at the Bodleian Library of Oxford University, and other letters of Ada's that came to light in scattered locations did indeed answer the first of my purposes, but made the others moot. New mysteries and questions about Ada's life and her involvement in the science of her time continually arose, the solution of one puzzle always leading to another.

I was fascinated, and my study continually widened. To read these documents is to explore a museum of the social, intellectual, and medical history of the early Victorian period, as well as to tread a tantalizingly parallel path along Ada's search for the truths that would free her of her mother's Truth. I had entered a world both self-examining and perplexing, eerily familiar and startlingly strange: Babbage obtains government grants to build his calculating engine and has a scheme to send messages along wires strung from the tops of church steeples; doctors diagnose their women patients as hysterics, and prescribe opium and bleeding as cures; the effects of hypnosis are attributed to suggestibility, and to a cosmic magnetic fluid; Lord Lovelace worries about overpopulation, unemployment, and the dangers of giving working people the vote; Lady Byron discusses the state of the American money market, and questions the wisdom of abolishing slavery; Ada and her husband skate on artificial ice, but must plan their journeys so as not to tire their horses; Ada swears unabashedly, gambles, and takes a lover, but feels constrained to buy her books and geometry models anonymously.

Nevertheless, a second biography within a decade, of a figure whose achievement turns out not to deserve the recognition accorded it, requires some justification. My study diverges from Mrs. Moore's in a number of ways. The areas she felt unable to explore—the mathematical, the scientific, and the medical—are central to my treatment. For this reason alone, I have been able to clear up a number of puzzles and misinterpretations about Ada's life and activities. To take one example among many: Ada's "curious letters" to Augustus De Morgan, "enquiring, speculating, arguing, filling pages with equations, problems, solutions, algebraic formulae, like a magician's cabalistic symbols,"[8] turn out to be a correspondence course in calculus, in which he was tutoring her. Another example is my conclusion that Ada's "self-exaltation" and "religious mania" were not caused, as Mrs. Moore suggests, by the opium she ingested; rather, they were the symptoms of a condition for which opium was employed as treatment.

I have been fortunate in the course of my work to have come upon several previously overlooked documents that have helped to resolve certain questions even as they raised others. Stumbling across Woronzow Greig's memoir of Ada's confidences, for example, revealed her vaguely-alluded-to adolescent "misconduct" as an attempted elopement with her tutor. A chance reference, by a mutual acquaintance, to a scholar working on the life of an obscure New Zealand missionary and naturalist started me on a course of detective work that produced a likely candidate for the identity of the tutor; later discoveries, however, have made it necessary to discard my hypothesis, and his identity remains a mystery. Again, close attention to Ada's essays and scribblings revealed not only the unsuspected range of her scientific and literary interests but also the complexity of her relationship to the mysterious John Crosse, about whom it was then necessary to discover a good deal more. But Mr. Crosse vanished, until I chanced to come across a reference to the sea-change his name had suffered.

More important than differences in subject matter, emphasis, and sources, however, are differences in interpretation. Ada's correspondence with her gambling partner, Richard Ford, for example, reveals to me less about the existence of a "mathematical system" for betting on horses than about her role as ringleader, rather than dupe, of her confederates. And again, my conclusions concerning the causes of the failure of Ada's aspirations to a scientific career are very different from the traditional ones that Mrs. Moore accepts.

Important differences in interpretation can arise on a surprisingly concrete level, as well as on the more general one where the reading of the evidence is agreed upon. For example, accepting the tradition that Ada was being blackmailed by "a group of unscrupulous racing men," Mrs. Moore interprets a letter from Lord Lovelace to Ada concerning her aborted meeting with an "extortioner" as relating to such blackmail and as demonstrating that her husband was aware of it long before the period when other references to her gambling begin to appear—in spite of the fact that this letter stands in unexplained isolation from any similar references. But if the blackmail tradition is rejected in the absence of any other evidence in its favor, it is easier to recall that, while Ada's survivors may have been blackmailed as a result of her activities, she herself was vulnerable to extortion as a result of her father's—and then this otherwise inexplicable letter slips into its context.

The most important lesson I learned in the course of my investigations is that there is no substitute for inspection of the original documents. In some cases, examination of the original can reveal what even a photocopy might tend to conceal. It sometimes happens, for example, that letters or other papers are dated not by the writer, or at least not at the time of the writing, but years later, and by the recipient or his heirs, very possibly inaccurately. A difference in the inks used, not revealed in a copy, will often show up on inspection of the original and provide the clue to this practice even when the handwritings are too similar to be distinguished. Thus some puzzling distortions in the course of events can be eliminated. A rarer and more striking instance of the extra information to be gained by handling the original documents occurred in the case of a letter that has been interpreted as a cryptic message from Ada to Babbage, perhaps pertaining to a gambling conspiracy. Inspection reveals that a page has been partially torn away, and a slightly newer blank piece of paper of similar color carefully glued in its place. (This is a frequent practice of the librarians at the British Library in preparing damaged documents for use.) It was the interruption of the text, caused by the missing piece, that made the message seem so strange.

All this is true even if published letters or extracts are accurate copies of the original texts, which is by no means always the case. Every student or scholar sometimes makes mistakes in transcribing, and some make more than others; every book has its share of printer's errors. Many of these mistakes may be trivial, but in at least one instance a mistaken transcription may have started a new tradition. In a pair of notes apparently written on 18 June 1846, Ada urged

Babbage to introduce her to a certain countess, whose name was reiterated several times in the course of the two letters. Mrs. Moore, thinking, as always, of references to Byron, reads this name as "Countess Italia-Italia" and surmises that it was the Countess Teresa Guiccioli, Byron's mistress during the last years of his life, whom Ada longed to meet. She hesitates only because there is no evidence that this countess was in England at the time. In his biography of Babbage, Anthony Hyman adopts Mrs. Moore's suggestion with little of her hesitation. Yet Ada's handwriting is quite clear, and strikingly distinctive to anyone who has studied any sample of it. What she has clearly written in this case is not "Italia-Italia" but "Halen-Halen."

Who was the Countess Halen-Halen? I don't know. There was a Spanish general named Juan van Halen, Count of Pericampos, who led a colorful career in the first half of the nineteenth century, and had a wife alive in 1846, but there is nothing else to indicate that she was the lady in question, and I know of no other likely candidates. Maboth Mosely, who had preoccupations of her own, read the name as "Countess Harley-Harley" and assumed it was a reference to Countess Harley Telecki, Babbage's friend in later life. This lady, however, could not have been more than ten years old at the time, and certainly not yet married to Count Telecki.

The second lesson I learned involved the unreliability of all manner of sources when taken in isolation, including first-hand reminiscences and personal memoirs. We tend to forget that our ancestors, like ourselves, could lie, forget, make mistakes, omit, mislead, and be constrained by kindness or politeness to the dead. "*De mortuis nil nisi bonum*," said Babbage, "appears to savour more of female weakness than of manly reason";[9] yet he himself was much influenced by it, especially where Ada was concerned. And the dead have been known deliberately to plant false contemporary documents with an eye on biographers of the future. Lady Byron in particular was a consummate practitioner of this art, but the memoirs of personalities as diverse as Mrs. Crosse, Mrs. Somerville, and Mrs. De Morgan, when checked against the surviving body of contemporary evidence, can all be seen to contain significant errors.

Perhaps the most disconcerting discovery was that standard reference works too, being at best only as good as their sources, can be shot through with false information. A useful check is to look in each for information already known from another source. *British Family Antiquity*, published in 1809, for example, and *Collins's Peerage of England*, of 1812, both fail to list William King, the future Lord Lovelace, among his father's children, although he was the eldest. The former lists Lord

King's brother as heir presumptive, while the second notes the birth of William's brother, six years his junior, in 1811, and designates him heir. The author of the first work, moreover, claims to have checked the manuscript of the entry for each family with its head in order to correct any errors it might contain. No wonder Lord Lovelace was convinced his father disliked him!

Nevertheless, standard reference works remain useful for checking, or sometimes bringing into question, information found in other places. For example, various writers, including Woronzow Greig, who conducted his own investigation, at different times and in different places have asserted Byron's descent from the Scottish royalty. According to *The Scots Peerage*, however, of the two princesses who have been named as forebears, one, Princess Jean, daughter of James II, did not exist. The other, Princes Annabella, daughter of James I, was the second wife of the Earl of Huntly, whereas Byron's ancestor, Huntly's third son, was almost certainly the offspring of his third wife, Elizabeth Hay. Thus, the claim of royal descent, for the purpose for which I wished to use it—that of tracing the possible transmission of a hereditary disease—must be treated as uncertain at best, even if this means that no satisfactory conclusion can be reached on this point.

The use and even the reliance on some secondary sources is of course unavoidable. I have tried to use them with caution and have confined their use as much as possible to background material. Of the secondary sources available, I have selected a few to rely on that present views compatible with those I have arrived at in interpreting the primary materials I worked with; and I have avoided discussing vexed questions that might be of great interest but to which I could contribute little.

Many of the letters and other documents used were undated, or only partially dated. For example, the year was often missing. In the frequent cases where the year could be guessed to within a decade or less by means of other evidence and the day of the week as well as the month and day were given, the year could be determined by using a perpetual calender. Surprisingly few writers have employed this device, with the result that many letters have been interpreted out of sequence. To be sure, the dating method just given is by no means infallible, since even where full dates are given, there is an occasional mismatch between the day of the week and the rest of the date. Consequently, I have tried to check all dates against internal evidence—references to public or private events, the use of names or addresses that changed at known dates. Sometimes, certain phrases or locutions provided clues, when they tended to be used frequently by the writer at certain periods but subsequently dropped out of use. An example is Ada's frequent reference to herself as a fairy during the mid-1840s.

Just how much to present of the evidence underlying my arguments and interpretations presented a problem. I have tried not to encumber the narrative excessively, at least on most points. Many of the steps in some of the chains of inference by which I have reached conclusions, made assertions, or accepted statements are omitted or relegated to the notes. Nevertheless, I have considered it important to present within the body of the text a general indication, and occasionally a detailed illustration, of the type of detective work upon which my construction of events and character is based.

In all the quoted material, the original spelling has been retained, although the punctuation has occasionally had to be altered slightly in the interests of intelligibility. It has been necessary to make many elisions to eliminate extraneous material that might obscure the point under discussion, or simply try the patience of the reader beyond endurance; many of the writers and correspondents were astonishingly verbose. A few nonessential phrases have been retained simply for flavor, however, and I have tried not to make elisions that in any way would alter the original purport of the quotation.

In organizing the exposition, I have been torn between the advantages of thematic and chronological presentation. While the former suits more closely my intention to center my study on particular issues, the latter conforms more naturally to the expectations of the reader of a biography, and obviates such confusion as might result from the practice of skipping back and forth in time. The major drawback of the chronological form is the expectation that all types of ground will be covered, including much that has been adequately dealt with elsewhere, to which I have nothing new to contribute; with the thematic form, on the other hand, one risks presenting a distorted and out-of-perspective view of the subject's life. I have chosen to compromise, with the result that each chapter overlaps others at least partially in time, while the focus of the narrative is shifted to another important aspect or issue. In the process I have tried to construct a biography that may be read independently of any other work, yet with an emphasis on areas that, while vital to the subject and crucial to the understanding of her life, have received inadequate treatment elsewhere.

Acknowledgments

I am indebted to the kindness and patience of many people, in both professional and private capacities, who shared their time, knowledge, and enthusiasm with me at every stage in the preparation of this book: Joan Abramson, who first urged me to write about the mystery of Lady Lovelace; the staffs of the libraries and archives of the British Library, the British Museum, the National Portrait Gallery, the Science Museum Library, the University of London Senate House Library, the University College Library, the Science Reference Library, the Institute for Historical Research, the Royal Institution, and the Royal Astronomical Society, all in London; the staffs of the Bodleian Library and the Museum of the History of Science, in Oxford, and of the Fitzwilliam Museum and the Peterhouse and Trinity College Libraries in Cambridge; Allan Bromley and Garry Tee, who patiently discussed my questions about Babbage's machines; Donna Haraway, J. Michael Dubbey, and David Kellogg, who made sobering and helpful criticisms of various portions of earlier drafts of the manuscript; Donna Haraway again, and Andrew Hodges, without whom this book would never have appeared; Peter Whitehead, Margaret Aldridge, Anthony Crosse, Major General and Mrs. John R. C. Hamilton, Audrey Mead, and Stephen Turner, who brought me closer to Ada's friends, lovers, and family.

Quotations from the Lovelace Papers appear with the permission of the Earl of Lytton; quotations from the Somerville Papers appear with the permission of the Fairfax-Lucys, and through the kindness of Gerard Turner; quotations from the Babbage Papers, the Wentworth Bequest, the Hobhouse Diaries, and the Egerton Papers appear with the permission of the British Library Board; other quotations appear with the permission of the Science Museum Library, the Nottingham Public Record Office, John Murray, Harcourt Brace Jovanovich, Routledge & Kegan Paul, and Chatto & Windus. Portions of chapters 2 and 3 appeared in somewhat different form in the autumn 1984 *Victorian Studies*. Individual acknowledgments accompany the illustrations.

Chronology

8 July 1835	Ada marries Lord King after a brief courtship.
12 May 1836	Birth of Byron Noel, later Viscount Ockham.
1 July 1836	Babbage decides to use punched cards with his engine.
22 September 1837	Birth of Anne Isabella King.
30 June 1838	Lord King created Earl of Lovelace.
2 July 1839	Birth of Ralph Gordon Noel King.
June 1840	Ada begins to study mathematics with Augustus De Morgan.
11 August 1840	Lord Lovelace is appointed Lord Lieutenant of Surrey.
August 1840	Babbage goes to Turin to discuss the Analytical Engine.
Autumn 1841	Ada suffers a more serious breakdown.
1842	De Morgan's *Treatise on the Differential and Integral Calculus* is published.
October 1842	L. F. Menabrea's memoir on the Analytical Engine is published.
August 1843	Ada's translation of the Menabrea memoir, with notes, is published.
November 1844	Ada meets John Crosse.
13 September 1845	Faraday succeeds in rotating the plane of polarized light.
Spring 1847	Lord Lovelace publishes "Theories of Population."
Summer 1848	The Lovelaces collaborate on a review of Gasparin's book on agricultural meteorology.
1851	Ada leads a gambling confederacy and suffers disastrous losses.
June 1851	Ada has a series of severe hemorrhages; her doctors eventually agree on a diagnosis of cervical cancer.
27 November 1852	Ada dies.
16 May 1860	Death of Lady Byron.
1862	Death of Byron, Lord Ockham.
18 October 1871	Death of Charles Babbage.
29 December 1893	Death of Lord Lovelace.

1

I Understand, Mamma

Nothing in Lady Byron's life became her like the leaving of her marriage. "The Separation," as it came to be called, shaped and directed the rest of her existence as nothing before or after. Then, too, within the context of her personal and social situation, to shed a husband while retaining position, reputation, fortune, and child was no mean accomplishment.

The wife of Byron was born in 1792 to Sir Ralph and Lady Milbanke after fifteen years of barren but hopeful conjugality; she was christened Anne Isabella and called Annabella. By the time she arrived, her mother, born Judith Noel, was principal heiress to the Wentworth family estates; the current incumbent, Judith's brother, Viscount Wentworth, had no legitimate offspring and little chance of any. The wait had been so long that only Lord Wentworth caviled at her sex, writing his sister early in 1792, "A nephew it must certainly be, for altho' you say you will be as well pleased with a girl I cannot in this respect pay a Compliment to the Sex at the expence of my veracity."[1] No matter; doted upon by a passionately maternal mother and a proud, indulgent father, little Annabella had every reason to believe she would always have her way, and not only within the family circle. She was of "high blood," as Byron would later put it, and the eventual heiress to a title more ancient than his own. Her parents were in debt, to be sure, and the estates her mother would inherit, which in due course would descend to her, were encumbered, but that was not unusual, and of little consequence while rents and income continued to flow.

By the beginning of the nineteenth century, most upper-class young men, in fact most who could properly be called "gentlemen," went routinely to a university to finish their education. This shared educational terminus endowed them with a certain uniformity of cultural background and social network, even if a good many of them managed to finish without having learned much or passed any examinations; for it was then possible to take a degree merely by satisfying the residence requirements. More important for their future social, political, and even professional life were the friends and contacts they made while fulfilling the terms of residence. For the more studious, of course, the opportunities, and again the peer group, were there to assist the aim of learning as well.

For women there was no such expected educational mold, no such common intellectual opportunities. The training of a young lady was completely up to her parents, and particularly her mother. It required special aptitude, unusual dedication, and indulgent or intellectual parents for a lady to become "learned." Learning had no value on the ever-looming and all-important marriage market, but it was not necessarily a disadvantage, if a girl was otherwise personable, well connected, and well dowered.

The vagaries to which the education of an upper-class girl were subject are memorably illustrated in the famous exchange between Lady Catherine and Elizabeth Bennet in *Pride and Prejudice*, a book that Annabella Milbanke herself pronounced "the *most probable* fiction I have ever read." "It depends," she wrote her mother, "not on any of the common resources of Novel writers. No drownings, nor conflagrations, nor runaway horses, nor lapdogs & parrots, nor chambermaids & milliners, nor rencontres and disguises. . . . It is not a crying book; but the interest is very strong."[2]

Annabella was not only opinionated and decisive but robust and precociously studious. Her mother's letters attest that she was admired and complimented wherever she went. It is true that Judith insisted on this admiration, but all the evidence indicates it was merited. She read widely and deeply, and filled her notebooks with solemn philosophical and moral comments on all she read and thought and observed. She became particularly addicted to writing "characters," short psychological sketches of persons she knew or would like to know.

Her education departed interestingly from the picture that Jane Austen's "fictions" have left us of the standard accomplishments of the genteel young lady of the time. She learned drawing and dancing, true; but unlike both her mother and her daughter, she seems to have had little interest in music. She read history, poetry, and literature. She studied French, Italian, Latin, and Greek. But she departed most from the commonalty of elegant females in her interest in philosophy, mathematics, and science. For her more abstruse studies she turned, as her daughter was later to do, to William Frend, a mathematician, scientific writer, and friend of her parents.

Mr. Frend had been a Cambridge cleric but had lost his orthodoxy and turned to Unitarianism. His independent, if rather eccentric, views appealed to the liberal, kindly Milbankes, and he influenced Annabella's later involvements in Unitarianism and reform, which sat so oddly with her strident religious pronouncements and punitive moralism. In letters written to her during her adolescence, Mr. Frend employed the mock-formal device of referring to both himself and his correspondent in the third person. When she was fourteen, one such letter began, "Mr. Frend is very much concerned that a variety of occupations has prevented him from noticing the receipt of Miss Milbanke's communications," and went on to inform her, "The manner in which the fourth book of Euclid was completed was highly satisfactory."[3] Three years later, "Mr. Frend" was "very much pleased with the elegant manner of solving the series of numbers to the fourth power."[4] Studies undertaken as a recreation from the real business of life could be pursued at an extremely leisurely pace, and, in fact, the extent to which she ever became a "mathematical Medea" was greatly exaggerated by Byron, who had difficulty in keeping his accounts.

Along with—in her case—series, the serious business of life commenced in earnest when a young woman reached seventeen and made her entrance into "the world." In later life Annabella pictured herself as having left her studious rural preoccupations with reluctance: "Arrived at the age of 17, I was anxious to postpone my entrance into the world, of which I had formed no pleasing conception, and I was too happy in my pursuits—drawing, book-collecting, verse-making—to wish for any other appropriation of my time. But my 'hour was come.' "[5] Other, more contemporary, evidence suggests that from her first visit to London at age eight she could be torn from its excitements only with difficulty. The lives of gently born females awaiting marriage were so constricted in the nineteenth century that she was far from alone in her eagerness for urban pleasures. But Annabella early showed signs of her lifelong addiction to self-righteousness and contorted reasoning when she explained her intention to leave her sick parents (who should properly have been her chaperons) and hurry to town for her third London season early in 1812. She was, she said, so terribly worried over the health of her friend Miss Montgomery that it would contribute to her parents' peace of mind as well as her own if she joined her friend in London. It was even an act of self-discipline on her part to take up their reluctant permission to depart. "It is also natural to me to be *less* disposed to take, in proportion as *more is given*, and it is really in opposition to this impulse that I do not refuse your offer," she concluded sagely.[6]

Despite her precociously developed moral overachievement, she received half a dozen offers of marriage during her first three London seasons, including one from Lord Byron. She refused them all while she composed a "character" for her ideal husband.

London early in the nineteenth century was organized for the pleasure and convenience of the wealthy and powerful. The "season" began rather slowly after the Christmas holidays, which the rich spent on their country estates. It lasted until August, when the stench of the Thames became overpowering and the year's crop of birds and game was waiting to be shot in the rural retreats. During the season an aristocrat was expected to maintain a house, or at least an establishment, in town when he had a daughter of marriageable age. Once there, it was necessary to attend an interminable, and with good fortune, productive round of parties, balls, dinners, plays, and concerts.

In addition to entertainments in private homes and gardens, there were many public pleasure gardens, theaters, and assembly rooms in which to admire the celebrities of the day and be seen in turn. The entertainments helped to mask the actual social, economic, and, increasingly, political business being transacted, that is, the arrangement of profitable marriages. A doggerel verse of the period lampooned this half-hidden preoccupation, with its thinly veiled commercial purposes rampant in a social group that still looked down upon "trade":

Then when Matrons speak of suppers small
"A few choice friends besides ourselves—that's all"
This language in plain truth they mean to hold,
"A girl by private contract to be sold."[7]

Although the "matrons" took a major role and interest in the forging of marriage alliances, F. M. L. Thompson, who quotes the above lines, goes on to acknowledge that

When momentous issues were at stake, and the rescue of an estate from ill-fortune was in question, the conduct of marriages was liable to be taken out of the hands of the matrons. . . . Necessity impelled an impoverished aristocrat to seek a bride of fortune. An heiress to a landed family was the most desirable solution . . . but one who was willing to unite such wealth to foundering gentility could not always be found, unless a large advance in rank was involved. In default, the occasion called for an infusion of mercantile wealth. . . . Lord Sefton's adviser, canvassing ways of clearing off a debt of some 40,000 pounds . . . came to the possibility of the heir, Lord Molyneux, marrying. "To marry a fine brought up Lady with little or no fortune would be to hurt the Estate. By the Estate he has a right to expect a large sum with a Lady, not to look at less than 60,000. . . . *many* a great and rich banker would be glad of an offer to give his daughter that fortune for her advancement and dignity."[8]

The sentiment expressed reflects the commensurability of status and money, as well as the lingering view that income derived from the rent of agricultural or urban land was socially superior to that proceeding solely from banking, commerce, or industry. Nevertheless, if inherited land could be made to yield added revenue from the coal or other minerals it contained or (later) from its value to the spreading railway network, no social status would be sacrificed by taking advantage of this added piece of good fortune. Money that supported women, heiresses or widows, however, was usually invested for them by the trustees of their estates and often in such safe receptacles as government securities, which tended to yield relatively low returns. The passage quoted also reveals the near-universal assumption that the wishes of a marriageable woman were a faithful reflection of those of her father, and that even the heir would do well to attend the advice of the family banker.

In theory, a young gentlewoman was free to accept or refuse any eligible man who proffered himself. In practice, her contacts were carefully restricted, before entering the world, to family members and family friends, and afterward to those who were admitted to occasions presided over by her appointed chaperons and social sponsors. Under these circumstances, women had very little of a personal nature to judge by in accepting or rejecting a prospective husband. Men, to be sure, often had even less, but relied on the assumption that if little was known, there was little to be known. Byron, according to his best friend, chose Miss Milbanke in the belief that she was rich, amiable, and "of the strictest purity." Most couples at marriage scarcely knew each other, unless they were related. (In the small, select circle in which Miss Milbanke and her future husband moved, cousin marriage was common and approved; it could help to keep the family wealth together. Its dangers were less well recognized.)

The men admitted to the great houses and to the functions sponsored by the matrons were carefully screened for their religious and class origins; there was also a tendency for prominent hostesses to have political preferences, and to make up their guest lists accordingly. For the aristocracy, entrance to the House of Lords was almost automatic for the holder of a title, while entry to the House of Commons for heirs and younger sons was smoothed by the control over local constituencies held by large landholders. The political aspect of social life became more important as the century wore on. As the direct predominance of large landholders waned, ambitious middle-class men sought heiresses to help finance their aspirations to a parliamentary career.

After a marriage offer had been made and accepted by the parties most nearly involved, it remained for others—solicitors, guardians, and trustees—to hammer out the financial arrangement, which was known as the "marriage settlement." Until this process was completed to the satisfaction of the bride's guardian and of the current owners of any property that the bride or groom hoped to inherit, and until trustees were appointed to protect the interests of both parties, no marriage could be celebrated. The details could take months to complete, but there were generally accepted notions about how much money could be expected to change hands—at the marriage, upon the deaths of the couple, and upon the marriages or maturation of their children.

Once more, the novels of Jane Austen provide a number of allusions to such matters. At the opening of *Mansfield Park*, for example, we are told that "Miss Maria Ward of Huntingdon, with only seven thousand pounds, had the good luck to captivate Sir Thomas Bertram . . . and her uncle, the lawyer, himself allowed her to be at least three thousand pounds short of any equitable claim to it." Thompson makes these guidelines even more specific: "In marriages between equals in aristocratic circles, portions of 10,000 to 30,000 pounds were normal, and the bride could expect a jointure of at least 10%."[9] The "portion" was the amount a daughter or younger son could expect to be given or to inherit out of the family wealth and property; the lion's share was reserved for the eldest son. A daughter generally received her share upon marriage—rather, her husband received it; a younger son was provided for when the time came to launch him on a career or profession.

If a woman did not marry, her portion was used to support her after her father died. In the more usual case where her portion became her dowry, a woman could expect to receive during her husband's lifetime an annual allowance known as "pin money"—usually several hundred pounds, or one percent of her dowry—which she could spend as she liked. After her husband's death she received an income for life from his estate, which was known as her "jointure." The dowry, then, could be considered a kind of insurance policy, intended to maintain a well-born woman in the style her family considered appropriate, or as an investment.

Normally, returns on investments, other than highly speculative ones, then ran around five percent, so, to make up the expected ten percent, a man's estate had to be encumbered to maintain his widow if she was of equal or higher birth than his own, often to the great annoyance of his heirs. An estate charged with the support of several dowagers (in contrast to one charged with the debts of a profligate incumbent) was considered heavily burdened indeed. The pin money, on the other hand, represented a less than normal return on investment; but this was supposed to be discretionary money, and a wife expected to receive her maintenance in addition. In reality, there was some vagueness over just what the pin money was intended to cover, and more than one well-dowered wife was reduced to maintaining herself and her children on her pin money, salvaged by her trustees, after a dissolute husband ran through her fortune and deserted her. Fortunately, the several hundred pounds of aristocratic pin money amounted to as much in many cases as a middle-class family income.

A classic illustration of such an outcome was the case of Catherine Byron, the mother of the poet, who had brought well over £20,000 to her husband in 1785 but had to subsist with her son, and the appurtenances of gentility, on £150 a year after she was deserted in 1790. During at least part of the next eight years, she lived on even less, while she was paying interest on a loan she raised for her husband's benefit before he died in 1791. Yet in 1798, when Byron inherited the family title and estates at the age of ten, his mother found that their financial troubles were only beginning.

From the sixteenth century it had been the custom among the aristocracy and other large landholders, who collectively constituted the gentry, to ensure the proper descent of important estates by means of a legal device known as the "strict settlement." It was also called a marriage settlement, since the head of the bridegroom's family, usually his father, found marriage a convenient occasion to place conditions upon his heir, who might otherwise inherit his patrimony with the freedom to sell or alienate it at his pleasure. By law, land could be tied up by "entail" for no more than three generations, so a practice grew up by which it was resettled in each generation. Generally, the heir about to be married was quite happy to agree that his future property would descend to his hypothetical eldest son, and content to exchange a few future freedoms for the assurance of an adequate income during his father's lifetime. The device of the strict settlement, the bridegroom's side of the marriage settlement, increased in importance when the profound changes in England's agricultural system during the eighteenth and early nineteenth centuries resulted in a smaller, but wealthier and more powerful, landed class, in possession of larger and more highly consolidated estates.

The patrimony inherited by Byron at the age of ten was not his father's, but had belonged to his great-uncle. This Lord Byron was known as "the Wicked Lord," mainly because he had killed his neighbor and kinsman in a dispute over game, for which crime he had been tried in the House of Lords. Like too many Byrons, the Wicked Lord was reckless with money in his youth. Later on, he deliberately despoiled his own lands and sold part of them (illegally, because they were entailed), just to spite his son, whose marriage he had disapproved. The son died in 1776, leaving as heir a son who died in 1794, so the final heir of this calamitous family, the future poet, inherited an estate free of entail but burdened with debt and litigation.

Byron's mother descended from the Scottish nobility and was thus considered somewhat inferior to the English aristocracy; she was ignored by her son's paternal relations, including the guardian formally appointed to surpervise his affairs. Alone, except for an extremely dilatory and possibly venal family solicitor, she had to learn the arts of managing a debt-ridden estate and a lawsuit for the illegally alienated land. Mrs. Byron was very frugal, but both mother and son were bound, in his case with enthusiasm, to keep up the munificent appearances appropriate to the nobility, regardless of actual penury. Appearances included not only fine clothes but an irreducible minimum of servants of both sexes, a carriage and sufficiency of horses, a ready supply of pocket money for possibly quixotic disbursements, and a suitably careless scrutiny of bills. Not required were the prompt payment of tradesmen's bills or of debts other than those incurred in gambling. The upshot of Byron's strict adherence to this gentlemanly code of honor was that by the time he reached his majority he was seriously in debt; after he came of age, his debts continued to increase.

His solicitor's remedy was that he should sell his estates, a solution he at first resisted. The hope of recapturing the disputed land, which contained rich mines and quarries, was always before him to soften the need for urgent action or even economy. But Byron was not really interested in property, except as it yielded income. The lawsuit was not finally settled, and the land in question legally sold (for far less than he had expected), until a few months before his death. His ancestral seat, Newstead Abbey, was sold years before then at a price that formed the bulk of his legacy.

His mother's solution, when he was in his early twenties, was to revert to Lord Sefton's adviser's advice: he must "mend his fortune in the *old* and usual way by marrying a Woman with two or three hundred thousand pounds.'"[10] She forgot the marital disaster this custom had brought upon herself. Byron's own first remedy when he came of age was to try to escape his creditors by borrowing more money and embarking on an extended tour of the Mediterranean, where lay adventure and freedom far exceeding the English gentleman's privileges with housemaids. He got no farther than Greece, but there he enjoyed with some youths relations that in England at the time were punishable by death.

After two years of wandering and frolicking with both sexes, Byron ran out of borrowed money. He returned to England intending to take up not only his debts but also a political career in the House of Lords, to which his title admitted him and for which his talents and flair for the dramatic seemed to augur success. Soon afterwards, too, the publication of the first two cantos of *Childe Harold's Pilgrimage*, a romanticized account of his grand tour, brought him instant celebrity. The fascination of his fame, youth, beauty, and peerage was only enhanced by the brooding melancholy he affected, perhaps over his debts, or his congenitally misshapen right foot, or the scandalous propensities of his family. He was also not averse to letting fall occasional hints of his own "crimes." All of this made him, as Miss Milbanke reported to her mother, "the object at present of universal attention," fit to enliven the longueurs of the 1812 mating season.

Byron shared with his future daughter, though not to such extremes, the wide swings of mood that caused him to pursue some interest or activity for a period with intense enthusiasm and dedication, only to abandon suddenly in disgust and boredom what had previously seemed so enthralling. Then he would become prey to deep depressions. His parliamentary career represents a case in point. He had been a member of the Whig party (the predecessor of the Liberal Party) since his Cambridge days, and on 27 February 1812, when he delivered his maiden speech in the House of Lords, he had chosen his subject in consultation with Lord Holland, a leading Whig. It was a defense of the stocking workers of Nottingham, who were protesting their displacement by labor-saving machinery called "stocking frames." The protests sometimes took the form of vandalism and breaking the frames, and a bill had been introduced to punish such behavior by death. Byron's speech, both sensible and compassionate, was considered a great success, though very radical. The bill passed nonetheless, after modification. His second speech favored Catholic emancipation, another liberal cause, but he made only one other. Soon he became disillusioned with the Lords and ceased to attend.

His time and attention were taken up with his new social and sexual successes, and his debts. He had become reconciled to selling his home in order to shake off his financial burdens and assure himself of an income, steady though modest by aristocratic standards. But why did he not marry and, at a stroke, mend both his fortunes and his character? Despite his glamour, it was not a simple matter. The women who now threw themselves at him most freely and engagingly were married already. Maidens were constrained to be far more demure, and Byron was surprisingly passive in sexual pursuit.

In the early nineteenth century the sexual freedom of married women in polite society was steadily eroded. In Byron's youth upper-class women, once wed, were permitted great latitude within their social stratum. The change was brought about, not by the influence and example of Queen Victoria and her prudish Prince Consort, nor by the rectitude of the previous consort, Queen Adelaide, but by the recognized desirability of controlling and regulating the assimilation of new wealth into the ranks of the politically powerful. In the early nineteenth century, because of its increased size and, even more, because of the number of claimants to membership, elite society became Society, more formal, rigid, and rule bound than it had previously been when the landed gentry knew each other and knew their political ascendancy supreme and unchallenged. The newly rich of the industrial revolution had to be admitted, but slowly, and after having fully adopted the standards prescribed by Society's arbiters. In this process of class formation, women played a crucial part in establishing and maintaining the rules both by their own behavior and by their censure of others.

Progressively deprived of their traditional economic and public pursuits, respectable women in the nineteenth century were increasingly required to perform only the essential female functions of defining and defending, through their lineage and their personal conduct, the identity of the intermarrying group to which they belonged. Thus, while Annabella could not appear in public unchaperoned, her mother campaigned actively for her father's parliamentary elections and even made speeches on his behalf. Her father's sister, Lady Melbourne, was an important Whig hostess, whose adulteries were widely known but confined within aristocratic circles. As Lady Byron, separated and widowed, Annabella could herself manage her estates, found experimental schools, and organize philanthropies. Her daughter, brilliant and blue blooded, as a married woman was excluded from the management of both her own and her husband's property, and had to exercise the greatest care not to have her name mentioned outside a close circle of friends and relatives.

Annabella's aunt, Lady Melbourne, was one of the most fascinating and congenial of the ladies who cultivated Byron. She was determined, she told him, to teach him "Friendship," but she was also the mother-in-law of Lady Caroline Lamb, another and much younger married woman who threw herself at him with such abandon that an open scandal was threatening. Lady Melbourne tried to avert the danger by encouraging him to propose to her niece, but he responded so half-heartedly, using the aunt as a go-between to convey his offer, that Annabella very sensibly refused. Then, as tired and disgusted by Society as he had been by the House of Lords, Byron prepared once more to go abroad, as soon as his estates and debts were settled.

While waiting, he turned easily to other women, including his own half-sister, Augusta Leigh, and Annabella began to repent her refusal. Less than a year after rejecting his proxy proposal, she daringly took the initiative in opening her own correspondence with him. It was a most improper thing to do, however she extenuated it; couples who were not engaged were not supposed to correspond at all, or even to call each other by their first names (a familiarity that some wives, including Lady Byron, never attained). Yet Annabella was at the same time writing to another young bachelor, the brother of her friend Miss Montgomery.

Piqued perhaps at receiving overtures from a girl who was otherwise so controlled and decorous (the mother of another rejected suitor had called her an icicle), Byron responded; the clandestine correspondence continued for over a year. It was not an entirely secret exchange, however, for against Annabella's express wishes Byron continued to discuss her and her letters with Lady Melbourne, and Annabella had her own confidants. With Lady Melbourne he also discussed his growing erotic attachment to Augusta, though she counseled him against this danger as earnestly as she had against that of eloping with Lady Caroline. Incest did not actually become a crime in England until 1906, but adverse publicity and social censure were punishments more feared in Society than legal penalties, particularly among women, and particularly among those who, like Augusta, had a court appointment and income to lose.

For a long time Byron ignored Annabella's many hints that she was in love with him, but at length Augusta too became urgent that he should marry. Her first choice, and his, was her good friend and relation by marriage, Lady Charlotte Leveson-Gower. When an offer in that direction was refused because Lady Charlotte's parents had other plans for her, he fell back upon an alternative plan and made a second proposal to Miss Milbanke. Evidently, he was more attractive to wealthy young ladies than to their guardians, but Annabella had long since bent her parents' wills to her own. The second offer was as half-hearted as the first—he was at the same time cheerfully planning to go abroad with a friend should she refuse—but this time she closed with it at once.

During the ensuing months, in which his as-always dilatory solicitor went leisurely about the business of drawing up the marriage settlement, Byron's nervousness over the prospect of marrying a woman of whom he knew so little (and not all of that congenial) intensified. A visit to improve the acquaintance was delayed; when it did take place, it was not a success. The Milbankes bored him. His attempts to familiarize himself with the young lady were, if anything, too successful. Roused to the point where she felt her propriety threatened, she cut short the visit and ordered him to leave. But despite his misgivings, he was now bound by the prevailing etiquette to complete the marriage; the rupture of an engagement was a woman's prerogative. Moreover, although the sale of his estate was in progress but stalled, with the purchaser unable to produce more than the initial £25,000 deposit, marriage itself still seemed advantageous. If his new wife did not succeed in domesticating him even as she eased his debts, she could still produce an heir, and he could still avail himself of the freedom permitted to husbands. If worst came to worst, he could always resume his travels.

In the marriage settlement he was prepared to be generous; or else the Milbanke solicitor drove a hard bargain. Annabella's dowry was to be £20,000, of which £16,000 would go toward the dowry of a daughter or the portion of a younger son. All of this was quite normal; however, he agreed to settle the income from £60,000 of his own property on her as a jointure, and this was £10,000 more than his advisers felt warranted. Furthermore, he did not insist that her father and uncle settle their property on her at the same time. To be sure, there was not much danger there: she was an only child, the obvious recipient of whatever her father and mother were free to leave her, and her mother stood to receive the bulk of her uncle's estate. Anabella's father was himself badly in debt, the result of lavish entertaining and the expenses of his political campaigns. (As a baronet, he was not a peer, and he had to campaign for election to the House of Commons.) The upshot was that most of her dowry was not paid during Byron's lifetime. Instead, Sir Ralph paid his son-in-law interest on the amount owed, and most of that was given to Annabella in the form of the pin money that was part of the marriage settlement. Marrying an heiress was not to prove an immediate solution to the problem of debt.

The marriage itself has been generally termed a disaster from the first by historians, a sentimental lot who tend to believe that the married state is inherently a happy one. Yet in the end both partners achieved what they could not have done unmarried. Byron had often spoken of marriage with contempt; he did not like to dine or to spend the night with a woman, and he was totally averse to the type of long-term cohabitation that marriage usually implies. The only women he liked were perfectly undemanding, adjusting themselves readily to his preferences for companionship. After the separation was achieved, he was once more free to travel abroad, still technically married and immune from other attempts and expectations along those lines. Moreover, he was no longer so flayed by guilt; he felt he had been punished enough.

As for Lady Byron, she had been accustomed to having her way from earliest childhood, but as a grown woman she could not expect to gratify her taste for universal domination outside her family circle, or even within it, in any situation so well as in that of a titled widow of independent means, with sole control over her child. Even before Byron's death, separation, with herself cast in the role of the suffering saint, brought to her many of the advantages of widowhood. It was as if she had had to pass through marriage to emerge on the other side.

Nevertheless, for a while both were miserable enough. As a husband Byron behaved so badly that his wife formed a theory that he might be insane, and Augusta, a witness to some of his outbursts, eagerly concurred. Yet he was shocked and outraged when Annabella left. True, he had talked of separation from the very beginning, but he had thought to set the time and the circumstances, if not to do the actual leaving. It was very upsetting: her sudden escape had been carefully though quietly prepared, and had been planned even before the birth of their child.

If the child had been a son, it cannot be doubted that Byron would have fought harder than he did to retain his parental rights. Even as it was, to leave her husband and retain her daughter meant that Annabella had to thread her way through a legal and social minefield. Two other famous marital disasters, though they occurred later in the century, are illuminating with respect to what she faced. Notably, both concerned sexual irregularities on the part of the husband.

The first was that of the John Ruskins. Effie's flight to her parents' house was as fearfully and stealthily planned as Annabella's own. Ruskin was impotent, and since Effie proved upon examination to be still a virgin after six years of marriage, she was granted an annulment.

Nevertheless, her action raised a storm of criticism, some ladies even suggesting she should have been grateful for her situation. Effie claimed it was John's unkindness that was her real reason for ending the marriage, but the law did not recognize unkindness. Still, her social position was affected; Queen Victoria refused to receive her for forty years, relenting at last only as a favor to her dying second husband, the president of the Royal Academy, John Millais.

The second case involved a distant cousin of Virginia Woolf who had married into the aristocracy, but whose husband was a flagrant homosexual. According to Woolf's biographer, "Lord Henry fled to Italy and there, in that land of Michelangelesque young men, lived happily ever after. His wife discovered that she had been guilty of an unformulated, but very heinous, crime: her name was connected with a scandal. Good society would have nothing more to do with her. She was obliged to retire from the world."[11]

Byron had demonstrated that he was anything but impotent. Their daughter was born on 10 December 1815, less than a year after the wedding. His brutality was inadequate as grounds for separation or divorce, although he had neglected, insulted, and humiliated her since their wedding day, both in private and before witnesses. Equally inadequate were his frequent drinking bouts and that, drunk or sober, he tried to frighten her by threats and vandalism. And also inadequate as justification for wifely desertion were his infidelities, or that he taunted her with them, as part of a catalogue of mental cruelties he repeatedly practiced on her. Finally, and so far ignored even by modern scholars, who take such practices as much for granted as did the societies of the past that they study, there is evidence that the sexual molestation begun before marriage continued afterwards and might on more than one occasion have amounted to rape—again, inadmissible as grounds of wifely complaint.

One of the lines most frequently quoted from Byron's own account of the marriage—burned soon after his death, though not before avid perusal by a number of people—was, "Had Lady Byron on the sofa before dinner." This was in reference to the wedding day, and to a time following a dismal drive in a freezing carriage, during which both were nervous and out of sorts. According to her later narratives, he had already begun his mysteriously menacing prophecies. As for Annabella, if she was not exactly "purity & innocence itself," as her daughter and the rest of the world came to believe, she was certainly inexperienced, and this preprandial consummation, when a servant announcing dinner was momentarily expected, must have been an unwelcome and unpleasant culmination of his premarital gropings.

As it began, so it seems to have ended. Byron's insistence that Annabella had "*lived* with him, as his wife, up to the day of her departure,"[12] while meant to imply an affectionate and pleasurable relationship during that period, actually suggests a painful and disagreeable experience for a woman so soon after childbirth. She had many reasons to wish to flee, but she had no recognized grounds for complaint, none at least that were not double edged. It took an unusually determined woman to make a successful escape, and the effort marked the rest of her life.

To achieve a separation while retaining her social acceptability, her financial independence, and the custody of her month-old child required that "the world" be made to believe that her husband was a monster of iniquity and she was a faultless saint; yet she could not be seen to be the one who did the showing. Around this objective Lady Byron organized her existence. She had obtained the expert advice of Dr. Stephen Lushington, a prominent lawyer who was to argue in favor of widow burning in India before the Privy Council, but who never failed to support his noble client against all her adversaries as long as she lived and afterward. As she explained to her mother, "I have been *perfectly* confidential [i.e., confiding] with Dr. Lushington and so far from thinking that the *suspicions* could do any good to me, he deprecates beyond anything the slightest intimation of them, as having the appearance of Malice—and altogether most injurious to *me* in a social view. The Misfortune of my case is that so little has passed before Witnesses—and the wife's deposition *unsupported* is of no avail."[13]

Lady Byron herself developed a taste and skill for legalistic ratiocination, in which she exceeded all her highly accomplished, expensive, and obliging solicitors. Byron too was of some assistance, at first praising her truthfulness, character, and conduct to her father and to his own friends, though later he changed his mind. Furthermore, it had been he himself who had dropped broad hints to his wife about his homosexual activities in Greece and his incestuous attachment to his sister. Yet she could prove nothing, and had reason to fear a court procedure even more than he. A whispering campaign of rumors, for which she would claim no responsibility, was much better suited to her purposes, and for this Lady Caroline Lamb, her cousin by marriage and his discarded mistress, eagerly offered her services.

There could be no going back. Historians and biographers who have deplored her inflexible refusal to consider Byron's boyishly earnest appeals for reconciliation overlook what was always so clearly before her: if she put herself once more in his power, she would appear to have condoned all his past conduct and to have denied the more shocking accusations. She could never again hope to escape. Furthermore, as even her otherwise hysterical and overwrought mother so sensibly pointed out, if much of his admitted unkindness and cruelty had been justified (by him) as revenge for the refusal of his first offer of marriage, what would his behavior now be if she returned to him?

Partly in return for a disavowal of responsibility for the most scandalous rumors, Byron, after a perhaps unflatteringly short struggle, agreed to a separation in which his wife's pin money was increased to £500 per year, the disposition of her inheritance was to be left to arbitration when the time came, and the rest of the marriage settlement was left unchanged. Immediately afterward he left England, never to return alive. When Lady Byron's mother died in 1822, she inherited the Wentworth estates, and both Byrons, as was required and customary in similar cases, added the Noel family name to their own. The arbitrators decided to split the Wentworth income evenly between them, but legal delays prevented Byron from receiving any of his share until over a year later, shortly before his own death.

No mention was made in the separation agreement of another vital issue, the custody of the the child Ada. This was because the legal situation at the time was so much in the father's favor that Annabella's lawyers decided not to raise the question. Had Byron chosen, or been provoked, to press his rights, he could certainly have gained possession of his daughter. He never attempted to do so, only issuing an occasional protest over what he considered an exceptionally high-handed instance of subversion of his paternal dignity, as when Ada was made a ward in chancery (a legal device for protecting the property settled on a minor) without consulting him. "A Girl," he thought, "is in all cases better with the mother." Nevertheless, much of Annabella's behavior during the separation battle and afterward was justified, if not determined, by anxiety over retention of her child. Thus, a variety of rumors, ranging from aggravated adultery to bigamy, homosexuality, and sodomy, after preliminary circulation, were allowed to die, while the whisper of incest was remorselessly, though intermittently and surreptitiously, revived.

Byron's half-sister Augusta was an obvious choice to be given in the care of the baby if it were taken from its mother, especially since he chose to travel in unhealthy and politically unstable foreign places. The suspicion of incest simultaneously attacked both his and her suitability. Annabella prepared documents in secret to lend plausibility to the charge, if it should become necessary, and mercilessly persecuted

Augusta under the pretext of reforming her character. When the time was ripe, she would reveal to her daughter the wickedness of the aunt whose namesake she was. Ada had been christened Augusta Ada and was actually referred to as "little Guss" by Augusta, Annabella, and Judith during the first weeks of her life. Byron himself, however, seems always to have called her Ada. It was, he explained, a family name, dating from the time of King John.

In addition to undermining Byron's and Augusta's reputations for parental fitness, Annabella strove mightily to build a case for her own. Considering the universal assumptions about the overwhelming force of maternal love, it is surprising to see how deliberately and self-consciously she went about this. Her passion was for control, not care; and there is much evidence that, whether by character or by circumstances, she found herself unable to love her child. Any private doubts she might have had over the inadequacy of her affection served only to strengthen her determination never to let others suspect any such failings. Motherhood succeeded wifehood as the name of her propensity for self-justification, only to be replaced by grandmotherhood.

Soon after the proceedings for separation were underway, Annabella become restless and dissatisfied with permitting her parents to act in her name in dealing with solicitors and well-wishers in London. Asserting that "the Child is weaned necessarily & without difficulty," she returned to town to take matters into her own hands. There she stayed until the bitter end, lingering until Byron was safely out of the country, but not without becoming concerned that her action might inspire adverse comment, which she took prudent steps to prevent. "As I am accused openly of total disregard of the Child's welfare," she wrote her mother, after Byron had urged reconciliation as her maternal duty, "I think it may be well to write you some letters on the subject *which may be kept*, and I shall begin tomorrow." Then, as she explained in a postscript, finding she had some time in hand, "I shall write a letter to-day *to be kept*."[14] So she proceeded to write some letters of instruction and inquiry about the baby's health and daily routine.

Yet she was nothing if not introspective, and there was none to whom she tried more assiduously to justify her feelings and conduct than her notebook. She reflected and decided that her inability to love her child stemmed from her uncertainty over custody. She expressed her sentiments in verse, for she too wrote poetry:

The Unnatural Mother, Dec 16, 1815

My Child! Forgive the seeming wrong—
   The heart with-held from thee
But owns its bondage doubly strong,
   Resolving to be free.
. . . .
And if already taught to feel
   She must not feel too far,—
Devoted once with fruitless zeal
   Her peace on earth to mar,
Then ere another passion rise
   In kindred with the first
She pauses o'er those tender ties,
   And sees them—formed to burst![15]

It was a nice touch; it was all his fault, even her maternal coldness. If dated correctly, the verses were composed only a few days after Ada's birth and show that she was then planning her flight and foreseeing its possible consequences. The feelings she expressed here persisted, and months later she penned more verses along the same lines, observing that "heart-wrung I could almost hate / The thing I may not love."

Time moved on; with Byron safely abroad, her anxiety over custody must have subsided. Yet she still found herself having deliberately to plan and justify her maternal behavior. A few days after Ada's first birthday she filed a position paper in her journal:

I will endeavour seriously to consider and diligently to execute the duties of a Mother, and to divest myself of wrong bias arising from my particular circumstances or morbid feelings. I think Ada has arrived at an age when a watchful & judicious superintendence may form the basis of good habits, & prevent the rise of evil ones. It is now, as it always has been, my opinion that a Mother should give this attention more systematically & unremittingly than is usually considered incumbent upon her. . . . I shall suffer from interfering powers, & want of sympathy with my views. . . . I shall thus dare to engage my affections—I *might* now with-hold them—I might spare myself the danger of loving—the fear of deprivation—the vexation of opposition—But all these I will meet and Thou—to whom it is known that I would do thy will as allotted to me do thou bless & confirm my humble pledge to be a good Mother.[16]

Though she flirted with Unitarianism, Annabella generally remained among those who called themselves "Christian, unattached"; nonetheless, she began early to amass a reputation for piety that became ever more insistent as she grew older. In the beginning, at least, it was quite as deliberately constructed as her maternal tenderness. "I have made a good impression," she wrote her mother in describing some new acquaintances, "but the funniest thing is that because I go to church very regularly & sometimes talk pye-house (pious) . . . it is supposed in the 'Assembly of the Saints' that I am on the high-road to Heaven."[17]

Her determination to acquire a command over Ada's habits needed no reinforcement or aid from divine sources. When Ada was three, she made another entry in her journal:

The cause of the ascendency of one mind over another is in general, not so much the superior strength of the governing character as the correspondence of certain of its qualities to the weakness of the governed — Therefore, if emancipation is desired, a resolute and unsparing investigation of our own infirmities, and the annihilation of every delusion of the Imagination is the only means of radical cure.[18]

Perhaps she was already forming her views of the special nature of her role in her daughter's life. Ada's psyche was endlessly analyzed, not only by her mother but by other "experts," friends, and teachers, who were invited to add their "characters" of Ada to those Lady Byron composed herself. One, unfortunately undated, that the anxious mother must have found particularly congenial began, "A desire to govern the minds of others is a leading feature in Miss Byron's character. She will gain ascendency over most of those with whom she comes in contact. The few whom she cannot govern will generally be those who might exercise almost a slavish control over her." The same sage attributed these deporable propensities to "the nature of her nervous system."[19] The victor in her clash of wills with Byron was certain she had sufficient steel to overcome any desire to govern on the part of a daughter whose character she later termed "so anomalous — so gifted & so defective." And so in contrast to her own. "My rule over the baser kind of spirit is so absolute," she concluded, "that I think I must have some qualifications for a Police officer, or Governor of Convicts."[20]

In view of his wife's determination to possess and dominate her child so absolutely, it is natural to ask what kind of parent Byron might have been. While Lady Byron was girding herself for governance, his fantasies were far more sentimental:

To aid thy mind's development, to watch
Thy dawn of little joys, to sit and see
Almost thy very growth, to see thee catch

Knowledge of objects, wonders yet to thee!
To hold thee lightly on a gentle knee,
And print on thy soft cheek a parent's kiss, —
This, it should seem, was not reserved for me;
Yet this was in my nature. . . .[21]

His nature had a fair chance to realize itself in connection with his illegitimate daughter Allegra, born in 1817, whose sole custody he took from her mother in infancy. Since illegitimate children and their fathers were more socially acceptable than their mothers, the arrangement seemed to offer Allegra something like the social and worldly advantages she might have had if she had been born in wedlock. But in the event, Byron kept her with him for only a short space of time. He was then living in Venice, and for some months she was placed in the house of the British consul there, who did not approve of her, passed her around to others, and finally delivered her back to her father. When she became spoiled and demanding as a result of the amused attention she received, Byron placed her in a convent.

Given the precarious political situation and the medically insalubrious climate in which Byron lived, there was some practical reason to shelter a young child in such a place. Once she was installed, however, he never visited her, though Shelley did. He even resented the child's demands for visits, sweets, and gifts, which he condemned as cupboard love. But when Allegra died in the convent, aged only a little over five years, Byron grieved extravagantly. He had her expensively embalmed and shipped back to England for burial at Harrow, his old public school.   If Allegra had survived, how would he have aided her mind's development? His plans for her included a Catholic upbringing and a proper Continental marriage. He heartily disliked women with intellectual pretensions, and his ridicule had often reduced even the complacent, self-righteous Annabella to tears. What effect would he have had on Ada, who gloried in her mind but retained a diffident sensitivity to criticism? How would she have flourished in the face of his sardonic wit?

In his famous farewell lines to his wife, Byron had asserted, "Even though unforgiving, never / 'Gainst thee shall my heart rebel." But before very long he was denouncing her bitterly to his sister, and then he pilloried her far more publicly, in a cutting picture of a pretentious bluestocking:

Her favourite science was the mathematical,
  Her noblest virtue was her magnanimity,
Her wit (she sometimes tried at wit) was Attic all,

> Her serious sayings darkened to sublimity;
> . . . .
>
> She knew the Latin—that is, "the Lord's prayer."
>   And Greek—the alphabet—I'm nearly sure;
> She read some French romances here and there,
>   Although her mode of speaking was not pure. . . .

Just in case anyone might not recognize the picture as drawn from life, he added a few snippets of unmistakable autobiography.

> For Inez called some druggists and physicians,
>   And tried to prove her loving lord was *mad*,
> But as he had some lucid intermissions,
>   She next decided he was only *bad*;
> . . . .
>
> She kept a journal, where his faults were noted,
>   And opened certain trunks of books and letters,
> All which might, if occasion served, be quoted;
> . . . .
>
>   Calmly she heard each calumny that rose,
> And saw *his* agonies with such sublimity,
> That all the world exclaim'd, "What magnanimity!"
>
> No doubt this patience, when the world is damning us,
>   Is philosophic in our former friends;
> 'Tis also pleasant to be deemed magnanimous,
>   The more so in obtaining our own ends;
> And what the lawyers call a "*malus animus*"
>   Conduct like this by no means comprehends:
> Revenge in person's certainly no virtue,
> But then 'tis not *my* fault, if *others* hurt you.[22]

To offset the image of implacable self-righteousness she had presented to the world during the separation contest and its aftermath, Lady Byron, for the rest of her life, strove to project herself as a being dominated by feeling. Even her God was "all-loving," she informed a friend. Her prevailing weakness, she told her journal, was that she "ascribed to the actions of others motives of a loftier or less worldly nature than really existed."[23] Only her ineluctable adherence to Truth, which she invariably judged most beneficial to the sinner she was hoping to forgive at the moment, prevented her from absolutely abandoning all principle, so overwhelming was her "passionate devotedness."

She had now achieved ascendancy over almost her entire circle of family and friends, and over none was her power more absolute than over her daughter. But the older Ada grew, the more delicate and difficult her task became, and the more subtle and adroit her efforts needed to be. With a daughter to rear, with the necessity as well as the inclination to lead a life of unblemished virtue, and with the fruits of her reflections on Byron's mismanaged youth, it was only to be expected that she should develop an interest in education and turn to good works.

The establishment of village schools was a favorite form of charity among the ladies of the gentry. In them poor and orphaned children were trained, usually in farm work if boys and in the domestic arts if girls. The schools thus provided not only evidence of upper-class responsibility but a pool of trained servants and laborers as well.

Lady Byron founded such a school in 1818, in the environs of her parents' house at Seaham. For advice about this project she turned to her old tutor Mr. Frend, and also mentioned her other educational undertaking. "My daughter is a happy and intelligent child, just beginning to learn her letters—I have given her this occupation, not so much for the sake of early acquirement, as to fix her attention, which from the activity of her imagination is rather difficult."[24]

Mr. Frend did not object to the hobbling of imagination, but brushed aside the subterfuge and responded to her boasting in his usual jocular manner. "I am glad to hear so good an account of yourself & your little one. As to the latter, do not be in any hurry. My eldest little girl gave alarming signs of being a prodigy, but I so effectually counteracted them that her mother began in her turn to be alarmed when she was between six and seven years old lest she should be backward in her learning."[25] He was referring to his daughter Sophia, who was to become one of Lady Byron's most slavishly devoted friends and a detractor of Ada's. His correspondent, however, was not to be deterred by his example.

At the recommendation of Lord Brougham, she became interested in the school run by Emanuel de Fellenberg at Hofwyl, Switzerland; she was so impressed that she wrote a paper on its history. The Fellenberg system of "learning by action" was an offshoot of Pestalozzi's techniques for schooling the vagabond orphans of the Napoleonic wars. Fellenberg, however, ran a two-tiered school. One was for poor boys, who learned not only agricultural work but various kinds of practical crafts as well, such as carpentry, mechanics, and leatherwork, all of which they practiced on the Fellenberg estate. The other, "higher" school was for the sons of the well-to-do. In their case, "action" was afforded by "military exercises, swimming, riding, pedestrian excursions, skating, gardening, turning and other mechanical operations." If yet more action seemed warranted, Lady Byron continued,

Pupils of the upper school who were found to require physical strengthening or, as was the case with many, bodily fatigue, were sent for a time to field-labour with the lower school, a proceeding which in both cases acted as a wholesome medicine; whilst by the boys themselves, getting up at three o'clock in the morning to earn a breakfast with a thrashing-flail was regarded as one of the greatest pleasures. The sons of the wealthy thus learnt to respect labour in the persons of the pupils of the poor school; whilst on the other hand the poor learnt to view their richer companions, not as enemies, but as sympathising friends.[26]

The regime was, she felt, the first step toward solving the problem of "how the leading classes of society, those who employ labour, could be trained to recognize the duty incumbent on them to educate the working-class and elevate them morally in the same degree as they avail themselves of their labour to increase their own property."[27]

The schools were for boys only, since for girls social considerations aways overrode educational ones, and upper-class girls were carefully guarded against any form of sexual or social mixing. Later, Lady Byron sent one of her grandsons to Hofwyl for a time, but he was not permitted to mix with even the boys of the "upper" school. She remained firmly convinced that boys' schools were breeding grounds for homosexuality and that Byron had been depraved at Harrow. The "industrial" school she founded near her home at Ealing Grove was to be for poor boys alone.

Of course there was never any question of sending Ada to a school, although some of the "industrial" techniques were adapted for her. She learned to sew, early and well. Later, in a gesture of affection and respect, she made one of those beribboned caps that nineteenth-century matrons wore indoors for her friend and mentor, Mary Somerville; and even as countess, she made her own petticoats.

In a notebook kept to record Ada's progress during her sixth year, written tellingly in Ada's name, Lady Byron declared her determination to teach Ada herself, with only occasional assistance, to avoid "the evil of Governesses." Another notebook, however, reveals the experiences of one governess who was hired during this same period and lasted only eight weeks. The task of poor Miss Lamont could not have been an easy one. At five and a half, Ada's schedule already covered "lessons in the morning in arithmetic, grammar, spelling, reading, music, each no more than a quarter of an hour long—after dinner, geography, drawing, french, music, reading, all performed with alacrity and docility."[28] At least at first. As the record continued, it appeared the little pupil sometimes showed signs of restlessness. This is scarcely surprising when it transpires that she was forced to spend part of the time reclining on a board, during which the lessons continued in the form of questions and answers. Outside of lessons, there were periods when she was required by her mother to lie perfectly still.

For good behavior and performance Ada was rewarded with "tickets," which might still be withdrawn for subsequent failure or disobedience. When she had accumulated a sufficiency of tickets, they were exchanged for some suitable prize, such as a book or picture. Nevertheless, the notebook kept on her behalf stresses that she was supposed to be working chiefly for the joy of pleasing her mother:

I want to please Mama very much, that she & I may be happy together. . . . Geography amuses me very much. . . . The French has not interested so much as some others—and one night I was rather foolish in saying that I did not like arithmetic & to learn figures, when I did—I was not thinking quite what I was about. The sums can be done better, if I tried, than they are. The lying down might be done better, & I might lay quite still & never move.[29]

Miss Lamont's journal confirms the stringency of the requirements to lie still and please Mama. If she so much as moved her fingers, her hands were encased in black bags.

. . . before Lady Byron she was immediately subdued—submitted to have the finger bags put on, and went into confinement into a closet for half an hour. . . . Lady Byron went to Leicester at 2 o'clock—during her absence, Ada never, I believe for a moment, lost sight of the stimulus of doing well that she might give pleasure to her mamma on her return by a good report.

It must be remembered that Miss Lamont's journal was written for inspection by Lady Byron, so her account of Ada's motives in her mother's absence need not be taken too seriously. More often, and distractingly, Mamma was present during the lessons.

In the evening, on occasion of being reproved for some slight shew of carelessness at her work, Ada, feeling some resentment arise, was going to reply—when, immediately checking herself, she went up to her mamma, and in a whisper said—"Give me some good advice."

On occasion Mamma could check exuberance as readily as resentment:

In the evening while her mamma was at tea, Ada amusing herself by singing, presently exclaimed, "I think mamma I have a very good voice I shall be able to sing better than you." . . . Lady Byron calling her said in an impressive manner "Ada did you give yourself your voice?" to which she replied "O I understand Mamma, we will talk of that when I am going to bed."[30]

Miss Lamont was dismissed, as Lady Byron explained to the lady who had recommended her, because she had not the strength and firmness to motivate her charge by only "a sense of duty, combined with the hope of approbation from those she loves."[31] Instead, the unfortunate governess often fell back on "complex methods," such as coaxing and persuasion.

It was right after the departure of Miss Lamont that Ada made the inevitable inquiry about her father, apparently for the first time. As Lady Byron noted the occasion, "Ada asked me today if Grandpapa & Papa were the same. I said no, that they were different kinds of relations. She replied, 'then mine's not a Papa?' I said I would explain to her more about that when she was older. Her mind did not appear to dwell on the subject."[32] By a lucky chance, Lady Byron's account may be compared with Ada's own version of this or a similar incident, which was recorded by Mary Somerville's son, Woronzow Greig, who claimed to have become Ada's most intimate confidant. "The confidence she reposed in me," he said, "was very much greater than a woman could safely repose in anyone, and thus my acquaintance with her secret history was greater than even if I had been her lover, as she told me many things which she would not have ventured to communicate to one who stood in that relation to her." According to Greig,

Adas feelings toward her mother were more akin to awe and admiration than love and affection. The familiarity of mother and daughter never subsisted between them, there was always a degree of repulsion and distrust altho they were proud of each other. . . . Moreover Ada once when she was very young while walking in the garden with her mother said "Mamma how is it that other little girls have got Papas and I have none." Lady Byron prohibited her daughter in such a fearfully stern and threatening manner from ever speaking to her again upon the subject, that the poor girl shrank within herself and as she more than once told me acquired a feeling of dread toward her mother that continued till the day of her death.[33]

It is somewhat surprising to find that, notwithstanding her mother's system of education, Ada retained a love of learning almost throughout her life; she even came to prefer mathematics to geography. It was most fortunate, too, because before long she was visited by the first of a series of incapacitating illnesses that only exacerbated the maternal and social restrictions already burdening her short existence.

Like Harriet Martineau, Florence Nightingale, and Elizabeth Barrett, Lady Byron was one of the great nineteenth-century invalids whose physical frailties, while often real enough, gained them sympathy and exempted them from so many of the tedious and bothersome duties expected of all women, yet miraculously left them able to pursue activities that posterity agreed were more valuable. Lady Byron's illnesses were of such a nature as not to prevent her traveling, or organizing and supervising both philanthropies and family concerns. Like Ada, Lady Byron had started out as a healthy and active child, but she was often and progressively ill from late adolescence. Indeed, to read her descriptions of her diet and the preparations with which she dosed herself is to wonder that she remained as resilient as she did and that she survived to the age of sixty-eight. Ada's illnesses were never to be so convenient.

Vegetables were not then a well-regarded source of nourishment for the upper classes, and fruit was actually considered harmful to children. At one point Lady Byron announced that she ate "*nothing but meat eggs and biscuits*"; her appetite for mutton was legendary. When headaches, indigestion, and "bilious attacks" followed, the medical men she consulted often prescribed preparations of metallic salts, such as antimony and zinc, that could be quite toxic. Both doctors and patients were much given to the use of emetics, laxatives, and purgatives, which Lady Byron referred to as "opening medicine." Letters were peppered with prescriptions, traded back and forth as freely as gossip. In truth, doctors knew little more than their patients, and the gentry used their authoritative status to dose their servants as well as themselves and each other. In one letter Lady Byron proudly announced that she had saved her maid's life "by a timely dose of Castor Oil when she was in danger of an inflammation in her bowels."[34]

Lady Byron, and the legions of physicians she consulted, believed even more firmly in the benefits of bleeding. This staple of "heroic" medicine had enjoyed waves of popularity for many centuries. Any condition that was accompanied by fever, swelling, or excitement was thought to be caused by an excess of blood, which carried impurities, either in the affected area or throughout the body. Bloodletting was considered an appropriate treatment even for hemorrhages; the logic of this apparently consisted in the belief that the hemorrhage was the body's attempt to rid itself of excess blood and poisons. From the doctor's point of veiw, too, bleeding had the desirable effect of "lowering" the patient, rendering him relaxed and quiet, and so confirming his expectations of improvement.

Bleeding could be accomplished in a number of ways: by lancing, cupping, or the use of leeches. Lady Byron was very fond of the latter form of treatment. In one letter she told Ada triumphantly, "I am rather better for a horrid *mouthful* of Leeches this morning."[35] They could be applied almost anywhere. Many conditions, when they afflicted women, were held to be somehow connected with the reproductive organs, and a number of derangements, both physical and mental, were attributed to sexual excitation, which, correctly, was thought to result in the sexual organs becoming engorged with blood. When Ada was four, for example, Lady Byron confided to her mother that her doctor was "positive that all my complaints are dependent upon a disorder of the womb, that has existed ever since Ada's birth. . . . The vessels in that region are in consequence overloaded and will require continual depletion by cupping and leeches."[36] The remedy was not pursued vigorously enough to meet her demanding physiology, and many years later, when Ada, now a married woman, commented on her mother's lack of physical exercise, she explained (asking Ada to burn her letter) that "in consequence of the frustration of *one* of the purposes of my existence, a congestion took place in one set of organs which made exercise most mischievous & likely to induce a fatal disease."[37] Sufficient bleeding at the right time, she thought, could have prevented this perilous condition.

Under the supervision of a mother who adhered to such dietary and medical regimes, it is not surprising that Ada early developed a delicate stomach, though this may have been unconnected with her later agonizing attacks of "gastritis." (For a discussion of Ada's lifelong health problems, see the Appendix.) Then, in her eighth year, she began to suffer from severe headaches that affected her eyesight, or at least hindered her reading, for several more months. Since headaches were attributed, once more correctly, to dilated blood vessels in the head, bleeding—from which even children were not exempted—was the treatment of choice. When Byron, then in Greece, received the news that his daughter was afflicted with "blood to the head," he at once made the usual connection, and wrote back:

Perhaps she will get quite well when she arrives at womanhood . . . if she is of so sanguine a habit, it is probable that she may attain to that period earlier than is usual in our colder climate;—in Italy and the East—it sometimes occurs at twelve—or even earlier—I knew an instance in a noble Italian house—at ten. . . . I cannot help thinking that the determination of blood to the head so early unassisted may have some connection with a similar tendency to earlier maturity.[38]

His use of the word "sanguine" and Lady Byron's references to her "bilious attacks" show that medical thought still bore traces of the ancient Greek humoral theory. This held that the body contained four fluids, or humors, that corresponded to the four elements of which the universe was composed. The humors—blood, phlegm, yellow bile, and black bile—determined by their relative proportions not only health and sickness but also the predominance of certain personality traits. The humoral theory was both a physical and a psychological system, an attempt to connect mind and body. There were other such theories to come into Ada's life.

Before the end of her childhood Ada was well acquainted with discomfort, pain, and physical restraint; another frequent visitor was death. Her effusive and affectionate grandmother died when she was six, and the mysterious Papa two years later (bled to death by his own physicians, as it happens). In the following year Grandpapa died. Ada's sadness and bewilderment in this period are revealed in two letters she wrote to a younger cousin, the son of the man who inherited her father's title. Calling the boy "Brother," she fantasized their loving and comforting each other when all the adults had departed.[39]

The deaths in the family, on the other hand, greatly enhanced Lady Byron's income and independence. With her ostentatiously unostentatious manner of living, she clearly did not need all of her revenues, so she handed her jointure to the new Lord Byron, who had received his peerage denuded of the family estates. Although Jane Austen taught us that ten thousand pounds a year was as good as a lord, many lords had to make do with considerably less. Still, it was considered shameful, and possibly degrading to his rank, for a nobleman to be unable to maintain a minimally aristocratic lifestyle. Byron had willed his money—what remained after the amount that yielded his widow's jointure—to Augusta, who, with her large and feckless family, needed it just as badly as the new Lord. But Annabella was finding out just how effective a financial obligation could be in securing devotedly loyal friends, and she harbored a jealousy and resentment toward Augusta that could barely be concealed by pious moralism.

The removal of her husband and ailing parents also freed her to make a grand tour of the Continent, taking Ada with her; they remained abroad for two years. Only a few months after they returned, Ada came down with measles, followed by serious complications. She was then thirteen, a significant age. Since her return she had been pursuing an interest in astronomy, corresponding with Mr. Frend and his daughter Sophia about it. On 27 May 1829, Mr. Frend wrote to inquire of Lady Byron:

How does Miss B. come on with her astronomy. The next month toward the end will exhibit Jupiter to her to great advantage & at a reasonable hour. I hope you have a good telescope & it will be an amusing exercise to sketch the planet with his moons & observe the variation of their positions in succeeding nights. She may be fortunate enough to witness a few eclipses & occultations but I would not consult books on the occasion. She may make tolerable guesses at the approaching phenomena & verify them by her own observations.[40]

But Ada was to enjoy no such starry amusements. On 29 June Lady Byron explained her delay in replying by "the serious anxiety which I have had reason to feel on Ada's account for the last two months. . . . Ada has been and still is in a perfectly helpless state; the loss of all power to walk or stand having followed other effects of the measles, and too rapid growth.—There is not, I am assured, any danger in her present disabled state, but as it deprives her of the pursuits of mind, as well as of active employment, my thoughts & time are more than usually occupied by her."[41]

There are a number of possible causes of temporary paralysis of the legs. Most of them, however, do not persist for more than a few months if recovery is eventually to be as complete as it was in Ada's case. The fact that her "disablement" stretched, with decreasing severity, over several years, ending only when she was considered of marriageable age, suggests that her recovery might actually have been delayed by the prolonged and stringent bed rest—which itself can weaken muscle tone—to which she was subjected in addition to the other debilitating treatments favored by Lady Byron.

A series of letters written by Ada to a friend of her mother's a year after the onset of her illness shows that she was permitted to sit up for only half an hour a day, a period that was increased to an hour toward the end of the summer. She admitted her "low spirits," but at least her schoolwork had been resumed—often again in a reclining position. By the autumn of 1831, however, she was walking on crutches and optimistically seeking advice on building a bridle path. At last, in September 1832, a letter from her mother's former maid, Mrs. Clermont, bore congratulations on her walking without "supporters," though oddly enough her mother had noted she was able to do this, as long as she had weights in her hands, some six months previously, indicating that her problem was at least partly one of balance. For some time after even this date, she was often weak and giddy.

During the period when she was still on crutches, the first reference to a new and fashionable interest of Lady Byron's appears in their correspondence. It occurs in response to a suggestion by Lady Byron that Ada should take the carriage into London, where her mother was then visiting a friend. Ada was curiously reluctant to quit her solitude and her studies, worrying over what might happen to her Latin verbs as a result of the interruption. Finally she asked, "have you a bedroom amongst your ground floor apartments? If not it might be rather awkward for me. — Having now stated all the fors and againsts which occur to my constructive organs, I leave it to your judgment."[42]

Lady Byron must have been an early convert to phrenology; but once converted, she ceased to be the plaything of fashion. Although one of its founders, Spurzheim, had crossed the Channel in 1814 to lecture to the benighted Britons, the London Phrenological Society was not established until 1824; it had taken a decade to catch on. Like the ancient humoral system, phrenology was an attempt to relate body—or in this case, brain—to mind. The "functions" of the brain were classified in terms of a set of behavioral "faculties," and attempts were made to relate these faculties to the structure of the brain as it appeared to the anatomists of the day. Some "organ" of the brain was supposed to give rise to each faculty, or behavioral disposition. Just how many faculties there were, and their exact locations in the head, were matters of some dispute. Among the faculties were included such "feelings and propensities" as "combativeness," "constructiveness," "destructiveness," and "acquisitiveness." Then there were faculties and associated organs for "sentiments," such as "veneration," "hope," "ideality," "conscientiousness." Still another broad class were the "knowing faculties," including "individuality," "form," "size," "weight," and "color." Finally there were "reflective faculties," such as "comparison," "wit," "causality," and "imitation." Only the less noble and desirable of the faculties were shared with the lower animals. The organs corresponding to these faculties in particular individuals could be large or small, giving rise to greater or smaller corresponding dispositions. The shape of the skull, being molded around the protuberances beneath, was therefore affected by the sizes of the various organs. Thus, a person with a large organ of veneration could be identified by phrenologists, not by his deeds of piety and devotion, but by the bump on his head over the location of the enlarged organ.

One of the many memoirists of the period recorded a conversation with a prominent phrenologist named Deville:

He told . . . of an anonymous lady whom he had to caution against sensitiveness to the opinion of others. Some years afterwards she came again and brought her daughter, who, when finished, was sent into another room, and the lady consulted him upon her own cranium. He found the sensitiveness so fearfully increased as almost to require medical treatment. He afterwards met her at a party, when she introduced herself as Lady Byron. Her third visit to him was made whilst Moore's Life of her husband was being published, and, in accordance with his prescription, she had not allowed herself to read it.[43]

Moore's *Life, Letters, and Journals of Lord Byron* was published in two volumes in 1830 and 1831, by which time Lady Byron had been consulting phrenologists for a number of years. Deville may have been correct in his conclusion, from measuring the bumps on her head, that she was sensitive to public opinion. He was wrong, however, about her not having read Moore's book; she even published a pamphlet to register her objections to it, though it was supposedly printed for private circulation only. Moore offered to include it with his second volume.

Most phrenologists were doctors; like psychoanalysis, however, it was a game that any amateur could play. Phrenology had social and religious, as well as medical and scientific, implications. Although the bumps and their underlying organs were innate, proper training and redirection could achieve compensatory enlargement or diminution within limits; hence, phrenology encouraged a compassionate sternness on the part of "governors": parents, teachers, employers, jailers, and madhouse keepers. Phrenology presented a kind of smorgasbord of progressive but not revolutionary ideas, from which so strong-minded and opinionated a woman as Lady Byron could pick and choose those she found congenial. It was perfectly suited to provide the final touch of authority to her judgments and pronouncements upon others.

The advice and exhortations in her letters to Ada were sprinkled with phrenological terms. "I want to see the Bird [Ada] to raise its bump of Self-esteem a little—I am sure it is morbidly sensitive," she wrote at one point.[44] At other times she felt Ada's self-esteem was entirely too high. Ada soon adopted the phrenological vocabulary, but her attitude toward the entire system was as fluctuating and ambivalent as her other associations with her mother. Her scientific bent led her not only to check phrenological beliefs against the opinions of the men of science of her acquaintance, but also to test the diagnostic powers of individual phrenologists. The skeptical Babbage was induced to undergo several phrenological examinations, the results of which Ada preserved. Her side of the correspondence became at times a running debate that challenged Lady Byron's certainty in one of the few areas Ada felt permitted to question. In February 1841, for example, she wrote asking if her mother's faith were at all shaken by a recently published critical book. Lady Byron replied in true form: "I may say I have read nothing to alter my conclusions about the *human* head—it is the application of the *same* principles to *Animals* which appears to me to be proved fallacious."[45]

The following month Ada wrote again, this time describing a visit that she, her husband, and a friend had paid that same Deville who had pronounced Lady Byron so morbidly sensitive years before. They had gone incognito, said Ada, and "it was very clear that he thought *much* the most highly of Sir G. Wilkinson, amongst the three." It was a sign of the success of their incognito, as well as a point against Deville's discernment, that he should have been more impressed with a disguised explorer and author than with an unknown earl and countess. Ada continued,

I think he failed with me in several points. He hit off one characteristic very cleverly & accurately, viz: my extreme *pain* & mortification at the slightest disparagement from others, & the tendency to *exaggerate & magnify* the circumstances to a remarkable extent—He dwelt very much on the predominance of the *Sentiments* over everything else in me. Now this is wrong. *Intellect* has at least an even share, if it does not carry the day, which I think it does. He said that Combativeness, Destructiveness, Self-Esteem, Hope, Order & Time, bear *no proportion* at all to the rest of the head; & that but for the Firmness, Conscientiousness & Causality the character would be a weak one.—Can we get phrenologised at Paris by the great man there?[46]

Apparently Deville had early discovered that he could invariably impress his female clients by capitalizing on the social insecurity, the vulnerabiity to gossip, and the heightened self-consciousness from which genteel women suffered so agonizingly. The myth that Ada's overriding mode of perception and response was intellectual—in contrast to her mother, who was "all feeling"—was already well established when this exchange took place, at a time when phrenology was becoming merged with mesmerism, which in due course became even more of a battlefield between them.

Ada had been declared far enough recovered to enter the world at the usual time—the first London season after she turned seventeen. It was a vital rite of passage for young ladies: as soon as they had made their bows at the Queen's Drawing Room, they were of an age to marry. Exercising her prominent organ of intellect, Ada also celebrated the occasion by drawing up a document in which she attempted to explain to her mother her own views on the freedoms that parents should permit their mature offspring. It is fascinating to compare it with Annabella's declaration of her reasons for leaving her parents to hasten to London on her own.

The principle point on which I differ from you is "Your being constituted my guardian by God *forever*." "Honour thy father & thy mother," is an injunction I never have considered to apply to an age beyond childhood or the first years of youth, in the sense at least of *obeying* them. Every year of a child's life, I consider that the claim of the parent to that child's *obedience*, diminishes. After a child grows up, I conceive the parent who has brought up that child to the best of their ability, to have a claim to his or her gratitude. . . . But I cannot consider that the parent has any right to direct the child or to expect obedience in such things as concern *the child only*. . . . I consider your only claim to my obedience to be that given *by law*, and that you have no *natural* right to expect it after childhood. . . . Till 21, the law gives you a power of enforcing obedience on all points; but at that time I consider your power and your claim to cease on all such points as concern *me alone*, though I conceive your claim to my attention, and consideration of *your* convenience & comfort, rather to increase than diminish with years. . . . I consider that the law gives you the power of enforcing it, beyond the age when you have a natural right to do so.[47]

If Ada hoped to clear the air and bring her mother to an understanding of her point of view in this way, she was bound to be disappointed. In addition to the moral pressure Lady Byron herself could bring to bear on any attempt at independence, she did not hesistate to confide in a circle of sympathizing friends, who in turn did not hesitate to scold and lecture Ada as soon as she exhibited any defect in veneration. It is no wonder that Ada despaired of "conversational litigation," as her mother called it (and who should know better?), to resort to a more active form of rebellion.

Because what happened was considered so shameful as possibly to affect Ada's marriageability, Lady Byron and her friends referred to the event only in the most oblique terms. The one explicit account that survives is Ada's own, and that at second hand. It appears in the memoir left by her confidant, Woronzow Greig, among his own family papers. After presenting her pedigree (a subject in which he took a special interest), his reminiscences become much more personal, and he offers a vivid picture of Ada in her teens.

My first recollection of Ada Byron about 1832 or 3 [1833 or 4 crossed out] is when as a young girl she was a visitor at the house of my mother at the Royal College Chelsea . . . and as she had even in those early years a decided taste for science which was much approved by Lady Noel Byron she took every opportunity of cultivating mothers acquaintance. Ada was then rather stout and inclined to be clumsy, without colour and in delicate health. She used to lie a great deal in a horizontal position, and she was subject to fits of giddiness when she looked down from any height. She seemed to be amiable and unaffected. As might be expected at this early period of life she had not much conversation. She was reserved and shy, with a good deal of pride and not a little selfishness which disclosed itself with her advancing years. Her moral courage was remarkable and her determination of character most pronounced.[48]

Though he mentions Ada's propensity to dizziness and fatigue, Greig makes no mention of her being on crutches, so the acquaintance is far more likely to have been formed in one of the later, crossed out, years than in 1832; probably, from the evidence of Ada's surviving letters, it was early in 1834. His description of her appearance agrees well with one left by her father's old friend, John Cam Hobhouse, who met her in February 1834 and recorded in his diary, "she is a large coarse-skinned young woman. . . . I was exceedingly disappointed." Greig attributed her early taciturnity to the demure behavior of the well-brought-up maiden, but his own account suggests another cause.

At this time Lady N. Byron was residing at Fordhook in Middlesex, and her most intimate friends were the late Miss Doyle, Miss Montgomery and Miss Carr the sister-in-law of Dr. Lushington and now living with them [that is, Miss Carr, at the time of writing, was living in Dr. Lushington's household]. These three ladies were constantly with Lady Byron who was entirely led by them, and as her daughter informed me they took the most unwarrantable liberties with Lady Byron and interfered in the most unjustifiable manner between mother and daughter. This annoyed Ada so much that she gave them the nickname of the three Furies which so far as appearances went was not unwarrantable as the ladies in question had all passed their premiere jeunesse and none of them was remarkable for good looks. . . .

As Ada grew older the interference of these ladies became more insufferable, but every attempt to resist it was repulsed by Lady Byron.

A short time before my family became acquainted with Lady Byron and her daughter, the former had engaged the services of a young man the son of some humble friend to come for a few hours daily to assist her daughters studies. As might have been foreseen a feeling of tenderness soon sprang up between these young people. It was not observed at first either by Lady Byron or her three friends. But Ada was reprimanded for chattering with her young master instead of attending to her studies. To this she paid no regard, and in consequence she was ordered to leave the room on one occasion by a "Fury." She did so unwillingly and in a state of high indignation. In the course of a few minutes she returned, and in pretence of carrying away some of the books, she managed to place in the young mans hands a slip of paper appointing an assignation at midnight in one of the outhouses.

The assignation took place and Ada informed me that matters went as far as they possible could without connexion being actually completed. My remark upon her telling me this was this youth must have been more or less than most.

In Greig's draft there is a carat after the word "without" in the last paragraph quoted, and the words "complete penetration" appear above the line, lightly crossed out. Perhaps he felt such minute specificity would convince his intended readers of his good information, but was already becoming uncertain of the tastefulness of the whole enterprise. The last sentence, conveying the Victorian gentleman's mixture of admiration and contempt for a man who had the opportunity of completing a seduction but refrained from doing so, is the first of a number of revealing intrusions into his narrative. Following this witty comment, his story plunges ahead:

After this her feelings toward the young man naturally became stronger and more uncontrollable. At length the mothers eyes were opened and the young mans visits were dicontinued. Driven to madness by disappointment and indignation at the conduct of the Furies, Ada fled from her mothers house to the arms of her lover who was residing at no great distance with his relations Lady B' humble friends. They received her with dismay and took the earliest opportunity of returning her to her mother before the escapade was known. The matter was hushed up, and the only persons cognizant of it besides the mother and friends was myself—who was informed of it by Ada, and Lovelace to whom Lady B. communicated the fact before her [Ada's] marriage. . . . To what extent she herself was cognizant or enlightened him I know not. But I suspect neither knew all the events.

Just when, for whom, and for what purpose this curious account was prepared can only be a matter of speculation. From the surviving correspondence between Ada and Greig, it does not seem likely that she made these confessions before the mid-1840s. From internal evidence, his account was written in the 1850s, possibly soon after her death—in other words, less than a decade after he heard the story. A regret expressed in another part of this memoir that he had failed to take down most of her early reminiscences suggests that he might even have made note of this one. In any case, it is clear from the unflattering comments on Ada and "Lady B." as well as the comments that reveal much about his own character, that he meant his account to be both frank and full.

The same cannot be said for another account that must refer to the same episode, that of Sophia De Morgan, Mr. Frend's daughter, which was written more than forty years after the events. Mrs. De Morgan was asserting the intimacy of her own friendship with Lady Byron:

After a visit paid to us at Stoke Newington with her daughter when the little girl was about fourteen I saw Lady Byron oftener and in the year 1832 went to stay with her at Fordhook. . . . I became acquainted with her anxieties on her daughter's account, & on one or two occasions had it in my power to prevent the consequences of Miss B's heedlessness & imprudence. I do not think this matter need be further entered into. There was I hope, no *real* misconduct at that time and an open scandal was prevented but it was very evident that the daughter who inherited many of her father's peculiarities also inherited his tendencies. . . . as I said before these occurred when Miss B was only fourteen or fifteen & were I believe simply imprudence.[49]

Mrs. De Morgan's memory is clearly failing here. Ada could not have been fourteen or fifteen when she attempted to elope, since at those ages she was flat on her back or on crutches, and she and her mother did not yet live at Fordhook, whither they moved in 1832. Nevertheless, she does seem to be referring to the same events as Greig when she hopes there was "no *real* misconduct," an unmistakable allusion to sexual activity. Now, Mrs. De Morgan, or Miss Frend, as she was then, was one of Lady Byron's confidants in the matter of Byron's incest; her statement here was given at the request of Lady Byron's grandson, who was busily gathering evidence upon just this point. She was not afterward in the circle of real intimates; a letter to Ada from her mother, many years later, mentions evading Mrs. De Morgan's prying questions on one occasion, perhaps after the discovery that she was in inveterate and malicious gossip where Ada's affairs were concerned. So the fact that she knew of Ada's elopement and might even have been involved in "preventing the consequences" suggests that she and her father might have been among the "humble friends" mentioned by Greig and that Ada's young lover was some connection of theirs. The Frends, if not exactly humble, were definitely middle-class.

In any case, Ada was returned, humiliated, and subjected to many lectures and scoldings from her mother, the Furies, Sophia Frend, and anyone else to whom Lady Byron cared to communicate her anxieties. Her spirit was broken, temporarily at least; she pronounced herself chastened and determined to mend her ways. Yet her season in London had already opened new doors through which she glimpsed new possibilities of freedom. She declared she would not marry until she had enjoyed the independence of coming of age; and she had already met Charles Babbage, whose influence, Greig declared, "eventually did her much harm."

2

The Much Desired Great Unknown

Charles Babbage is gradually becoming recognized as one of the most important scientific figures of the nineteenth century. He was also the possessor, as he was well aware, of one of the most fascinating intellects of all time. Like Ada, he was much given to observing and recording his own thought processes.

He was born in London in 1791, the only son of a banker, which made him decidedly a gentleman, though not an aristocrat. He supplemented the relatively meager instruction of teachers and tutors with a private study of mathematics so broad and deep that, arriving at Cambridge, he found he knew more than most of the dons. Gathering a group of like-minded students around him, he set out to reform the state of mathematics in England, which at that time was far behind the Continent, particularly France, partly because of the great veneration in which Newton was held. Consequently, the first campaign waged by Babbage and his friends was to persuade the university to adopt Leibniz's superior "d" notation for the differential calculus in place of Newton's outmoded "dots." The subject of notation, the symbols or tools of mathematical entities or processes, was always an important interest of Babbage's. He understood that, in intellectual work just as in manufacturing, the tools used could advance or retard the process.

To accomplish their aim, Babbage and his friends formed a society whose purpose was to translate a small book on the differential and integral calculus by Lacroix. It was published in 1816, the joint work of Babbage, John Herschel, and George Peacock. In recognition of the almost sacrilegious nature of the step they were suggesting, Babbage facetiously proposed to title the work "The Principles of pure D-ism in opposition to the Dot-age of the University."[1] Then, to persuade the languid dons to adopt it, they published a companion volume of problems with their solutions, all in the new notation.

The campaign was successful, but on leaving Cambridge, Babbage was unable to secure a position he considered worthy of his talents; one by one, such appointments fell to men of lesser acquirements. Cambridge University, he noted in his autobiography, finally forgave his irreverence by electing him to the Lucasian Chair of Mathematics, Newton's old post. But the honor came so late that he almost declined it; his father, in whose eyes his refusal to accept lesser posts would have been vindicated by the appointment, was dead. Then, too, he was already deep in his work on his first calculating engine. Before he plunged into calculating engines, however, he had produced a series of papers in pure mathematics that made highly original and important contributions to the development of modern algebra. A recent assessment of his work in this area concludes that it was a thousand pities he turned to computers in 1820 and aborted the more theoretical phase of his career.[2]

According to Babbage's accounts, which vary somewhat, the impetus behind the invention of his first machine, which he called the Difference Engine, was the need to produce error-free numerical tables of various kinds, a task that seemed beyond the capabilities of fallible human beings. As skillful and inventive in practical matters as in theory, by 1822 he had constructed a small model that worked well enough to prompt him to write an open letter to Sir Humphry Davy, then president of the Royal Society, in which he also mentioned that he had thought of several other types of machine as well, suited for different types of mathematical computation.

To accumulate and store the numbers, Babbage adopted the device that had been used by Pascal and Leibniz: tall, vertical shafts or axes on each of which a large number of disks were stacked, by means of the holes in their centers. The disks did not actually touch each other, but each could be independently turned around its axis by means of an attached toothed gear wheel. Each wheel had ten teeth, corresponding to the ten divisions marked on the edges of the disks, one for each digit 0 through 9. The digits on each disk had values one decimal place higher than those on the disk below, so that an ordinary decimal number could be read off each column by reading down a vertical line from the top. The number on any level could be changed independently by turning the disk. The number on one column could be added to that on another by connecting the two columns by means of gear wheels and turning the disks of one until all the digits in the chosen vertical line were zeros. The second column then held the sum of the two original numbers.

A slight complication was the occasional need to carry from one disk level to the one above when the sum of two digits on the first level was greater than 9. Babbage worked out several clever mechanical schemes to accomplish this, of which he was very proud. The overall operation of the Difference Engine depended on the fact that many mathematical functions can be approximated by several terms of a power series. The successive values of the powers of any number, and the sums of power series, can be arrived at by repeated additions of several orders of differences between terms. Thus the Difference Engine worked by connecting a number of columns of disks in such a way that the successive numbers on each were added to the column next in line. The column on the end, containing the highest order of difference used, always had the same constant number on it. Unlike previous mechanical calculators, once the Difference Engine was set up for a particular task, it would proceed through all the necessary steps without further intervention. All the human operator had to do was turn a handle at the top of the model.

After a leisurely delay of almost a year the recently formed Astronomical Society awarded Babbage a gold medal for his invention and a grant of £1,500 to launch him on the construction of a full-sized engine. Many of his friends, able scientists, were enthusiastic about the project. In keeping with his principle that a theoretically feasible plan should not be frustrated by practical deficiencies, Babbage determined to build his engine of the finest available materials, by the most advanced techniques, in the hands of the most expert workmen. At the recommendation of Marc Brunel, who pioneered the making of machines with interchangeable parts, he hired Joseph Clement, who had been trained at the firm of Henry Maudslay, the nursery of the most highly developed mechanical engineering practices of the time. Still, the Difference Engine project required that Babbage himself design many new and ingenious tools to make the parts he needed. These tools by law belonged to Clement, the maker, instead of Babbage, the designer and employer.

In the end the Government contributed a total of £17,000 towards the construction, which proved a much longer and more involved task than first estimated. The support, however, was erratic and uncertain from the beginning (as government payments often were), and Babbage had often to advance money to Clement out of his own pocket. In 1833, in the course of moving the work to a safer location, Babbage had a demonstration portion of the machine put together; it worked perfectly. Shortly thereafter a dispute over payments broke out between Babbage and Clement. The latter not only stopped work and laid off his assistants, but appropriated all the drawings and parts of the engine, except for the portion that had been assembled. That had been removed to Babbage's drawing room, to enable him to display its workings to admiring visitors.

It was almost two years before he regained his plans and parts, and nine years before the changing Governments could be brought to a definite decision regarding future support for the project. In the meantime his restless mind had continued to evolve the possibilities of a calculating machine that would transcend the limitations of the Difference Engine: an Analytical Engine, capable of multiplication and division as well as addition and subtraction, hence able to perform any numerical calculation, and needing no constant order of difference. It was not long before he thought of instructing such a machine by means of punched cards, such as were then used to control the weaving of patterns in materials by power-driven looms.

Babbage met Ada Byron at a party in June 1833, during her first London season, soon after he had set up his demonstration engine. Ada, wrote Lady Byron to her friends Dr. and Mrs. King, was delighted with him. He responded to her delight in his customary fashion, by issuing an invitation to view his Difference Engine. Sophia De Morgan records the scene as follows:

I well remember accompanying her to see Mr. Babbage's wonderful analytical engine. While other visitors gazed at the working of this beautiful instrument with the sort of expression, and I dare say the sort of feeling, that some savages are said to have shown on first seeing a looking-glass or hearing a gun—if, indeed, they had as strong an idea of its marvelousness—Miss Byron, young as she was, understood its working, and saw the great beauty of the invention. She had read the Differential Calculus to some extent, and after her marriage she pursued the study and translated a small work of the Italian Mathematician Menabrea, in which the mathematical principles of its construction are explained.[3]

The engine that Ada and her mother went to view was not, of course, the Analytical Engine, no model or portion of which existed until long after Ada's death, but the Difference Engine. Ada did see its beauty, but as to an understanding of its workings, contemporary evidence reveals only Lady Byron's:

We both went to see the *thinking* machine (for such it seems) last Monday. It raised several Nos. to the 2nd & 3rd powers, and extracted the root of a Quadratic Equation. —I had but faint glimpses of the principles by which it worked—Babbage said it had given him notions with respect to general laws which were never before presented to his mind—For instance, the machine would go on counting regularly, 1, 2, 3, 4, &c—to 10,000—and then pursue its calculation according to a new ratio, which was, I *think*, 10,002, 10,005, 10,009—but I am only certain that the numbers were no longer successive ones, and that their differences were neither in arithmetical nor Geometrical ratio, as far as I could apprehend. —If this occult principle of change existed in the law according to which the machine was constructed, (for Babbage discovered it to be latent in the mathematical formula originally applied by him) it may be consistent with the *general* laws of our solar system that the sun shall not rise tomorrow. —He said, indeed, that the *exceptions* which took place in the operation of his Machine, & which were not to be accounted for by any errors or derangement of structure, would follow a greater number of uniform experiences than the world has known of days & nights. —There was a sublimity in the views thus opened of the ultimate results of intellectual power.[4]

This letter reveals that Babbage had already been struck by the cosmological implications of his machine that he later elaborated in the *Ninth Bridgewater Treatise*, his contribution to natural theology. He was very fond of mystifying his guests with the apparently miraculous behavior of the engine, but it was a simple matter for him to arrange that one or more additional number columns be coupled into the operation after a set number of turns.

As delighted as Ada might have been, and as interested in science, she knew very little mathematics at the time. It was not until March of the following year, 1834, that she turned to the same Dr. King, repentent over her attempt to elope with her tutor, and asked him for help with a rather elementary course of study:

How far I am *really* and *permanently* awakening to a sense of religious duty and religious obligation, time alone can prove. I cannot but feel very distrustful of myself. . . . You said much on the subject of controlling our imaginations and our thoughts. I often think of this now, as I cannot but perceive that it is a paramount duty for one in my circumstances to exercise this sort of self-government. . . .
I must cease to think of living for pleasure or self gratification; and there is but one sort of excitement, if indeed it can be called by that name, which I think allowable for me at present, viz: that of study and intellectual improvement. I find that nothing but very *close & intense*

application to subjects of a scientific nature now seems at all to keep my imagination from running wild, or to stop up the void which seems to be left in my mind from a want of excitement. I am most thankful that this strong source of interest does seem to be supplied to me now almost providentially, & think it is a duty vigorously to use the resources thus as it were pointed out to me. If you will do me so great a favour as to give me the benefit of your advice and suggestions as to the *plan* of study most advisable for me to follow, I shall be most grateful.—I may say that I have *time* at my command, & that I am willing to take *any* trouble. It appears to me that the first thing is to go through a course of Mathematics—that is to say—Euclid, and Arithmetic & Algebra; and as I am not entirely a beginner in this subjects [*sic*], I do not anticipate any serious difficulties, particularly if I may be allowed to apply to you in any extreme case. My wish is to make myself well acquainted with Astronomy, Optics &c; but I find that I cannot study these satisfactorily, for want of a thorough ac- quaintance with the elementary parts of Mathematics. . . . In short, here I am, ready to be directed! I really want some hard work for a certain number of hours every day.[5]

Dr. King agreed that mathematics kept sexual longings at bay:

My dearest Ada
You are quite right in supposing that your chief resource and safeguard at present, is in a course of severe intellectual study. For this purpose there is no subject to be compared to Mathematics and Natural Phi- losophy: Because 1° they require individual attention to comprehend them, 2° they have a natural sequence of ideas, which the mind can work out of itself, when the train of thought is once suggested. 3° they have no connexion with the *feelings* of life: They cannot by any possibility lead to objectionable thoughts.
In early life I pursued that study under these impressions as well as others. I considered them a moral discipline, tending to control the imagination, and give one mental self command. I should recommend you a complete Cambridge course.[6]

He then went on to recommend both a set of books from which to study and a method of learning that would not leave her thoughts free to stray even on her daily walks. "If you continue this course of study for a year you will find it answer it's purpose, in all ways," he assured her. The rest of the letter is concerned with his eagerness to help her, the wildness and self-assurance of youth, the wretchedness into which her impatience and love of pleasure were leading her. "So it is with most of us in early life," ran the solemn conclusion; "we begin with self indulgence: then creep into doubtful improprieties: and end in crime and misery."

If for no other reason, this letter is interesting for what it reveals about the scope and quality of mathematical education in the early nineteenth century. Dr. King had left Cambridge in 1809, the year before Babbage entered it. The relative decline that had characterized British mathematics for a century was striking at Cambridge: here mathematics was so strongly stressed that it formed a major part of the Senate House Examination, and yet, in contrast to the situation abroad, there were few opportunities for a career in science. Cambridge graduates mostly went on to become gentlemen of leisure, clergymen, doctors, and lawyers. Mathematics was considered good mental exercise, good moral discipline, but of little use in itself.

The letter also sheds a good deal of light on why Ada had found Babbage so delightful. Babbage too considered mathematics, and problem solving in general, good mental exerise, but was totally free of the moral concerns that haunted Dr. King. Nevertheless, Ada on this occasion responded eagerly to Dr. King's proposed regimen; nine days later she was already demonstrating that she could soon push her guide to the limits of his Euclid:

I usually do four *new* propositions a day, and go over some of the old ones. I expect now to finish the 1st book in less than a week. I use Lardner's Euclid, which is the one now I believe most approved by mathematicians [so much for Simpson's, which Dr. King had recommended]. It has notes, which I also read. . . .[7]

The common response to Ada's intellectual pretensions on the part of the numerous moralizers who surrounded her was to redouble the sermonizing. Dr. King fell back on this device, sending her some religious texts to chasten her dangerously rising spirits. She thanked him for the extracts, and redirected his attention to geometry, incidentally revealing that it had taken her longer than she expected to finish the first book:

Will you answer me the following questions? Can it be proved by means of propositions & deductions from the *1st* book *only* that *equilateral* triangles being constructed on the sides of a *right* angled triangle, and also one on the *hypothenuse*, the sum of the triangles on the sides is equal to the triangle on the hypothenuse? I think I have heard that this *is* capable of proof by the 1st book, but that the proof is a difficult one. It strikes me that it ought to be as demonstrable as when the figures are *four*-sided & equilateral. —Mama and I are reading Whewell's Bridgewater Treatise. How interesting it is![8]

He answered, now panting noticeably:

My dear Ada
You will soon puzzle me in your studies. When I was at College we
had few problems deduced from Euclid. We *got up* a set of books and
seldom went out of them, except the high men, i.e., the first 4 or 6
Wranglers, i.e., the men of the first class. I imagine all *similar* figures,
on the sides of a right angled triangle, have the same property. That
on the hypothenuse being equal to the other two. I sat down to think
of it the other day but had not time to make it out. . . . I think Whewell
is a book quite worth studying like Euclid and should advise you to
do it. I do not say at this moment—But loose reading does no good,
especially in early life. . . . You must *trammel* your mind in these things
for a year before we can judge of the effect. Some day I shall point
out to you a system of Logic, & Morals, but it would be now premature.[9]

Impatient Ada was already dipping into as yet forbidden knowledge.
In 1829 the eighth Earl of Bridgewater had died, leaving £8,000 at
the disposal of the president of the Royal Society, to be distributed
among persons selected to write and publish at least a thousand copies
of a treatise "On the Power, Wisdom and Goodness of God, as man-
infested in the Creation." Whewell's was the third treatise to be commissioned, and he chose to write on "Astronomy and General Physics
considered with reference to Natural Theology." In it he denied that
"mechanical philosophers" and mathematicians had any special authority to pronounce on such matters as the administration of the
universe. Babbage, a sharp critic of the Royal Society, was not among
the chosen authors, but Whewell's treatise inspired in him a strong
urge to rebuttal. Undeterred whenever he felt he had a contribution
to make, in 1837 he was to bring out the *Ninth Bridgewater Treatise* on
his own, largely to argue against Whewell. It was to make a profound
impression upon Ada.

Despite Dr. King's revelation that he was not among the "high
men," and despite the deadening way in which he presented the
subject, Ada persisted with her studies, and by the following September
we learn that she had managed to acquire two hapless pupils of her
own.

Dear Dr. King
. . . We are now at Buxton in Derbyshire. We came to see our friend
Lady Gosford, who is here for her health. The society of so excellent
& so delightful a person can hardly fail I think to be beneficial to a

young person like myself. She is one of those who certainly make religion the rule of life. —Her daughters too are doing me great good I believe, though in a very different way. They are amiable young women, with good natural abilities, but of rather indolent habits. I am trying to excite & rouse them to various objects of study & interest, and so much as possible to make my little talents, such as they are, of use to my young friends, whom I would gladly serve, were it only for their Mother's sake. I teach three fourths of the day at least, and find that I myself gain more perhaps than they do. I am endeavouring to induce one of them to take up mathematics, but I have rather a difficult task there; however, I do not despair.[10]

In the same letter Ada revealed that she and her mother had also taken a tour of the industrialized Midlands over the summer. Like many of the progressive upper class, they managed to remain oblivious to the wretched condition of the working poor. Instead, Ada was able to use the trip to preen herself a bit on her first-hand acquaintance with Babbage's calculating machine:

A tour of friends and of natural beauties too, is the very perfection of a tour, so far as enjoyment is concerned, except perhaps that *I* could wish to add that it was also a tour of manufactures and machinery. This indeed our's has been *partly*, but only partly, from unavoidable circumstances. This Machinery reminds me of Babbage and his gem of all mechanism. At the beginning of the *last* Edinburgh Review, there is a very clever article on this Machine, which you should read. I can hardly judge whether it will be perfectly intelligible to one who has never seen the original, or models, but I should think it would to you. At all events a great part would. Pray get it.

Ada persisted in her efforts with the amiable young women and after returning home began an instructive correspondence with them. A note written in November, designated "For Livy [Olivia]" and signed "Yours ever musically," serves up one portion of her "little talents." But the major part of her attention was clearly directed to Lady Gosford's other daughter, Annabella Acheson, Lady Byron's godchild and namesake. Enclosing a proof in geometry, she urges Annabella to persevere and to consult her in every difficulty. She closes, optimistically and self-consciously, "So this you see is the commencement of 'A Sentimental Mathematical Correspondence carried on for — — — years between two Young Ladies of Rank' to be hereinafter published no doubt for the edification of womankind." She signs it, symmetrically, "Ever yours mathematically."[11] Her mention of "the edification of womankind" is a humorous reference to the practice of women who presumed to publish work of an instructive or abstruse nature, explaining in the dedication that their intention was the uplift of other women. They did not presume to instruct men.

In succeeding dispatches Ada continued to lecture and correct her friend. Taking a leaf from Dr. King's book, she scolded Annabella for having read material before it was assigned. To prevent such unruliness, she planned to write her own geometry text. She cautioned the frustrated Annabella to make haste slowly—how many times she would hear that phrase! Sometimes she would report complacently on her own progress, with the assurance of her intention to place all her learning at her pupil's disposal.

At last the amiable young woman struck back, and in the usual way. She was, after all, her mother's daughter and Lady Byron's namesake. In a letter dated 29 June 1835, she complained that Ada was too little aware of the evil existing in all human hearts, including her own, and scolded her teacher for "reason" and unbelief. But by that time all Ada's need for penance had ended. The year of trammeling had had its desired effect. She had so far awakened to her religious and filial duty as to have made a most suitable match for herself. For in the intervals of mortifying her flesh with a course of "severe intellectual study," she had been attending the operas, balls, and dinner parties of the London season, with the object of attracting a marriage mate of whom her mother could approve.

In a letter to Annabella Acheson late in 1834 Ada remarked, "Mr. Babbage & Mrs. Somerville are very kind indeed to me. The latter generally inquires with interest 'how my pupil is going on?' "[12] Ada had the previous spring acquired another friend-tutor in Mary Somerville. There was every reason for Ada to look to Mrs. Somerville as a guide and model and for Mrs. Somerville to regard Ada with sympathy and interest, and to befriend her, despite the thirty-five-year difference in their ages.

Like Ada, Mary Somerville (1780–1872) had obtained most of her mathematical and scientific instruction by importuning friends and acquaintances. In this Mrs. Somerville had been perhaps the less fortunate, for her father, Vice-Admiral Sir William Fairfax, had been home from sea at least periodically during her childhood and adolescence, and at those times had done everything he could to discourage her studies. "I was annoyed," she wrote in her autobiography, "that my turn for reading was so much disapproved, and thought it unjust to give women a desire for knowledge if it were wrong to acquire it."[13] She had been almost thirty years old, a widow with a competence, before she had both the freedom and the means seriously to address her interests. "The only thing in which I was determined and inflexible," she said, "was in the prosecution of my studies; they were perpetually interrupted, but resolutely resumed at the first opportunity."[14]

At her father's insistence she had been sent at the age of ten to a fashionable boarding school for the education appropriate to a well-connected young lady. There, learning, such as it was, was impeded by concern for the development of her figure:

A few days after my arrival, though perfectly well made, I was inclosed in stiff stays with a steel busk in front, and, above my dress, bands drew my shoulders back till the shoulder blades met; then a steel rod with a semicircle that went under the chin was clasped to the steel busk in my stays. In this constrained state I and most of the younger girls had to prepare our lessons. The chief thing I had to do was to learn by heart a page of Johnston's [sic] dictionary, not only to spell the words, give their parts of speech and meaning, but as an exercise of the memory to learn their order of succession.[15]

She was allowed to leave this school after one year, which she did with great joy but little education. Despite the drill with the dictionary, or because of it, she remained a poor speller: when she and playwright Joanna Baillie compared notes on the matter in later life, they agreed that, when in doubt, they tended to choose, not the best word for their purposes, but the one they could spell. And she also bemoaned her want of memory.

In her early teens Mary Fairfax discovered a subject for which her aptitude was remarkable. Seeing some strange symbols in a ladies' fashion magazine, she inquired what they meant and was told that they were "algebra." (A writer, ironically enough reviewing Laplace's *Traité de Mécanique Céleste*, in 1807 lamented the state of mathematics in Britain by pointing to the competent mathematical talent being wasted in concocting and solving mathematical problems in women's magazines.) Eventually she learned the nature of the subject from her younger brother's tutor, whom she then persuaded to buy her a copy of Euclid and an algebra text. These she studied in private, over the opposition of most of her family; but, though determined, she was shy and shrank from open rebellion: "I did not assume my place in society in my younger days; I lost dignity by it, and in an argument I was instantly put to silence though I knew and could have proven that I was in the right."[16]

In 1804 she married Samuel Greig, a cousin who held a commission in the Russian navy. They were transferred from Scotland, where she had grown up, to London, where, alone for much of the time, she continued her studies although her husband too did not really approve. In 1807 he died, leaving her with two sons, of whom the elder, Woronzow, was named for one of his father's Russian connections. Returning to Scotland, she felt free for the first time to pursue her interests openly. In 1812 she married Dr. William Somerville, another cousin, who strongly supported her studies from the first.

In 1816, again as the result of a husband's appointment, she removed to London. Both there and on the Continent, though not wealthy, the Somervilles became friendly with many well-known scientists and writers including Babbage, whose salons they frequently attended. Dr. Somerville joined the Royal Society and in 1826 communicated to it the results of a series of experiments his wife had carried out on the magnetizing effects of sunlight (which was widely believed in for a time). Then, in 1827, Henry Brougham, an old friend, asked Dr. Somerville to persaude her to translate Laplace's *Mécanique Céleste* for the Society for the Diffusion of Useful Knowledge. She agreed somewhat reluctantly, setting the condition that the project was to remain a secret until completed and that the manuscript was to be destroyed, and the secret never revealed, should it be found unsatisfactory.

For over four years she worked surreptitiously, ready to thrust her books and papers out of sight if a visitor was shown into the drawing room of a morning. In the evenings she and her husband kept up the moderate social schedule they had established. She neglected neither her household duties nor the instruction of her daughters. The book appeared in 1831 and was acclaimed, but even then her response was measured:

In the climax of my great success, the approbation of some of the first scientific men of the age, and of the public in general, I was highly gratified, but much less elated than might have been expected, for although I had recorded in a clear point of view some of the most refined and difficult analytical processes and astronomical discoveries, I was conscious that I had never made a discovery myself, that I had no originality.[17]

Above all, she feared to make a fool of herself. When her experiments on the magnetizing effects of sunlight were later falsified, she bitterly regretted having written and published her paper. To ward off criticism, she assumed a style of modesty and self-effacement. Her autobiography is full of little self-deprecating touches, like the comment on her difficulties with spelling. Her assessment of her own capacities contrasted sharply with the soaring aspirations Ada was to develop. But she had written and published a successful book, and in the process had acquired a taste for the enterprise.

Mrs. Somerville immediately set to work upon another book, a study of the interrelationship among the physical sciences, which appeared in 1834 and was even more successful. It was appropriately dedicated to Queen Adelaide, consort of William IV, with the modest hope that "if I have succeeded in my endeavour to make the laws by which the material world is governed more familiar to my countrywomen, I shall have the gratification of thinking that the gracious permission to dedicate my book to your majesty has not been misplaced."[18] Her work was nonetheless used for the instruction of men, including classes at Cambridge. Yet her own experiences had left her with strong views on female education, as she wrote in her autobiography:

It was the fashion of a set of ladies such as Mrs. Hannah Moore [*sic*], Mrs. Elizabeth Hamilton, & Mrs. Grant of Logan to write on female education. I detested their books for they imposed such restraints & duties that they seemed to be written to please the men; even my friend Mrs. Robert Napiers book . . . I never read through.[19]

She saw to it that her daughters did not lack the opportunities she had wanted; but she was never quite able to shake the belief that she had contributed to the early death of her precocious eldest daughter because she had "strained her young mind too much."

I have perseverance and intelligence but no genius, that spark from heaven is not granted to the [female] sex, we are of the earth, earthy, whether higher powers be allotted to us in another state of existence God knows, original genius in science at least is hopeless in this, at all events it has not yet appeared in the higher branches of science.[20]

This ambivalence colored Mrs. Somerville's acquaintance with Ada; it is revealed in their correspondence, though not in her autobiography:

All the time we were at Chelsea we held constant intercourse with Lady Noel Byron and Ada who lived at Esher and when we went to reside in Italy I kept up a correspondence with both as long as they lived. Ada was much attached to me and often came to stay with us. It was by my advice that she studied mathematics. She always wrote to me for an explanation when she met with any difficulty; among my papers lately I found many notes asking mathematical questions, and I heard after I went to Italy that she was very successful as a mathematician.[21]

Mary Somerville's autobiography was written when she was nearly ninety, many years after these events, and must be checked against other sources for details and dates. The Somervilles moved to Chelsea in 1824, while Lady Byron did not live at Esher until after her daughter's marriage. I have found no evidence that the acquaintance was formed before the spring of 1834; Ada, then, had already appealed to Dr. King for help in mathematics when she met Mrs. Somerville. On 7 June of that year Mrs. Somerville wrote to Ada offering to visit, although she was very busy preparing a second edition of her book.

The friendship ripened throughout the summer and fall; then, in February 1835, Mrs. Somerville received a disturbing letter from her young admirer:

I cannot deny that I was shattered when I left you, but then I am for some unaccountable reason in a weak state. Altogether now, & at this moment can hardly hold my pen from the shaking of my hand, though I cannot complain of being what people call *ill.* . . . When I am weak, I am always so exceedingly terrified at *nobody knows what,* that I can hardly help having an agitated look & manner; & this was the case when I left you.[22]

That these nervous fits or "weak states" were severe, of some duration, and attributed at least in part to excessive study is shown by a letter written on the following 4 April from Brighton, where Lady Byron had taken her to recover. A pattern had been established that was to be repeated many times over the years.

I assure you there is no pleasure in the way of exercise equal to that of feeling one's horse flying under one. It is even better than waltzing. I recommend it too as a nervous medicine for weak patients. I am very well able now to read Mathematics provided I do not go on too long at a time; & as I have made up my mind not to care at present about making any particular progress, but to take it very quietly & as much as possible merely for the sake of the improvement to my own mind at the time, I think I am less likely to be immoderate.[23]

Before the end of June Ada wrote again to hint of her engagement, and it was not until November, three months after her marriage, that she wrote with a specific mathematical query.

Mrs. Somerville's son Woronzow (1805–1865), who considered himself something of an insider, did not learn of the match much before his mother. He was a friend and former classmate of Ada's intended groom. His confidential relationship with Ada did not develop until some years later; however, many of the details of his recollections of this period are echoed in other sources. Of Ada's activities during the London season of 1835 he reports, "She constantly accompanied my mother and sisters to Babbages evening parties which were then the fashion in London, and thus she commenced an acquaintance with a man who from his scientific pursuits and reputation acquired an influence over her which eventually did her much harm." According to Greig the still sickly, clumsy young girl, whose task at this point was to attract a suitable husband—a project that had recently come close to disaster in her mother's eyes—was vulnerable to "many notorious fortune hunters," including at least one clergyman. "The Revd Charles Murray, the Treasurer of the Ecclesiastical commission who eventually decamped with £9000 of the monies of the commission after having incurred debts to the amount of £25000 . . . made a very daring attempt to inveigle her . . . into a marriage because she was an heiress." Lady Byron's reaction upon learning what was afoot was once more humiliating to poor Ada, whose defense, as so much else in this early period, would be echoed at the end of her career: "When Lady Byron discovered it and taxed her with her baseness, she said it was all a mistake."[24]

Meanwhile Greig had had an inspiration:

During the Spring of 1835 I suggested to my friend Lord Lovelace, then Lord King, that she would suit him as a wife. He and I had been to college together (Trinity Cambridge) and were and have continued through life on the most inimate terms. Even to me in whom he confides more than in any other person living, he is not very communicative. Accordingly he received my suggestion without remark and he did not mention the subject to me until the 12 June 1835 when he wrote to ask me to dine with him alone in St James Square as he had something particular to tell me. At dinner he surprised me by announcing his engagement to marry Miss Byron, as I was not even aware that he had been paying his addresses to her. But they had met at Sir Geo. Philip's place in Warwickshire and the courtship was concluded in a very short time. Ada afterwards told me that her mother having noticed Lord Kings attention thought it right to put her daughter on her guard and commenced doing so in a rather circuitous manner lest her communication should be too startling. But

Ada interrupted her by saying oh Mamma I know what you mean—you think Lord King intends to propose to me. I think so too, and if he does I have made up my mind to accept him.[25]

That Lady Byron, too, later confided in Greig concerning intimate family matters is explainable on the basis that he was an attorney and had developed a professional as well as personal relationship with both Lord Lovelace and his mother-in-law. That he did not himself become one of Ada's suitors is attributable to the great difference in wealth and social position between them; he did fall in love with her, afterwards, if not at this time. His matchmaking between the nervous, susceptible teenager and his titled, taciturn friend of thirty was either very imaginative or completely worldly.

William King (1805–1893) was no relation to Dr. William King, the Brighton physician who was Lady Byron's friend and Ada's mathematical guide. He was a descendant of Peter King, the son of an Exeter grocer who became first a lawyer, then a Member of Parliament, and finally Lord Chancellor of England for George I. With the purchase of an estate at Ockham in Surrey in 1707, this ancestral King had followed faithfully the tradition of Founding a Family, and in 1725 he was created the first Baron King of Ockham. At a time when any peerage over a century old was denominated "ancient," William, who had succeeded as eighth Lord on his father's death in 1833, was the holder of such a title. He was, moreover, a collateral descendant of John Locke, the philosopher, and possessed estates with an annual rent roll of £8,000, though these were encumbered by his mother's jointure.

After leaving Trinity, he had begun a diplomatic career, serving as secretary to the Lord Commissioner of the Ionian Islands. He had traveled extensively, in the Middle East as well as Greece. He spoke several languages, notably Greek, French, Italian, and Spanish. He had serious interests in architecture, history, economics, education, literature, and agronomy. His father's death had diverted his sober industriousness from diplomacy to national and county politics and the consolidation, expansion, and improvement of the family estates. In his politics he followed his family's liberal Whig traditions, which matched those of Byron and Sir Ralph Milbanke in favoring Catholic emancipation and opposing the Corn Laws. Thus, to his match with Ada he brought qualities of intellect and lineage that, if not equal to those of the Byrons and the Noels, were solid and substantial, and qualities of steadiness and responsibility for which the Byrons had never been known.[26]

He was also very much in love. The association with the Byron glamour and the favorable regard of Lord Melbourne—the prime minister and leader of his party, and Lady Byron's cousin—to say nothing of his bride's dowry and future inheritance, brought prospects exceeding his expectations. As for Ada, newly recovered from a nervous breakdown and basking at last in her mother's approval, her feelings were conveyed to her lover with suitable maidenly reserve: "I do not know when I have been in so calm and peaceful, and I hope I may add with truth, so grateful a state of mind."[27]

She was shortly to have even more reason for gratitude, as Greig tells us:

During the engagement and before the marriage, Lady B.' stern sense of justice and right, which existed however on some points only, caused her, as I learned from herself in one of those mysterious indefinite communications which she is in the habit of making even to those she most trusts, to communicate to Lovelace the escapade before mentioned of her daughter with the young tutor.[28]

Lord King chivalrously passed the unnecessary test, and Ada's gratuitous disgrace was averted, for Lady Byron was able to report to Sophia Frend: "He knows *all*, and is most anxious for the marriage."[29]

Lady Byron must have been at least as delighted by the match as the two principals, for she made a financial arrangement with her prospective son-in-law that both she and Ada, who of course had no say in it whatever, were later to regret. In addition to the £16,000 that had been settled on Ada by her own marriage agreement, Lady Byron added another £14,000 to be turned over to the happy bridegroom at once. (Later she would add another £10,000 in two installments.) The Wentworth estates were entailed, but Ada and her descendants were almost certain to inherit them, since it was most unlikely that either her mother or her mother's uncle or aunt, both in their fifties, would produce a male heir. If Ada inherited while her husband was alive, he and not she would be the life tenant of the property. If she inherited after he died, her jointure of £2,000 per year would cease. While he lived, she would receive pin money amounting to £300 a year, which, it later transpired, was even intended to cover such items as ball gowns. This was the same amount her mother had received. Lovelace never considered these terms unfair, but Ada did.[30]

At that time, however, there was too much else to think about. The marriage took place on 8 July 1835, only a month after the engagement, and within a fortnight Lady Byron, who had written her son-in-law transferring her guardianship of Ada to him, was exhorting her daughter, "May you prove the Mrs. Somerville of Housekeepers! I shall expect you to produce a Treatise 'On the Connexion of the Domestic Sciences,'—But I really think that you have all the practical qualifications for excelling in that sort of management, if you will attend to *Order*."[31]

Ada was more interested in emulating Mrs. Somerville's other accomplishments. In October, recovering from the discomforts of early pregnancy, Ada wrote to her mother to complain about Mrs. Margaret Carpenter, who was painting her portrait: "I conclude she is bent on displaying the whole expanse of my capacious jaw bone, upon which I think the word Mathematics should be written."[32] She was returning determinedly to her studies, and a few days later wrote to Mrs. Somerville herself:

I now read Mathematics every day & am occupied in Trigonometry & in preliminaries to Cubic & Biquadratic Equations. So you see that matrimony has by no means lessened my taste for those pursuits, nor my determination to carry them on, although it has necessarily diminished the time I have at command. But I suspect it is no bad thing to be limited in that respect.

By the bye I have a mathematical question to ask you, which I hope is not too trifling to be beneath your notice. . . .

$$\sin(a + b) = \frac{\sin a \cos b + \sin b \cos a}{R}$$

$$\cos(a + b) = \frac{\cos a \cos b - \sin a \sin b}{R}$$

The proposition these formulas relate to is, "The sines & cosines of two arcs *a* & *b* being given, the sine and cosine of the sum or difference of these arcs may be found by the following formulas &c. —["] The expressions for the sine & cosine of the sum, as above, are then demonstrated by a figure; after which occurs the following passage: "The values for the sine & cosine of *the differences* (*a* − *b*) may easily be *deduced from these two formulas*." Now however easily deduced, I have not succeeded in doing it. The required values are given in the book, though not the *method* of deducing them from the above. They are, sin (*a* − *b*) = ...

I suppose the deduction depends on some principle which I have overlooked, but I do not perceive how from consideration of the first two formulas, the last two can be deduced.

I am sorry to find that your copy of Legendre's Geometry is still in my possession. I thought I had returned it & my only excuse is—my marriage.[33]

This letter is rather startling, for it reveals that Ada had not progressed very far; despite hard work, skill and ingenuity in the manipulation of symbols did not come easily to her. The length and detail with which she described the problem reveal the extent of her uncertainty. Mrs. Somerville replied as tactfully as possible, pointing out that the formulas held for all values of the arcs, including negative ones, and gently reminding her that the half-forgotten Legendre contained the necessary substitutions. The sought-after "principle" was simply making the appropriate substitution for the variables of the formulas. It was a principle that would continue to elude her.

That this was not just a momentary lapse on Ada's part is shown by her next letter, written just two and a half weeks later.

I have another trigonometrical question to ask you, & am encouraged by your kindness to trouble you with these things, which is almost a shame too, when you are so busy. I believe this question is as simple to answer as the last, & am afraid you must think me a great dunce not to make out such easy things.

"Since the arc a is formed from the sum of the two arcs $(a - b)$ and b, by the preceding formulas we shall have:

$$R \sin a = \sin (a - b) \cos b + \cos (a - b) \sin b$$
$$R \cos a = \cos (a - b) \cos b - \sin (a - b) \sin b$$

and from these we find

$$R \sin (a - b) = \sin a \cos b - \sin b \cos a$$
$$R \cos (a - b) = \cos a \cos b + \sin a \sin b \ [\text{"}]\text{*}$$

My query is, how the latter two formulas are deduced from the former?[34]

Mrs. Somerville patiently responded ten days later, excusing her tardiness on the grounds that she had been busy: "When you are more in the habit of using sines and cosines you will readily make out all the transformations but till that time I shall have great pleasure in giving you what aid I can. The formulae proposed are...."[35] Mrs. Somerville then proceeded to make the required deductions by suitable multiplication, addition, and substitution. Her letter concludes with the words, "I rejoice to hear you go on so well, indeed I never saw you so strong and I have no doubt that your health will now be perfectly established." The unpleasantness of early pregnancy was past.

---

*Here R, the radius, must equal 1.

Now Ada pressed on with both mathematics and harp lessons. By the end of the following March she had gone onto solid geometry and was having some difficulties there.

Can you tell me if any *solid* models have ever been made for illustrating some of the Propositions of *Spherical* Geometry, and if so *where* such things are best to be had. Next to this, some extremely good plates on the subject would be a great help. The kind of propositions I refer to are those on the intersections of Circles of the Sphere; for instance the following, which I take from the Spherical Geometry which precedes Lardner's Spherical Trigonometry:
If three great circles intersect, they will divide the Sphere into eight spherical triangles. Each of the hemispheres into which any one of the circles divides the sphere will be divided into four spherical triangles which will be respectively equal side for side and angle for angle with the four triangles of the other hemisphere. Every two of these triangles have an angle in one equal to an angle in the other, and the sides opposite to these angles respectively equal, while the remaining sides and angles are supplemental.
Also here is another. . . .[36]

The interpreter of Laplace to the students of Cambridge and connector of the physical sciences for her countrywomen readily assured Ada that she would try to find the desired teaching aid with Babbage's help. Apparently such models were not as easy to come by as they had been in the past. Then she added, "Pray don't let the circles turn you crazy till we meet, for I am sure I can explain them to your satisfaction viva voce, though I doubt of my talents that way on paper."[37] In her reply Ada asked that the search be carried on with her own identity concealed. The curiosity she inspired as Byron's daughter, added to the reserve required of all respectable ladies, required extreme precautions against publicity.

I had not imagined I was forgotten, Mr. Babbage having mentioned in a letter to Lord King that he was making enquiries by your desire. I only hope the person you mention may be forthcoming, and should of course be too happy to remunerate him . . . I should not wish *my name* to be mentioned to him, if the enquiry was made. . . . I shall soon be able to have some mathematical talk with you. . . . I am very well satisfied with the progress I have made in the last 6 months, though I daresay it would appear to you small. How much I should like to have a mathematical child, and only think what pleasure I should have in teaching it, and how capable I might hope to be too by the time it was old enough (for I should not begin I think the 1st year).[38]

Obtaining the models proved almost as stubborn a problem as those they were intended to solve; in December 1836 we find Ada still negotiating for them. She thanked Mrs. Somerville for a list she had sent, but insisted on ordering through her, as she still did not wish the maker to know for whom they were intended. The following month she still had not received her order, for she wrote Mrs. Somerville that she was not in immediate want of them and would rather wait until that lady could bring them in person. It is strange that Ada, who had received the customary instruction in drawing and painting, did not think of attempting to construct models or draw the figures herself.

After that, Ada's pursuit of mathematics seems to have been superseded for a while by her interest in the harp. In June she reported that she was going "at a very snail's pace in mathematics. I should be devoting some hours to it now, but that I am at present a condemned slave to my *harp*, no easy task master either. . . . I play 4 or 5 hours generally, & *never* less than 3."[39] At this time her first child was just a year old and she was already well along in her second pregnancy, which she had hoped to postpone for a while.

The truth is, Ada found Mary Somerville quite daunting. "Do you know," she had once written, "I dare not disobey you for the world?"[40] Their intercourse, though frequent over a period of years, never attained the ease and confidence she was to enjoy with Babbage and with Mrs. Somerville's son. Except for her passion for learning which for a long time she kept as discreet as possible, Mary Somerville was a very conventional woman. Her autobiography reveals that she was proud of her ability to cook and sew, of her taste and appearance. Her career as a scientific writer and her subsequent fame had come upon her unsought and unexpected.

Moreover, much as she had encouraged Ada's scientific and mathematical interests, if one takes the bulk of the surviving correspondence into account, she was rather more encouraging along the ordinary lines of feminine activity, including some that became quite distasteful. Even in 1834, in a letter that commended Ada on her enthusiasm and industry in copying papers on the steam engine, she told Ada that a cap the latter had made for her pleased her "the more as it shows me a mathematician can do other things besides studying x's and y's."[41] After Ada's marriage, pregnancies and babies claimed a large share of epistolary attention, to which Ada dutifully responded. "You manage these matters so well," Mrs. Somerville wrote her at one point, "that you are just the person to have a large family and I congratulate you on being in the fair way for it."[42] Mary Somerville had herself borne six children, of whom three had not survived childhood. Ada did not dare tell her how she felt about having a large family.

If Ada had hoped that marriage would bring some of the freedom she had looked forward to when she should come of age, it was not long before her hopes were dashed. The guardianship that Lady Byron claimed to be relinquishing was actually drawn tighter by the affectionate, almost rapturous alliance that at once developed between mother and husband. William had been on bad terms with his father, was on bad terms with his mother and younger brother. Lady Byron became at once his mother and his partner in the managing of their estates and in the parenting of Ada. The baby-talk style that had characterized Lady Byron's letters to her husband during their brief year of cohabitation was now revived in corrrespondence between Ada and her two guardians. As Byron had been "Duck," and his sister Augusta "Goose," so now Lady Byron became "Hen." The younger generation at least escaped the barnyard: William became "Crow," and Ada "Thrush," "Bird," or "Avis." On a visit to her mother during her first year of marriage, Ada felt constrained to ask her husband for permission to accompany her mother to church in these terms: " . . . there is here a famous Unitarian preacher, whom the Hen thinks of going to hear next Sunday. Would you object to my going? Mama says that I may go separately if I please, but that *in your absence*, she cannot be so far responsible for my proceedings as to take me with her."[43] (Lady Byron was making an exaggerated display of the transfer of guardianship; although neither ever joined the church, Lord Lovelace, like Lady Byron, leaned toward Unitarianism at this time.)

For several years Ada seems to have tried faithfully to model herself after her mother: to earn a reputation for great learning and ability but to use her talents in the manner of an executive or administrator rather than attempting a career of her own. In 1836 Lord King opened an experimental school on their family estate at Ockham, in imitation of the one Lady Byron had established at Ealing Grove. Ada wrote her mother that she wished to commission the writing of four textbooks:

1. A good & amusing work on geography. —
2. Some simple book calculated to explain to *parents* the advantages of education for their children. . . .
3. Such a book as Combe's [i.e., phrenological], but adapted to young persons. . . .

4. A book instructive in the arts of economical *cooking* & cottage household management. Cobbitt's Cottage Economy . . . is too mischievous & absurd in many things. . . .
Our school is doing *so well*, that I am very anxious it should do *much better.*—There is a paradox.[44]

There were within Ada's purview, and even in her acquaintance, several women whom she considered capable of writing such texts as she needed. In her next letter she contemplated asking Sophia De Morgan to write one or two of them "since she had 2 medical connections." Miss Harriet Martineau, she thought, could write the book for parents "if she would avoid any of her queer theories." The celebrated journalist and popularizer of political economy was also a friend of Lady Byron's. The "queer theories" to which Ada might have been referring here are suggested by a letter Lady Byron wrote to another friend at about this time:

I lament some of her [Martineau's] theories, and none more than her female emancipation tenets. I consider them injurious to those she desires to befriend for these reasons.
1. That they lead women to undervalue the *privilege* of being exempted from political responsibility.—
2. That by stimulating them to a degree of exertion, intellectually, for which their frames are not calculated, the ability to perform private duties is impaired—Mind is transferred from its legitimate objects.
3. That by mixing imaginary with real grievances, the redress of the latter is retarded.—
I would however allow that some women, like Miss M herself, may be fitted for a wider sphere of action and influence, and her career shews that it may be obtained by them. If she cannot speak in the Senate, she no doubt influences Senators by her works. If she cannot vote, she effects more by instructing others in the duties of the elective franchise.[45]

Harriet Martineau was born in 1802; by the date of Lady Byron's letter she had been instructing her countrymen in the duties of the elective francise and much else for nearly a decade. She was the sixth of eight children of a Norwich manufacturer of Unitarian persuasion. Her career shows more about what was exceptionally than about what was regularly obtained by women. Her childhood, as she described it in her autobiography, shows that Ada's was by no means extreme or unusual as illness, despotism, and misery in genteel child rearing went. "Cheerful tenderness," she asserted, "was in those days thought bad for children," and added, "My moral discernment was almost wholly obscured by fear and mortification."[46]

Like Ada, she was not physically beaten, but her mother believed in ridicule and disparagement as a means of inculcating proper humility and obedience in the young. The numerous older children were also permitted to tyrannize the younger ones. Reflecting on her wretchedness, Harriet early developed a passion for justice, not only for herself but for other children and even for the servants. One of her earlier remembered miseries was that of being sent to deliver insulting and humiliating messages to the family maids. Later in life she became an energetic champion of the poor and oppressed, and a pioneer in advocating kindess to children for their own sake and not simply as an efficient means of control.

Educationally, Harriet was fortunate. When the local school for middle-class Unitarian children was threatened by a shortage of boys, Harriet and her sister were among several girls admitted to take up the empty places. The deafness that developed in her early teens, largely depriving her of the pleasures of music, fed her habit of voracious reading. Much reading, as frequently happens, led to writing, and by 1829, when the family fortunes foundered irrecoverably, she had already begun her forays into didactic and admonitory "authorship" that then blossomed into a widely acclaimed and remarkable journalistic career. Of the catastrophe that might have driven her, like so many gently reared young women, reluctantly into governessing — from which her deafness barred her — she alone could later say with satisfaction, "I, who had been obliged to write before breakfast, or in some private way, had henceforth liberty to do my own work in my own way; for we had lost our gentility."[47]

Before the period of the family financial debacle, a neighbor had made a fateful loan of a copy of Mrs. Marcet's *Conversations on Political Economy* to Harriet's sister:

I took up the book, chiefly to see what Political Economy precisely was; and great was my surprise to find that I had been teaching it unawares, in my stories about Machinery and Wages. It struck me at once that the principles of the whole science might be advantageously conveyed in the same way, — not by being smothered up in a story, but by being exhibited in their natural workings in selected passages of social life.[48]

Having discovered that she had been speaking prose unaware, she set out to do so consciously and deliberately. The result, *Illustrations of Political Economy*, when published as a series beginning in 1831, made her famous. In these simple tales she promoted the classical economics of free trade and laissez faire at a point when Britain was far enough ahead of the rest of the world in industrialization to enjoy the competitive advantages to be gained by dismantling the structure of protective legislation erected at an earlier time. Popular interest in the subject ran high.

Harriet Martineau was fortunate that many of her later enthusiasms also resonated to peaks of public interest. For forty years she continued to instruct, inform, admonish, and improve her fellow citizens, reaping the satisfaction of a useful life conducted according to her always moralistic, often shrewd, sometimes eccentric or controversial views — even as these developed and changed over time. In due course, at one point or another, she championed the abolition of slavery, female suffrage, socialism, religious free-thinking, and mesmerism. But although she was much lionized, had great confidence in her own decided opinions, and actually achieved the fame that Ada longed for, her view of her own ability, expressed in the obituary she thoughtfully provided for herself, seems singularly free of "queer theories":

Her original power was nothing more than was due to earnestness and intellectual clearness within a certain range. With small imaginative and suggestive powers, and therefore nothing approaching to genius, she could see clearly what she did see, and give clear expression to what she had to say. In short, she could popularize, while she could neither discover nor invent.[49]

Whether or not Ada really shared all her mother's views in 1837, it was only reluctantly that she considered writing her own text, and, as in the case of the geometry book, it was a project she did not actually carry out. Interestingly enough, a decade later Miss Martineau did publish a book for parents, though perhaps not the one that Ada had in mind.

The school at Ockham was an elaborate one.[50] It included three and one half acres of agricultural land and its own experimental forest. (To the end of his life William retained a love of exotic trees so great that his tenants complained their crops suffered from the excessive shade resulting from his enthusiastic tree planting.) In addition to practical work, the curriculum covered grammar, English composition, simple mathematics, linear drawing, history, geography, and the theory of music, together with a bit of natural philosophy, as science was then called. The school had workshops, and, not to neglect the pupils' physical development, a gymnasium was added. Ada's attempts to supervise its construction and design were rather unfortunate, for William wrote to her rather testily at one point:

The ropes are all useless—far too short. My dearest when will you trust my judgment & knowledge a little more—This Gymnasium has cost a good deal in all—some thought—some money & more in vexation & disappointment—our own workpeople have been taken off their business at times they could ill be spared, & yet here we are on Aug 9th as far from the use of it as the day it was put up—besides with the exception of one or two of the ropes there is really nothing in their composition different, better, or safer than I should have provided myself—it a little *chills* one, dearest, to be told that these little matters I am quite unfit to interfere in & that they are to be entrusted to others who end by disappointing us.[51]

Administration rather than direct achievement would be a position she would revert to from time to time; but after the birth of her third child, the irritating differences between her mother's position and her own left her increasingly restless and dissatisfied. Lady Byron had been effectively a widow after the first year of her marriage, in independent possession and control of her person, movements, and child. After the death of her mother, and still more after Byron's death, she had been in control of the income from a large fortune. These circumstances were in no small part a result of her iron will, which had been encouraged from her earliest days by her doting, almost worshipful, parents. Ada, on the contrary, had been ruled from her earliest days by that same iron will. Marriage, after the first euphoria, had proved to be an increase of restriction. Her husband and mother, joining their advantages in law, money, age, and custom—and for many years acting in complete harmony on all matters, great and small—intimidated her and made her attempts at domestic management seem feeble and somewhat ridiculous, even to herself. Early in 1838, while their town house was in the process of redecoration, she wrote to him, "I am afraid you will say I do much better about arranging & ordering *without* you than *with* you & perhaps I *am* then less confused and hurried from having no one to depend on."[52]

Of even greater importance, perhaps, was the mounting loss of control over her own body that she was experiencing. Her health, good only for short periods of time since childhood, had been deteriorating for some time. A severe illness, possibly cholera, after the birth of her daughter in September 1837, had forced her to wean the child early. In March 1838 she was feeling so weak and ill that she decided not to go to London that year, to avoid society. The problem seems to have been menstrual difficulties (she was given to exhaustingly heavy periods), and Dr. Locock, her family physician, in whom her implicit faith was seriously misplaced, advised still another baby as a cure. The experiment was tried, but without the hoped-for results, and at twenty-three she found herself the mother of three children, born within four years, and still in a weakened physical state.

Her one avenue of escape seemed to be that, the precocious child of precocious and gifted parents, she was permitted, encouraged, almost required to be a genius, and it was to the development of that genius that she now turned. There were to be no more children. That this was a deliberate decision is certain, and though no direct mention of the subject is made, either in correspondence or in Lovelace's writings on demography, it is almost equally certain that some form of contraception was employed.

"I am not naturally or originally fond of children," Ada wrote, "& tho' I wished for *heirs*, certainly should never have desired a child."[53] William would never express himself quite so bluntly, but it appeared that he too found the time spent with their children often trying and intrusive on his other interests and activities. We know, from the letter earlier alluded to, that Ada and her husband discussed the question of deliberately conceiving a third child, which implies that they might also have forestalled such a conception. We know that sexual relations, at least occasionally, also continued between them. Some years later, Ada closed a letter to Sophia De Morgan with the words, "I am so thankful at the alarm about *adding to my family* being *a false one*; that I don't in the least mind all I have suffered. I think *anything* better than *that*."[54] Still later, she wrote to her husband, in connection with the discovery of an abscess in her vagina, "Some light has been thrown on my *sexual* inabilities by the present circumstances."[55] More than separate bedrooms was available to upper-class couples interested in family limitation in the 1830s.

The choice of mathematics as a first object of serious study was almost inevitable. Except for the singular circumstance of being the daughter of Byron, it was her interest in mathematics that most distinguished her from other accomplished young women of high birth. Both Ada's governors approved her resolve. They were, surprisingly, willing to make almost any accommodations in the service of her studies. She was able to command long hours uninterrupted by the children's demands, to remain in town for considerable periods while William attended to the business of his country estates and to his duties as Lord Lieutenant of Surrey. She could make herself inaccessible to those she did not wish to see, including, sometimes, her mother. Complacently, Lady Byron wrote to her son-in-law, "Your account of the Avis delights me—In the studies which you so kindly & wisely encourage, she will find a balance to the excitable parts of her brain—Whilst the performance of a Mother's practical duties will prevent her from becoming a mere abstract quantity."[56]

The first person to whom she applied for instruction was of course her old and much-admired friend, Charles Babbage. But there had been a gap in the correspondence between them for a period of over a year and a half, and she knew that he was very much taken up with his calculating machine. So, rather timidly and by indirection, she wrote him in November 1839:

[I have] quite made up my mind to have some instruction next year in Town, but the difficulty is to find the *man*. *I* have a peculiar *way* of *learning* & I think it must be a peculiar man to teach me successfully.

Do not reckon me conceited, for I am sure I am the very last person to think over-highly of myself; but I believe I have the *power* of going as far as I like in such pursuits, & where there is so very decided a taste, I should almost say a *passion*, as I have for them, I question if there is not always some portion of natural genius even.—At any rate the taste is such that it *must* be gratified.—I mention all this to you because I think you are or may be in the way of meeting with the right sort of person, & I am sure you have at any rate the *will* to give me any assistance in your power.[57]

In his reply, dated 29 November 1839, Babbage chose not to take the hint. He told her how busy he was, and then: "I think your taste for mathematics is so decided that it ought not to be checked. I have been making enquiry but cannot find at present any one at all to recommend to assist you. I will however not forget the search."[58]

Ada continued to advert to her request over the next few months, perhaps hoping to change his mind. In February she wrote, "I hope you are bearing me in mind, I mean my mathematical interests. You know this is the greatest favour anyone can do me. Perhaps none of us can estimate *how* great. Who can calculate to *what* it *might* lead; if we look beyond the present condition especially?" After her signature she added her customary caution: "I have always forgotten to tell you, what perhaps you may have already grasped, that in any inquiries for a mathematical Instructor, I do not wish my name to be mentioned."[59]

Of those within her intimate circle to whom she had turned for help in mathematics in the past, none now remained accessible and congenial. Mr. Frend, her mother's old mathematical mentor, to whom she had occasionally turned with scientific questions in the past, was now in his eighties and very frail. Besides, his mathematics was so old-fashioned, he did not believe in negative numbers. Dr. King's knowledge too she had readily outdistanced years before; and his piety and preachiness at this point would be anything but welcome. Mary Somerville was by then living abroad. In 1838, Dr. Somerville's health had suffered a breakdown, and almost at the same time the Somervilles' capital was lost through imprudent investment. The combined circumstances led them to travel on the Continent and to settle in Italy, where many of the English middle class found they could live more economically than at home.

Even if Mrs. Somerville had remained in London, it is doubtful that Ada would have turned to her in 1839 for mathematical instruction. She had by then realized that the normal domestic and social excitements, even supplemented by study, were not enough to "stop up the void." The pleasure she had anticipated in teaching her children was not to materialize. Her tastes were too different, as she wrote to a prospective governess the following year:

I am admirable as an *organizer, director* of *other superintendants.* . . . How many moments there are when their presence must be irritating & intolerable to me. . . . Add to this my total deficiency in all natural *love* of children . . . and an exceedingly delicate and irritable nervous sytem . . . and you will not wonder that I begin to feel them occasionally (to speak plainly but truly) a real nuisance. . . . I believe I am fit to *educate*, with proper aids. But . . . as the *Chief*, the general.[60]

It began to look as if the search for "the desired Mathematical Man, the *Great Unknown*," might be as protracted as the search for the wooden models had been. Ada's husband, who had been created Earl of Lovelace in Queen Victoria's coronation honors list the previous year, had suggestions for alleviating her restlessness that were more in line with his own interests. "W[illiam] wants to set me about an Essay on Planting," she wrote her mother, almost in despair, in March 1840. Yet by June of that year "the man" had been found. It was Ada's fate to be the most fortunate and unfortunate of creatures, and no circumstance of her life was luckier or happier than that which secured her the services of Augustus De Morgan to be her mathematical instructor, at a time when he himself was in the process of making fresh contributions to a revitalized subject.

Most probably the arrangement was effected through the offices of Mrs. De Morgan, Lady Byron's lifelong idolizer and Mr. Frend's daughter. In her memoir of her husband she mentions that although she and her father were friendly with Mr. De Morgan for ten years before her marriage, in the course of which he met most of her father's acquaintance, she did not introduce him to Lady Byron until 1837. "My husband," she then says, "afterwards gave her daughter, Lady Lovelace, then Lady King much help in her mathematical studies."[61] Actually, Lord King became Lord Lovelace two years before Mr. De Morgan became his Lady's tutor. Mrs. De Morgan was writing forty years after the event.

It is possible that the connection was made by Babbage, though not likely since it came so long after Ada had made her initial request of him. The two men were well acquainted with each other, having overlapped mathematical interests and society memberships. They corresponded but were never really friends, although Babbage may have been unaware that De Morgan actually disliked him. As early as 1834 De Morgan had written Mr. Frend:

I was very sorry to find when I came home that Mr. Woolgar had been very uncourteously received by B– with my note. That unfortunate man will never rest until he succeeds in getting nobody's good word. He calculates very wrong (for a calculating machine maker) if he thinks such a thrower of stones as himself can stand alone in the world. It takes all his analysis and his machine to boot to induce me to say I will ever have any communication with him again, which nothing should induce me to do except the consideration that men of real knowledge should have more allowance made for them than some charlatans I know.[62]

They were often friendly with people on opposite sides of the disputes in which they became entangled. George Biddell Airy, who had declared the calculating machine a humbug and had been Babbage's successful rival for the Lucasian professorship in 1826, was De Morgan's teacher and lifelong friend. In the quarrel between the Reverend Richard Sheepshanks and Sir James South (over the mounting of a telescope), Babbage remained loyal to his friend Sir James, while De Morgan sided with Sheepshanks. The unkindest cut probably came in De Morgan's article in the *Penny Cyclopaedia* entitled "On Helps to Calculation." Dated 1846, it began: "[O]f all disgusting drudgery, numerical calculation is the worst: a combination of all the worry of activity with all the tediousness of monotony and all the fear of failure." De Morgan went on to examine a variety of devices, both mental and mechanical, to ease the drudgery: slide rules, tables, lightning calculators (that is, persons trained to perform mental arithmetic with great rapidity), even decimal coinage. No mention was made of Babbage's calculating engines. Considering how eager Babbage was to keep his inventions before the public eye, the omission can only have been deliberate.

There are striking parallels in the lives and characters of the two men, as well as interesting contrasts in the ways in which some of the same themes were played out. Both consciously organized their lives around positions of principle, to which both sacrificed much. But it was Babbage who became embittered, while De Morgan, who lost more and died broken, was known for his sweetness of temper. Both were generous men, who warmly recognized and forwarded those who succeeded in enterprises in which they themselves had been less fortunate. Both were of that upper-middle-class extraction that caused them to stand fiercely upon the status of "gentlemen," but De Morgan was a relatively poor man.

On his mother's side De Morgan was descended from James Dodson, a noted mathematician in his day and author of a work called the *Anti-logarithmic Canon*. It is a mark of the social snobbery of the early nineteenth century that De Morgan recalled once asking his aunt who James Dodson was and being told, "We never cry stinking fish." Dodson had been a teacher at Christ's Hospital, a charitable institution, though one renowned for its academic excellence.

De Morgan was born in 1806 in Madura, India, where his father was a colonel in the East India Company's service. Soon after his birth he lost the sight of his right eye to a tropical infection called "sore eye." This proved a blessing in disguise, for it exempted him (in his mother's eyes) from following the military profession of his forebears, which would not have been to his taste. His father died when he was ten, and his mother hoped he would become a clergyman, but the length and boredom of the hours spent in church during his childhood made him incapable even of listening to sermons as an adult. He steadfastly opposed all his mother's plans for a religious career, while obediently reading all the books on the subject she sent him. He refused to join any church while he lived, and saved an open confession of faith for his will, "because," as he declared there, "in my time such confession has always been the way up in the world."

Like Babbage and Mary Somerville, he had discovered mathematics almost by accident, but his predilection was not at first hindered, and he soon "read Algebra like a novel." He entered Trinity College, Cambridge, at age sixteen, intending to study classics, which he soon slighted in favor of mathematics and "loose reading." As a result of the latter he was graduated only fourth wrangler, although he was acknowledged by his teachers to be the best student of his year. His mother was disappointed, and he developed criticisms of the education and examination systems that were to color his career as a teacher.

Unlike Babbage, who had considered becoming a clergyman despite his scorn of piety and the established church, De Morgan felt he could not take a master's degree or expect a Cambridge appointment because he could not conscientiously subscribe to the required Thirty-nine Articles. It then seemed that he had only medicine and law left to choose between, and despite the qualms he felt over the conflicts of interest that could arise for the practitioners of either, he decided that law was in the long run less likely to trouble his already overdeveloped sense of honor. Accordingly, he began to read law, but soon after had the opportunity to apply for the post of mathematics professor at the newly opened University College, then called London University. It was an institution funded by subscription, mostly by well-to-do business men, whose purpose was to provide a first-class education without reference to religious affiliation; it offered an unprecedented opportunity to Jews and dissenters, to whom the doors of Oxford and Cambridge were closed. It did not even exclude women, although no women students appeared. Even without women the nondenominational character of the new university aroused such horror among conservatives that King's College was then founded to afford an alternative education in London to true believers.

In thus changing his profession, De Morgan once more disappointed his friends—a term used to designate those who are presumed to have a concern for one's welfare, an obligation to forward one's interests, and a right to meddle in one's decisions. With relief he turned his back on a promised brilliant and lucrative legal career, determined to "keep to the Sciences so long as they will feed me."

Because the new university was governed largely by a group of business men rather than academics, touchy questions respecting the autonomy and dignity of the faculty soon arose. When, in 1831, Charles Bell, the professor of anatomy, one of the foremost men in his field, was dismissed on complaint from some of the students, but for no misconduct or professional deficiency, De Morgan, representing the faculty, wrote a letter of protest to the board of governors and resigned. He then spent five years in the wilderness, supporting himself by tutoring private pupils and by actuarial consultancies. In 1836 his successor at the university was accidentally drowned, and De Morgan offered himself as a temporary replacement; he was the only available substitute who could take over the dead man's classes on such short notice. He soon convinced himself that the issue over which he had resigned had been resolved, and was reinstated on a permanent basis.

At this point he apparently felt himself well enough established to take a wife, and in 1837 he married the ex-prodigy whom Mr. Frend believed he had reduced to suitable mediocrity. Seven "young philosophers," as Babbage termed them, were produced with alarming rapidity. De Morgan's income from teaching never exceeded £500 and later declined to £300; it depended on the size of his classes. He was obliged to supplement it by taking private pupils and by writing, of which he did a prodigious amount. He contributed over 850 articles to the first edition of the *Penny Cyclopaedia* (a publication of the Society for the Diffusion of Useful Knowledge, with which he was actively involved); as later listed by his widow, these ranged from "Abacus" to "Young, Thomas." He wrote biographies of Newton and Halley. He contributed serious articles to the *Dublin Review* and satirical reviews to the *Athenaeum*. He wrote 33 articles for the *Quarterly Journal of Education*, and was an indefatigable obituarist and commentator. All this and much more he did in addition to his formal writings on mathematics and logic. Fortunately, his notion of relaxation included letter writing.

When not actually seated at his desk, De Morgan, like Babbage, was an inveterate club man, a founder and officer of scientific societies. Unlike Babbage, he never seems to have quit in disgust when these organizations failed to live up to their original purposes and high standards. Instead he remained to shoulder much of the organizational drudgery, content to free the titular head of such duties. When, for example, he was offered the presidency of the Royal Astronomical Society, he declined it to retain his post as secretary, ostensibly on the ground that his sightless right eye prevented him from becoming a practicing astronomer. The candidate he favored was Sir John Herschel, whose activities kept him on long voyages and out of England much of the time. "The President," he said, "must be a man of brass, a micrometer-monger, a telescope-twiddler, a star-stringer, a planet-poker and a nebula-nabber."[63] But of course the secretary set the agenda.

Like Babbage, he was remembered for his wit; but to judge from the samples of each that have come down to us, De Morgan's was by far the more labored and pedantic, though perhaps less arch. While Babbage was something of a courtier, and much in society, De Morgan was above all a tireless teacher. Even his satirical reviews for the *Athenae-um*, whose ostensible purpose was to poke fun at the ignorant dabblers in science who presumed to fancy that they had succeeded in solving age-old problems in the learned disciplines, were actually weekly lessons for the public in critical and logical thinking on these issues.

As a teacher of mathematics De Morgan paid careful attention to method and to the logical, linguistic, and contextual aspects of the subject. He despised what he considered the hypocritical pretense that an anonymous examination (easily got around, should the professor be so minded) was the fairest method of judging the quality and promise of a student. He felt the system relieved the instructor of responsibility without really protecting the student against favoritism. Instead he favored forthright evaluations in explicitly stated context:

The truth is, that examination in mental merits, by help of words which are to be construed independently of all knowledge of the mind they came from, is an absurdity. It would be absurd to read Locke or Paley in this way. An author, to be understood, requires appliances [applications] to be derived with all his works: and requires more, but more cannot be got. And if this be the case with the well considered works of matured minds how much more is it wanted in the case of the half formed language of beginners, whose minds are so often charged with more than they can adequately express—more often that not.[64]

These views were carried into his work as a mathematician, which contrasted interestingly in method with that of Babbage. Babbage's work was characterized by bold inventiveness and virtuosity in the manipulation of symbols, which encouraged him to generalize his results beyond those domains rigorously covered by the assumptions and definitions on which the results were based and to borrow devices from other sources, only half suggested by the problem at hand. De Morgan, in contrast, was concerned to clarify and establish on a rigorous logical foundation procedures that had for some time been in common use before extending and generalizing them in new domains. He wrote textbooks on arithmetic, algebra, and trigonometry. His first publication in logic, a pamphlet entitled *First Notions of Logic*, appeared in 1839 and was intended to assist his students in the principles of sound mathematical reasoning. When he began to guide Ada's studies he was in the course of completing his *Differential and Integral Calculus*, which was published in 1842. As customary in his texts, De Morgan there discussed the fundamental principles on which the subject was based. His text included the first precise analytical definition of a "limit," the concept from which the calculus, for the first time, was there rigorously derived. The same work also included a discussion of infinite series and his own test to determine convergence (that is, whether an infinite series has a finite sum) when the ordinary tests fail.

The remaining correspondence between Ada and De Morgan is clearly incomplete, full of gaps, and sometimes very difficult to order. Many of the letters are not dated, but those that are fall in a period between July 1840 and December 1842. They demonstrate that Ada enjoyed the rare privilege of querying an author about his work when the subject matter was still fresh in his mind, when the subject itself was still relatively unsettled and exciting, and when the author was among those engaged in developing it. They also show that De Morgan was a gifted and patient teacher and Ada an enthusiastic, active, and adventurous student, even though they do not take her past the first steps in the differential and integral calculus.

In what seems to be the earliest letter in the series, De Morgan responded to a query of Ada's concerning the nature of zero:

Zero is *something*, though not some *quantity*, which is what you here mean by *thing*. Some writers in the differential calculus use not 0, but 0/0, in an absolute sense, as standing for a quantity. . . . [De Morgan here argues that this is incorrect, because 0/0 cannot be evaluated without additional procedures.] Absolute modes of speaking, which are false, are continually used in abbreviation of circumlocutions.[65]

In this passage he called her attention to the assumptions being made simply by the use of a particular term. Shortly thereafter he sent her the proofs of his calculus text, with the remark that just at the beginning the reader is expected to learn differentiation "by rule for the present, as an exercise in algebraical work." This of course would not do for Ada, who immediately had questions about the nature of differential and integral calculus, to which he willingly responded:

The Differential and Integral Calculus deal in the same elements but the former separates one element from the mass and examines it, the latter puts together the different elements to make the whole map. . . . It is thought that Newton and Leibnitz had some remarkable new conception of principles, which is not true. Archimedes and others had a differential and integral calculus, but not an algebraical system of sufficient power to express very general truths.

Many persons before Newton knew, for instance, that if $\frac{(x + h)^n - x^n}{h}$ could be developed [evaluated when h approaches zero] for any value of n, the tangents of a great many curves could be drawn and they knew this upon principles precisely the same as Newton and Leibnitz knew it. But Newton *did* develope $\dfrac{(x + h)^n - x^n}{h}$ and *did* that which they could not *do*. It was the addition to the powers of algebra in the seventeenth century, and not any new conceptions of quantity which made it worth while to attempt that organization which has been called the differential calculus.[66]

Here he makes the point that the infinitesimal calculus was developed on a pragmatic basis before it could be rigorously justified. (De Morgan was candid with her in discussing the vexed problems of divergence, on which mathematicians had not yet reached consensus: "It is fair to tell you that the use of divergent series [i.e., series whose terms do not clearly become smaller and smaller as the series progress] is condemned altogether by some modern names of very great note. For myself I am fully satisfied that they have an *algebraical* truth wholly independent of arithmetical considerations.")[67] De Morgan was an antiquarian and historian of mathematics as well as a mathematician, and glad enough to go into these matters for her; however, he continued, she must not skip over essential stages in her haste to understand:

I should recommend you decidedly continuing the Differential Calculus, warning you that you will have long digressions to make in algebra and trigonometry. . . . In the meanwhile, as mechanical expertness in differentiation is of the utmost consequence, and as it is the most valuable exercise in algebraical manipulation which you can possibly have, I should recommend you thoroughness and keeping up the chapter you are now upon.

The fact is that Ada had been experiencing some of her old difficulties in symbol substitution and in the mechanical working out of the results. At one point she commented,

It had not struck me that, calling $(x + \theta) = v$, the form $\dfrac{(x + \theta)^n - x^n}{\theta}$ becomes $\dfrac{v^n - x^n}{v - x}$. And by the bye, I may remark that the curious *transformations* many formulae can undergo, the unsuspected & to a beginner apparently *impossible identity* of forms exceedingly *dissimilar* at first sight, is I think one of the chief difficulties in the early part of mathematical studies. I am often reminded of certain sprites and fairies one reads of, who are at one's elbow in *one* shape now, & the next minute in a form most dissimilar; and uncommonly deceptive, troublesome and tantalizing are the mathematical sprites & fairies sometimes; like the types I have found for them in the world of fiction.[68]

Her attempts to diagnose her problems are revealing of her penchant for introspection and her satisfaction with what she found:

I cannot help here remarking a circumstance which I think is almost invariably true respecting *all* my difficulties & confusions in studying. They are without *any* exception that I can recall, from misapprehension of the *meaning* of some symbol, or of some phrase or definition; & on no occasion from within any *error in my reasoning*, or from any *difficulty* in carrying on chains of deductions correctly, however complicated or profound or lengthy these may be.[69]

Her professor's views were somewhat different, but at this point he contented himself with remarking that "perhaps your *ultimate difficulties* will not be altogether the same as those you commence with." He then returned to her an exercise in which she had calculated some values of the variable x^2 (1, 9/4, 4, 25/4, 9, and so forth), drawn vertical lines proportional to these values, and connected the upper ends of these lines with a curve. A note in her hand says that the vertical lines represent the function x^2, but she cannot see their relation to the curve, or why he says that the *curve* represents x^2. A note in his hand says that this curve "and no other belongs to $y = x^2$." Armed with her new insight into her own thought processes, she wrote back,

I am afraid I do not understand what you were kind enough to write about the Curve; and I think for this reason,—that I do not know what the term *equation to a curve* means.—Probably with some study, I should deduce the meaning myself; but having plenty else to attend to of more immediate consequence, I do not like to give my time to a mere digression of this sort.—I should like much at some future period, (when I have got rid of the common algebra & Trigonometry which at present detain me), to attend particularly to this subject.— At present, you will observe, I have four distinct things to carry on at the same time:—the Algebra;—Trigonometry;—Chapter 2nd of the Differential Calculus;—& the mere practice in Differentiation.[70]

Again and again the difficulty of assimilating the answers to her abstract questions in concrete form and specific applications was to plague her. In his answer De Morgan cautioned her that she must "distinguish in algebra questions of *quantity* from questions of *form*." He then went on to define the equation of a curve, presumably satisfactorily, since this particular question does not recur, although on several occasions he was to remind her not to confuse quantity and form. He strongly recommended Peacock's book on algebra.

Peacock's *Treatise on Algebra*, published in 1830, was the first book on modern algebra, in that the subject was treated in its own right and not simply as arithmetic with unknown quantities. Many of the philosophical ideas contained in it appeared first in an unpublished manuscript by Babbage, which the latter called *The Philosophy of Analysis*, apparently written before calculating engines displaced his work in pure mathematics.[71] George Peacock had been, next to John Herschel, Babbage's closest friend at Cambridge, and a collaborator with him in his schemes and publications to reform and improve the standards of British mathematics. De Morgan had been a student of Peacock's at Cambridge in the mid-1820s, when the subject of algebra was as fresh for Peacock as that of the calculus was for De Morgan in 1840. De Morgan adopted Peacock's view of algebra and extended it in his own work.

In the same letter in which she demanded a definition of *equation to a curve*, Ada also complained of impatience with the progress she was making:

I wish I went on quicker. That is—I wish a human head, or *my* head at all events, could take in a great deal *more* & a great deal more *rapidly* than is the case;—and if *I* had made up my own head, I would have proportioned it's wishes & ambition a little more to it's capacity.—When I sit down to study, I generally feel as if I could *never* be tired;—as if I could go on forever.—I say to myself constantly, "Now today I will get through so & so"; and it is very disappointing to find oneself after an hour or two quite wearied & having accomplished perhaps one twentieth part of one's intentions—perhaps not that.—When I compare the *very* little I *do*, with the *very* much—the infinite I may say—that there is to *be done*;—I can only hope that hereafter in some future state, we shall be cleverer than we are now.

He comforted her with the same *festina lente* that she had once used to admonish Annabella Acheson, and added a spot of history:

Festina lente, and above all never estimate progress by the number of pages. What you say about comparison of what you do with what you see can be done was equally said by Newton . . . and the last words of Laplace were "Ce que nous connaissons est peu de chose; ce que nous ignorons est immense" [What we know is little; what we know not is immense]. So that you have respectable authority for supposing that you will never get rid of that feeling; and it is no use trying to catch the horizon.[72]

Soon her anxiety over her rate of progress abated, and she was able to write, "I am much pleased to find how very well I stand *work*, & how my powers of attention and continued effort *increase*. I am never so happy as when I am really engaged in good earnest; & it makes me most wonderfully cheerful & merry at *other* times, which is curious & very satisfactory." She even became philosophical about her errors: "I now see exactly my mistake. . . . I used once to regret these sorts of errors, & to speak of *time lost* over them. But I have materially altered my mind on this subject. I often gain more from the discovery of a mistake of this sort, than from 10 acquisitions made at once & without any kind of difficulty."[73] She sometimes thought out loud in writing to him:

I object to the necessity involved of supposing x to be *diminished without limits*, a supposition obviously quite necessary to the completion of the *Demonstration*. It has struck me that though this supposition leaves the Demonstration & Conclusions perfect for the cases in which x *is* supposed to *diminish without limits*, yet it makes it valueless for the many in which x may be anything which does *not* diminish. —No— by the bye, I think I begin to see it now; I am sure I do. It is as follows: —the supposition of x diminishing without limits is merely a *parenthetical one*, by means of which a limit for a certain expression $\frac{a^x - 1}{x}$ is deduced under those circumstances; & then the argument proceeds, that having already obtained in another place, a different limit for this same expression under the same circumstances, we at once deduce the *equality of the two limits*, from whence follows &c, &c. Thus this supposition of x diminishing without limits, is not a portion of the *main argument*, but only a totally independent and parenthetical hypothesis introduced in order to prove something else which *is* a part of the *main argument*. —Yes—this is it, I am sure. I had had the same objection to the Demonstration in *Bourdon* to which I have had the curiousity to refer. [The 1828 translation of Bourdon's *Elements of Arithmetic* was De Morgan's first publication.] I am sometimes very much interested in seeing *how* the *same* conclusions are arrived at in *different ways* by different people; and I happen to have been inclined to compare you & Bourdon in this case . . . and very amusing has it been to me to see *him begin* exactly where *you end*, &c.[74]

Ada refused simply to follow a derivation by rule but insisted on trying her own chain of inferences. Occasionally she was bold enough to take exception to parts of his calculus text and even to offer improvements:

I shall be exceedingly obliged if you will also tell me whether a little Demonstration I enclose as to the Differential Coefficient of x^n is valid. It appears to me perfectly so; & if it is, I think I prefer it to yours in page 55. It strikes me as having the advantage in simplicity, & in referring to fewer requisite *previous* propositions.[75]

His inelegant presentation was duly explained:

Your proof of the diff. co. of x^n is correct, *but it assumes the binomial theorem.* Now I endeavour to establish the diff. calc. without any assumption of an infinite series, in order that the theory of series may be established upon the differential calculus.

Besides, if you take the common proof of the binomial theorem, you are reasoning in a circle, for that proof requires that it should be shown that $\dfrac{v^n - w^n}{v - w}$ has the limit nv^{n-1} as w approaches v. This is precisely the proposition which you have deduced *from* the binomial theorem.[76]

On another occasion she termed one of his passages "awkward and inconvenient" and demanded to know why certain things were taken as assumptions while others were rigorously proved.

But these inconsistencies have always struck me occasionally, and are perhaps only in reality the inconsistencies in a beginner's mind; & which long experience & practice are required to do away with . . . I fancy a great proficiency in Mathematical Studies is best attained by *time;* — *constantly* & *continually* doing a little. — If this is so, surely then the University examining system must be very prejudicial to a real progress in the long run, particularly when one considers how *very very* little School-boys are generally prepared on first going to the Universities, with anything like distinct mathematical or even arithmetical notions of the most elementary kind.[77]

Wherever she drew these observations from, they were surely most congenial to his own views. Her studies with De Morgan could not have been in sharper contrast to her lessons with Dr. King. She wrote to her mother, "No two people ever suited better," adding that she could never repay him for his kindness.[78]

Although Professor De Morgan did take private pupils and needed the income to support his expanding family, there is no suggestion that he ever received more in payment for his services to Ada than an occasional gift of game from Lord Lovelace (a courtesy connoting social parity). He was also able to enroll Lovelace in one or two of his pet projects, including a society for the printing and exchange of scientific manuscripts, which, he explained, would be of use to Ada in her studies.

In another letter De Morgan spoke of including Lovelace among the members of the "Historical Society." William of Occam, the medieval theologian, took his name from the same place as the Lovelaces' elder son, he added; did Lord Lovelace have any papers relating to him? This was in reference to the fact that the renowned Schoolman was born in the same Surrey village where the King family seat was located. Now that Ada's husband was an earl, her elder son was known by courtesy as Viscount Ockham. "The Old Ockham," mused De Morgan, "will be a good example for the young one, though he was a monk, as I suppose. I would have been nothing else had I lived in his day."[79]

Did he already regret his marriage and his increasing family responsibilities? Sophia De Morgan outlived both Ada and her husband, and left memoirs of both. In her account of her husband she omitted almost all family correspondence and domestic details. What remains, however, suggests that there were many matters on which they disagreed and not many interests and activities they shared. De Morgan soon refused even to accompany his family to the country for summer holidays. A confirmed Londoner, he considered Blackheath "a miserable scene of desolation." Ada, too, increasingly wished to escape her husband's rural seats for the more stimulating atmosphere of London. Mrs. De Morgan's reminiscences of Ada were rather malicious, despite her reverence for Lady Byron, but it was some time before Ada suspected her dislike, and by then she had made her the recipient of some rather incautious confidences.

That De Morgan was, like Dr. King and Mrs. Somerville, an unpaid friend-tutor is also indicated by the repeated apologies with which Ada made claims upon his time and attention, despite the audacity with which she presented her ideas.

I *have another* enquiry to make, respecting something that has lately occurred to me as to the Demonstration of the Logarithmic & Exponential Series in your algebra, but the real truth is I am quite *ashamed* to send any more; so will at least defer this—I am afraid you will indeed say that the office of my Mathematical Counsellor or Prime

Minister is no joke. . . . What *will* you say when you open this packet? —
Pray do not be *very* angry and exclaim that it really is too bad.[80]

Her attitude toward the tutors, masters, and even the artists she employed was quite different. When Margaret Carpenter, for example,
was painting the portrait that was later exhibited at the Royal Academy,
Ada complained to her mother that Mrs. Carpenter was prone to rest
in the library; she ought to confine herself to her room when not
working. Only doctors and lawyers, particularly the latter, seemed
able regularly to submit bills without losing caste.

Of course De Morgan responded to her, "Pray send your point
about the exponential theorem." And the inquiries kept coming. What
is peculiar about Naperian logarithms? How do they differ from others
in simplicity or use? Where does convergence begin? Has he heard
of a science called descriptive geometry? She ranged from "Practical
Application" to stunning speculation:

You know I always have so many metaphyiscal enquiries & speculations
which intrude themselves, that I never am really satisifed that I un-
dertand *anything*; because, understand it as well as I may, my com-
prehension *can* only be an infinitesimal fraction of all I want to
understand about the many connexions & relations which occur to
me, *how* the matter in question was first thought of or arrived at, &c.,
&c.
I am particularly curious about this wonderful Theorem.

Here then suggests itself to me the question: "Then are there certain
truths & conclusions which can be arrived at by *pure analysis*, & *in no
other way?*" And also, how far *abstract analytical expressions must* express
and mean *something real*, or not. In short, it has suggested to me a
good deal of enquiry, which I am desirous of being put in the way
of satisfying.

It cannot help striking me that *this* extension of Algebra ought to lead
to a *further extension* similar in nature, to *Geometry of Three Dimensions*;
& that again perhaps to a further extension into some unknown region,
& so on ad-infinitum possibly. And that it is especially the consideration
of *an angle* $= \sqrt{-1}$ which should lead to this; a symbol, which when
it appears, seems to me in no way more satisfactorily accounted for
& explained than was formerly the appearances which Bombelle in
some degree cleared up by showing that at any rate they (tho' in
themselves unintelligible) led to intelligible and true results. You do hint
in parts of page 136 at the possibility of something of this sort.[81]

The issue Ada was grappling with here was the latest in a series of
problems in the treatment of mathematical concepts that have no

immediate counterparts in everyday experience (such as negative numbers). This time it was imaginary and complex numbers. Imaginary numbers are the square roots of negative numbers (real numbers, both positive and negative, are the square roots of positive numbers). Complex numbers are complexes of real numbers plus imaginary numbers (which latter can always be represented as the square root of minus one multiplied by a real number). "Bombelle" (Bombelli) was a sixteenth-century mathematician who worked with complex numbers as solutions of cubic equations, though he did not geometrize them as he did the real numbers, which can be visualized as lengths along a line, in one direction (called positive) or the other (negative). Around 1800 it was realized that complex numbers can be visualized as if they were plotted on a two-dimensional plane or graph, with the horizontal coordinate representing the "real" part and the vertical coordinate the "imaginary" part.

De Morgan called his own treatment of the subject "double algebra," and he must have been delighted at Ada's grasping the idea that a two-dimensional treatment could be generalized into three dimensions or more. His attempt to extend the geometrical interpretation into three dimensions with a "triple algebra" was not successful, because he was unable to devise a consistent set of rules for such a system. That achievement was finally effected by his friend William Rowan Hamilton, who was forced to use rules of multiplication different from those of conventional algebra. His system, which he called "quaternions" (he actually wound up with four dimensions), was first announced in 1843.

Ada's mathematical studies, intense and enthralling as they often were, seem always to have been rather sporadic, as she periodically lost interest and returned to them. In April 1841 a major interruption occurred when she went off to Paris to visit her mother there. Her letters attest to the difficulty of "getting my *head in* again" after such interruptions. Then a letter to Sophia De Morgan revealed a disruption of a more ominous kind.

You will I know be interested in how I have gone on. I have been *very ill indeed*, since I saw you. . . . But you shall hear (when we meet) what an extraordinary illness I have had; & what infinite trouble with myself; for there has been no end of the manias & whims I have been subject to. . . .

I should tell you that my Mother knows *nothing whatever* of all this; & that she never must. She has been spared much uneasiness &

anxiety; & now the thing is over; & never likely to recur; for I have gained much in knowledge & experience of my constitution and temperament. *Many causes* have contributed to produce the past derangements; & I shall in the future avoid them. *One* ingredient, — (but only one among many) has been *too much Mathematics.*

I need hardly say that since I returned here I have been *utterly unable* to *think* even of my Studies. I yesterday resumed them; but for some time I must give only from 1/2 an hour to an hour a day. Pray tell Mr. De Morgan all this; (he must wonder at not having heard from me).—But say that I hope to have at least *something* to trouble him with when I go to Town next week; & that I don't despair of getting on very *creditably* this winter, with an hour a day regularly given.[82]

The date on this letter is Tuesday, 21 December, which places it in 1841. Ada later recalled that she had experienced "singular states of brain & nerves in the autumn & winter" of that year, and came to suspect that she had been the victim of "mesmeric experiments," which had caused these "unnatural feelings mental & bodily." In any case, the "derangements" did recur; for several years her moods followed the swings between grandiose exaltation and deep, paralyzing depression that mark a manic-depressive disorder. (This and other possible diagnoses are discussed in the Appendix.) Eventually, it seems, Mrs. De Morgan communicated her growing suspicions not just to her husband but to Lady Byron as well. Mr. De Morgan sent an anxious letter of inquiry to Ada, but her mother, far from showing anxiety, acted at once to deny and suppress any rumors that might have been put into circulation. To Ada she wrote,

You are right, I believe, in your construction of the Professor's very kind letter, and I only hope you will see him & "do justice to yourself," by not appearing less right and reasonable than you are *au fond*—& by not reflecting any of the various diseases which may have been imputed to you—No one who saw you as you are could fancy a *discontinuance* of Mathematics necessary.[83]

To De Morgan she wrote,

Dear Mr. De Morgan
I find both from your letter to Lady Lovelace & from Mrs. De M's today, that you are most likely anxious to prevent Lady L. from injuring herself by mathematical study. But I feel apprehensive that this caution may be carried too far—I have at all times observed that she was the better for pursuits of that description—& if she would

but attend to her stomach, her brain would be capable even of more than she has ever imposed on it. . . . After the weeks which I passed with her at Bristol it is impossible for me to share in the extent of apprehension entertained by some friends who judge from report instead of observation. The consciousness of making progress in science seems to me an essential element in her happiness, & appears not less desirable to Lord Lovelace than to myself.[84]

De Morgan answered immediately:

I have received your note and should have answered no further than that I was very glad to find my apprehension (of being a party to doing mischief if I assisted Lady Lovelace's studies without any caution) is unfounded in the opinion of yourself and Lord Lovelace, who *must* be better judges than I am, on every point of the case but one, and *may* be on that one. But at the same time it is very necessary that the one point should be properly stated.

I have never expressed to Lady Lovelace my opinion of her as a student of these matters. I always feared that it might promote an application to them which might be injurious to a person whose bodily health is not strong. I have therefore contented myself with very good, quite right, and so on. But I feel bound to tell you that the power of thinking on these matters which Lady L. has always shown from the beginning of my correspondence with her, has been something so utterly out of the common way for any beginner, man or woman, that this power must be duly considered by her friends, with reference to the question whether they should urge or check her obvious determination to try not only to reach but to get beyond, the present bounds of knowledge. . . . There is easily to be seen the desire of distinction in Lady L.'s character; but the mathematical turn is one which opportunity must have made her take independently of that.

Had any young beginner, about to go to Cambridge, shown the same power, I should have prophesied first that his aptitude at grasping the strong points and the real difficulties of first *principles* would have very much lowered his chance of being senior wrangler; secondly, that they would have certainly made him an original mathematical investigator, perhaps of first-rate eminence.

The tract about Babbage's machine is a pretty thing enough, but I could I think produce a series of extracts, out of Lady Lovelace's first queries upon new subjects, which would make a mathematician see that it was no criterion of what might be expected from her.

All women who have published mathematics hitherto have shown knowledge, and the power of getting it, but no one, except perhaps (I speak doubtfully) Maria Agnesi,* has wrestled with difficulties and

---

*An eighteenth-century mathematical writer, appointed in 1750 to fill her father's chair as professor of mathematics at the University of Bologna upon his illness.

shown a man's strength in getting over them. The reason is obvious: the very great tension of mind which they require is beyond the strength of a woman's physical power of application. Lady L. has unquestionably as much power as would require all the strength of a man's constitution to bear the fatigue of thought to which it will unquestionably lead her. . . . Perhaps you think that Lady L. will, like Mrs. Somerville, go on in a course of regulated study, duly mixed with the enjoyment of society, the ordinary cares of life, &c., &c. But Mrs. Somerville's mind never led her into other than the *details* of mathematical work; Lady L. will take quite a different route. It makes me smile to think of Mrs. Somerville's quiet acquiescence in ignorance of the nature of force, saying "it is $\frac{dy}{dt}$" (a mathl. formula for it) "and that is all we know about the matter" — and to imagine Lady L. reading this, much less writing it.

Having now I think quite explained that you must consider Lady L's case as a peculiar one I will leave it to your better judgment, supplied with facts, only begging that this note may be confidential.[85]

It is important to examine this letter carefully as evidence of Ada's mathematical ability, particularly in view of the writer's belief that he was supplying "facts" (which included the "obvious" constitutional weakness of women) and his position that student evaluations are best made by assessing the mind rather than the productions of the student. First, it should be pointed out that his judgment was unfair to Mary Somerville; the woman who interpreted Laplace to the English-speaking world was no complacent dilettante. Well aware that her work did not constitute an original contribution, she nevertheless sought and appreciated the underlying unity and significance of physical and mathematical phenomena. This she demonstrated clearly in her correspondence with Faraday (the last physicist of substance who was able to disregard mathematical formulations), among others, as well as in her second book, *On the Connexion of the Physical Sciences*, which had been published ten years before De Morgan, in the passage above, quoted her statement from her first book, the Laplace interpretation.

In his comparison of their views of the nature of "force" (actually the formula is for acceleration), De Morgan presumably meant that Ada would have demanded a physical explanation, while Mrs. Somerville was content with a mathematical description or representation, which he considered less meaningful. But Mrs. Somerville's position at the time was at least a respectable one, having been adopted by Galileo and Newton, and succeeding developments have shown it to be even better. As modern physics has grown more abstruse and come to deal with concepts and entities farther removed from ordinary experience, mathematical description has proved more necessary and fruitful than the verbal descriptions and explanations that constitute the familiar forms of physical understanding;[86] ironically, De Morgan's own work revealed this tendency too.

A further irony was that it was Ada's inability to grasp this mathematical approach, her continued insistence on verbal explanations, that in the end limited her understanding of science and, as we shall see, marred her "pretty tract" on Babbage's machine. Unable to assimilate the symbolic processes with which alone highly complex and abstract matters may be rigorously treated, she remained vulnerable to the mystical leaps and flights, the striking analogies and picturesque conceits with which commonsense reasoning attempts to surmount the difficulties of mechanistic solutions.

In assessing Ada's questing, metaphysical turn of mind, it is evident that De Morgan identified it with his own. His speculation that her grasp of "first principles" would have lessened her chances of becoming senior wrangler was surely a reference to his own case. This identification permitted him to forget or overlook what the correspondence clearly reveals: that Ada had great difficulty getting beyond her probing "first queries" and acquiring a firm grasp of mathematical practice, and that, far from saying "very good," and "quite right," he was more often compelled to repeat the same admonitions, to clear up the same confusions, again and again. At twenty-eight, in a field where important original contributions are (still) often made before the age of thirty, and after ten years of intermittent but sometimes intensive study, Ada was still a promising "young beginner."

3

This First Child of Mine

It is quite possible that Ada had decided on a career as a writer, translator, and associate of Babbage even before she began her serious mathematical studies with De Morgan. Her correspondence with her mother and Babbage in 1840 and 1841 is riddled with hints about her future plans. In March 1840, while she was still hoping to persuade him to be her instructor, she told the latter, "Should there be no chance after I go to Town, of the much desired *great unknown* being found for me, I have some idea of having instead for this season some German lessons. I know a little of it already and have always intended to know more. Indirectly I think it would bear on some of my objects."[1] Translation was a good way to begin, whether or not original contributions were to follow. Mary Somerville, De Morgan, and Babbage himself had all begun their published careers as translators. There was no reason why she could not proceed on a course at least as successful and rewarding as those of Mary Somerville and her mother's friends Harriet Martineau and Anna Jameson.

To her mother, however, her ambitions were revealed in a different light. In November she confided that she hoped to use her mathematics to solve some of the world's mysteries. It was just at this time that Lady Byron slowly began to unveil to her daughter and son-in-law the more sensational aspects of Byron's wickedness.

During a lengthy stay in France Lady Byron had met and taken under her protection Medora Leigh, a daughter of the aunt Augusta whom Ada had been raised to consider so evil that she must avoid any contact with her. Medora had run away with the husband of her own elder sister and had borne a child out of wedlock to him. The pair, having exhausted whatever money they could lay their hands on, were living in penury in a small French town. Despairing of extracting more from her debt-ridden mother, Medora appealed to Lady Byron to help her leave her lover. Lady Byron invited the young woman to join her, invited Ada and Lovelace to visit them in Paris the following spring, and revived the old supposition that Medora was really the child of Byron's incest with Augusta. Soon she had dropped enough epistolary hints for Ada to have begun to suspect the depth of her father's "crimes," though it was not until February 1841 that an explicit announcement was made.[2]

Ada responded to the revelation of her father's iniquity with a thrill of recognition and identification. Flushed with pleasure over her increasing ability to concentrate on her studies, she soon began to add the redemption of her father's sins to her other inducements to a scientific career. Once more religion, mathematics, and sex became mystically and contrapuntally intertwined, but now it was the salvation of her father's soul for which she strove. In January she announced to her mother that her imagination had been so stimulated by her mathematical studies, through which she hoped to increase the glory of God, that if she continued she would be a poet. The following month she became even more explicit:

I believe myself to possess a most singular combination of qualities exactly fitted to make me *pre-eminently* a discoverer of the *hidden realities* of nature . . . the belief has been *forced* upon me. . . .
Firstly: Owing to some peculiarity in my nervous system, I have *perceptions* of some things, which no one else has . . . an *intuitive* perception of . . . things hidden from eyes, ears, & ordinary senses. . . .
Secondly: my immense reasoning faculties;
Thirdly: my concentration faculty, by which I mean the power not only of throwing my whole energy & existence into whatever I choose, but also of bringing to bear on any one subject or idea, a vast apparatus from all sorts of apparently irrelevant & extraneous sources. . . .
Well, here I have written what most people would call a remarkably *mad* letter; & yet certainly one of the most logical, sober-minded, cool, pieces of composition, (I believe), that I ever framed.[3]

To the actual announcement of Medora's "true" paternity she responded that she felt similar to her father in her defiance of authority but regretted that he had not possessed a little of her "real philosophical turn." She believed that she had derived from him the flower of his characteristics and thus ought to do greater things; "but there is less *flash* about me & much more *depth*," and "in all probability my reign (if ever I have one) over mankind will be chiefly *after my death*."[4] She had a duty toward him to live out truths, nevertheless.

With Babbage, during the same period, she was much more circumspect and even, initially, humble:

I am very anxious to talk to you. I will give you a hint on *what*. It strikes me that at some future time . . . my *head* may be made by you subservient to some of *your* purposes & plans. If so, *if* ever I could be worthy or capable of being *used* by you, my head will be yours . . . though I scarcely dare so exalt myself as to hope however humbly, that I can ever be intellectually worthy to attempt serving *you*.[5]

Toward the end of February, just as Lady Byron dropped the other shoe, she wrote him to say that she was more determined than ever in her future plans and thinking much about the "future connexion between *us*." But there is no evidence in the surviving correspondence on either side to suggest that he ever encouraged her wish for "connexion" until she laid before him a *fait accompli*. Her opportunity would not appear for another two years.

Babbage was certain that, as he put it in his autobiography, "As soon as an Analytical Engine exists, it will necessarily guide the future course of science." Many of his contemporaries, however, lacked his vision, including some, such as the Astronomer Royal, who were in a position to advise the Government. Though he might not want collaborators, Babbage certainly needed friends and promoters to keep the educated public apprised of new developments, to keep his engines alive in their eyes during the long weary years when he still hoped for official backing. Although he himself was energetic in his efforts to publicize his inventions, he preferred, whenever possible, to have the discussion and praise appear over the names of others, particularly prominent others, even if sometimes he had to do the actual writing himself.

In 1840 Babbage went to Turin to hold forth in a series of lectures and discussions explaining the proposed Analytical Engine to a group of Italian philosophers and men of science. He had hoped to have the most eminent of them, Baron Plana, write a report or article on the subject, but Plana pleaded ill health. In the end he had to content himself with the services of a young military engineer, Captain Luigi Menabrea (who eventually became Prime Minister of Italy). Babbage accepted the substitution with less than consummate grace. From Florence, over a year later, he drafted a letter to Plana:

If you had made a report on the subject to the Academy of Turin during the last year it might have been of special service to me in the discussion of the question with my own government. As it is I must be content with the description drawn up by M. Menabrea with which I am well satisfied because he seems to have penetrated completely the principles in which it rests.[6]

From this it may be seen that Babbage had reviewed Menabrea's article over a year before it was published. Correspondence from Menabrea indicates that it was rewritten to incorporate Babbage's suggestions. It appeared, in French, in the *Bibliothèque Universelle de Genève* in October 1842. Shortly thereafter Ada translated it into English, at the suggestion and under the supervision of Charles Wheatstone, who solicited translations for *Taylor's Scientific Memoirs*, a recently founded journal devoted to making available significant scientific and technical papers from abroad.

Wheatstone was a family friend who took a great deal of sympathetic interest in Ada's plans, constituting himself a sort of career adviser. The inventor of the concertina, and a scientist before science was elaborately subdivided, Wheatstone is best remembered as the developer of the electric telegraph and of a device combining balanced electrical elements known as the "Wheatstone Bridge." Although the latter was actually created by Christy, Wheatstone made use of it to accurately measure voltages, a use still current in introductory physics laboratories. Wheatstone and Babbage had much in common. Both were cryptologists of remarkable skill; both held academic appointments but gave no lectures. In Babbage's case the reason was presumably preoccupation with his engines; in Wheatstone's it was severe stage fright. (In 1848, however, Wheatstone, who was Professor of Experimental Philosophy at King's College, planned a series of public lectures on electricity. The Bishop of London forbade the attendance of women, on the grounds that they had previously "congregated too abundantly" to hear the geologist Charles Lyell. Wheatstone was so angry at this exclusion that he resigned his professorship.)[7]

Babbage suffered a serious illness in the autumn of 1842, which may partly explain his being unaware of Ada's translation until after it was completed, and why her contribution took the form it did. His role in the publication developed only later in the proceedings. Because of her sex and her social position, Ada could not make direct contact with the publisher, at least not at first; a letter from Wheatstone makes it clear that he was the original intermediary.[8] Thus it appears that the translation of Menabrea's piece, rather casually undertaken and fairly expeditiously accomplished, had been approved by Wheatstone and almost delivered to the publisher by the time (probably early in 1843) Ada mentioned to Babbage what she had been up to.

Some time after . . . the late Countess of Lovelace informed me that she had translated the memoir of Menabrea. I asked why she had not herself written an original paper on a subject with which she was so intimately acquainted? To this Lady Lovelace replied that the thought had not occurred to her. I then suggested that she should add some notes to Menabrea's memoir; an idea which was immediately adopted.

We discussed together the various illustrations that might be introduced: I suggested several, but the selection was entirely her own. So also was the algebraic working out of the different problems, except, indeed, that relating to the numbers of Bernouilli [sic],* which I had offered to do to save Lady Lovelace the trouble. This she sent back to me for an amendment, having detected a grave mistake which I had made in the process.[9]

Babbage's references to Ada were always kind, deferential, and flattering. But his autobiography was written twenty years after the events referred to here, and ten years after Ada's agonizing and scandal-clouded death. The autobiography was itself a polemic on behalf of his machines, as he explains there, and he had every inducement to value the most important piece of propaganda in favor of his masterpiece. He went on to say, "These two memoirs [Menabrea's article and Ada's Notes] taken together furnish, to those who are capable of understanding the reasoning, a complete demonstration—*That the whole of the developments and operations of analysis are now capable of being executed by machinery.*" Babbage's immense and growing reputation, the awe inspired by his abilities, has tended to induce acceptance of any statement he has made about his engines. When these passages are compared with the contemporary correspondence and with the publication itself, however, a very different picture emerges, a picture explainable in terms of the background and story of its writing.

To begin with, how well was Ada prepared to write on a mathematical subject? And how intimate was her acquaintance with Babbage's

---

*The Bernoulli numbers are the numerical multipliers of the terms in the power series expansion of $\dfrac{x}{e^x - 1}$:

$$1 - \frac{x}{2} + B_1 \frac{x^2}{2} + B_3 \frac{x^4}{2 \cdot 3 \cdot 4} + B_5 \frac{x^6}{2 \cdot 3 \cdot 4 \cdot 5 \cdot 6} + \cdots,$$

where the B's represent the successive odd Bernoulli numbers.

as yet immaterial Analytical Engine? The last surviving letters in Ada's mathematical correspondence with De Morgan are dated 16 and 27 November 1842 (hence shortly before she translated the Menabrea memoir). In them we find her wrestling with an elementary problem in functional equations. (The problem was: Show that $f(x + y) + f(x - y) = 2f(x)f(y)$ is satisfied by $f(x) = (a^x + a^{-x})/2$.) She was still unable to take a mathematical expression and substitute it back into the given equation. It was the same "principle" that had plagued her in her correspondence with Mary Somerville and in earlier letters to De Morgan. On 27 November, after having struggled for at least eleven days, she sighed,

I do not know when I have been so tantalized by anything, & should be ashamed to say *how* much time I have spent upon it, in vain. These functional Equations are complete Will-o-the-Wisps to me. The moment I fancy I have really at last got hold of something tangible & substantial, it all recedes further & further & vanishes again in thin air. . . . .I *believe* I have left no method untried.[10]

The evidence of the tenuousness with which she grasped the subject of mathematics would be difficult to credit about one who succeeded in gaining a contemporary and posthumous reputation as a mathematical talent, if there were not so much of it; perhaps the most telling and consequential appears in her translation of Menabrea. In one passage the author was considering a mathematical expression* that "becomes equal to the ratio of the circumference to the diameter" (that is, to π) when n, the number of factors, becomes infinite.[11] The translated passage then continues,

Nevertheless, when the cos of $n = \infty$ has been foreseen, a card may immediately order the substitution of the value of π. . . .

The passage as it stands is nonsense, and made nonsense by the phrase "cos of $n = \infty$." Where did this phrase come from? In the original paper the passage reads, "Cependent, lorsque le cos. de $n = \infty$ a été prévu . . . ," and Ada has made a literal translation of this. But a moment's consideration makes it clear that the "cos." is the result of a printer's error and that the phrase intended by Menabrea was "le cas de $n = \infty$" (the case of $n = \infty$), which makes perfect sense in

---

*The expression is known as Wallis's product (multiplied by 2). The cosine of n as n approaches infinity continues to fluctuate in value between $+1$ and -1.

the context. Ada had translated a printer's error; surprisingly enough, her mistake has been reprinted several times.[12]

Ada's understanding of the proposed mechanical and logical operations of the Analytical Engine, early in 1843, is at least equally dubious. She had been fascinated by Babbage's machines ever since her first glimpse of the "gem of all mechanism," as she called the Difference Engine, and had borrowed drawings of it in 1834. But her letters show that during the early years of marriage and motherhood she had not kept *au courant* of developments in the Analytical Engine, which would not have been easy in any case. Not only were Babbage's plans continually changing, but the direction of change was not always toward completion. On 29 November 1839, for example, he had written her, "I have just arrived at an improvement which will throw back all my drawings full six months unless I succeed in carrying out new views which may shorten the labor."[13]

An even more telling example of the machine's vagaries occurs in the Menabrea translation, where Ada observed in a footnote that she had altered a sentence "in order to express more exactly the present state of the engine." In the original the sentence in question reads (in French),

All the parts, all the wheels that constitute that immense mechanism have been placed together, their action has been studied, but it has not yet been possible to assemble them.[14]

In the translation we read instead,

The plans have been arranged for all the various parts, and for all the wheel-work, which compose this immense apparatus, and their action studied; but these have not yet been fully combined together in the drawings and mechanical notation.[15]

In the time between the writing of the first statement and the writing of the second, the engine seems actually to have dematerialized. In the early 1840s only Babbage can be said to have been intimately acquainted with his engines; and although both of his interpreters were relative novices, completely dependent on him for information and claims about the Analytical Engine, Ada, from friendship and close association during the writing period, must have been the more persuadable.

The *Edinburgh Review* article on Babbage's Difference Engine that Ada had recommended to Dr. King as possibly within his compre-

hension had been written by Dr. Dionysius Lardner, a science popularizer, also under Babbage's close supervision. Along with urging
the practical need for a machine that could produce infallibly accurate
numerical tables and discussing the then current status of Babbage's
negotiations with the Government, Lardner's article had included a
description of the physical structure and mechanical operation. In
keeping with the more general nature and immaterial status of the
Analytical Engine, Menabrea's account dealt little with mechanical
details. Instead he described the functional organization and mathematical operation of this more flexible and powerful invention. To
illustrate its capabilities, he presented several charts or tables of the
steps through which the machine would be directed to go in performing
calculations and finding numerical solutions to algebraic equations.
These steps were the instructions the engine's operator would punch
in coded form on cards to be fed into the machine; hence, the charts
constituted the first computer programs. Menabrea's charts were taken
from those Babbage brought to Turin to illustrate his talks there.[16]

Babbage had suggested several more "illustrations" to be included
in Ada's Notes, that is, several more of the charts, with attendant
discussion of the means of preparing the problems for machine processing. The examples she used, except for that of the Bernoulli numbers, were also among those that Babbage had prepared some years
earlier—a circumstance that explains the absence of discussion of these
other examples in the correspondence that flew between Ada and
Babbage during her preparation of the notes.

If Menabrea's account of the Analytical Engine was often on a more
general and abstract plane than Lardner's description of the Difference
Engine, Ada's Notes frequently moved to a level yet more rarefied
and detached from the physical embodiment of a machine, to expatiate
on the metaphysical implications and latent powers of an entire mental
industry. Here she seemed to be in her element. With her usual
painstaking thoroughness she had set out to remedy her ignorance
by commandeering from Babbage all of his relevant papers. A very
early letter in this series announces,

I have read your papers over with great attention; but I want you to
answer me the following question by return of post. The day I called
on you, you wrote on a scrap of paper (which I have unluckily lost),
that the *Difference* Engine would do
    Δ (something or other) but that the *Analytical* Engine would do
    Δ (something else that is absolutely general)

Be kind enough to write this out properly for me; & then I think I can make some *very* good Notes.[17]

As they are presented in the published version, Ada's Notes appear to be comments on specific points of Menabrea's memoir, clarifying, elaborating, extending, and occasionally correcting the translated article. The notes are unified and set off from the translation, however, by a set of recurrent themes appearing in a variety of contexts and on different levels of abstraction. One frequently repeated theme was the contrast between Babbage's two engines. Where Menabrea had remarked that the Difference Engine "gave rise to the idea" of the Analytical Engine (which was both chronologically and logically true), Ada denied the relationship. She insisted that there was no necessary temporal or conceptual relationship between the two inventions; instead she stressed the metaphysical differences. The Difference Engine, she went on to say, could do nothing but add and tabulate; hence, its entire significance rested in the numerical data it processed. The Analytical Engine, on the other hand, maintained a strict separation between numerical data and operations; hence, it transcended mere number. Perhaps the Government's final decision in the previous year to abandon the Difference Engine made Babbage in 1843 wish to focus public attention on his newer, superior, and still hopeful project.

For the modern reader the important distinction between the two machines is that the Difference Engine followed an unvarying computational path (except for the parlor games Babbage played for the benefit of visitors such as Lady Byron), while the Analytical Engine was to be truly programmable and capable of changing its path according to the results of intermediate calculations or processes. Yet this was a feature mentioned by Ada only in passing. Thus, modern interest in Ada's Notes has centered on the prototype programs they contained—a century before the actual construction of a programmable computer. But Ada's focus was quite different.

The ideas set forth, the themes stressed, and the way they are presented by Ada show that her Notes were inspired by views that were much more the expression of the period in which they were written than the anachronism implied by modern interests. For Ada— and for Babbage—the calculating machine was a metaphor as well as a harbinger of economic and scientific progress. For example, the theme stressed in presenting the "illustrations" or programs was that they demonstrate how certain lengthy, laborious, and complex calculations can be most efficiently executed—from the point of view of the time and effort of the mathematician—by organizing the operations to be performed by the machine into recurrent cyclical groups. The same set of operations can then be repeated over and over by the engine, with only the starting and stopping places indicated by the instructions. To describe this efficiency, Ada used an interesting analogy taken from economic theory:

In the case of the Analytical Engine we have undoubtedly to lay out a certain capital of analytical labour in one particular line; but this is in order that the engine may bring us in a much larger return in another line.[18]

Where did this idea come from? Ada had referred to knowledge as capital early in 1841, when discussing her career plans with Sophia De Morgan:

Now the philosophy, the training, the instruments in short, I am as you know gaining. . . .I shall be years before I have the necessary quantity of what I call *capital*. . . .I am enterprising; & the greater & harder the work, the greater my spirit of enterprise.[19]

In making her statement in the Notes, Ada was comparing the Analytical Engine with the Difference Engine, to the disadvantage of the latter. But the full development of the calculating engine as a metaphor in economics had appeared in 1832 in Babbage's analysis of the industrial system, *On the Economy of Machinery and Manufactures*. There Babbage had reflected to the full his laissez-faire and Utilitarian views. He elaborated in particularly loving detail the principle of the division of labor that had been illustrated by Adam Smith in *The Wealth of Nations*.

The benefits derived from the division of labor, Babbage explained, included not only the time saved and skill attained by each worker when restricted to the repetition of a small set of simple operations, and the time saved in training, but, most important, the money saved by the hiring of less skilled labor, such as that of women and children, for jobs requiring little strength or skill. A supply of cheap and readily available labor is also assured by this means:

Again, the facility of acquiring skill in a single process, and the early period of life at which it can be made a source of profit, will induce a greater number of parents to bring up their children to it; and from this circumstance also, the number of workmen being increased, the wages would soon fall.[20]

Babbage's language is so matter-of-fact and complacent as to leave no doubt of his approval. But the example of the division of labor that intrigued him most was one of the division of mental labor—also, as it happens, inspired by Smith's pin factory. Devised by Baron Gaspard de Prony, it was adverted to (presumably with Babbage's consent) by Menabrea as a preliminary to his discussion of the Analytical Engine. De Prony was director of the Ecole des Ponts et Chaussées and had been commissioned by the French government to supervise the preparation of the new mathematical tables needed by a newly decimalized nation. While considering how such a prodigious task was to be organized, he chanced upon a copy of *The Wealth of Nations*. At once he determined to manufacture his logarithms like pins.

He set up two workshops that would perform the same calculations and thus serve as a mutual check on accuracy. Above them were two other sections of mental operatives. The first section consisted of five or six of the most eminent mathematicians in France, charged with deciding which formulae would be best to use for step-by-step calculation of the functions to be tabulated. (They performed the programmer's task.) These formulae were then passed to the second section, consisting of seven or eight competent mathematicians, who would substitute numbers into the formulae and then pass them to the third section. (They performed the operator's task.) The second section also received back the finished calculations, and compared and coordinated the results.

The third section consisted of sixty to eighty persons who performed most of the numerical work, using only addition and subtraction. Babbage, in the *Economy of Machinery and Manufactures* immediately made the connection between this third section and his Difference Engine, and between the organization of calculations and that of a factory, now in terms of class as well as task:

From that part executed by the third class, which may almost be termed mechanical, requiring the least knowledge and by far the greatest labour, the first class were entirely exempt. Such labour can always be purchased at an easy rate.

As to the work of the second section, that was

in some measure relieved by the higher interest naturally felt in these more difficult operations . . . but when the completion of a calculating engine shall have produced a substitute for the whole of the third section of computers, the attention of analysts will naturally be directed to . . . a new discussion of the methods.[21]

While it replaced the laboring class, the calculating engine would only produce fresh and more challenging work for the best trained and most gifted men. That the divisions refer to class in two senses is clear, since Babbage's exposition went on directly to state that

the proceeding of M. Prony . . . much resembles that of a skilful person about to construct a cotton or silk-mill, or any similar establishment. Having, by his own genius, or through the aid of his friends, found that some improved machinery may be successfully applied to his pursuit, he makes drawings . . . and may himself be considered as constituting the first section. He next requires the assistance of operative engineers . . . and these constitute his second section.

Once more, an easily replaceable third section would actually perform most of the work, at least at first. Like Leibniz, Babbage deplored the waste of gifted, educated men in routine, boring drudgery and strongly recommended the use of the uneducated for such purposes.

It is remarkable that nine-tenths of this class had no knowledge of arithmetic beyond its first two rules which they were thus called upon to exercise, and that these persons were usually found more correct in their calculations, than those who possessed a more extensive knowledge of the subject.

When convenient, however, he saw no objection to replacing them by yet more accurate or efficient machinery. He did not at first consider the displacement of the second section in this fashion. But once his work on the Analytical Engine began, he realized that it could be assigned far more sophisticated tasks than the third section and that the work of the second could be greatly simplified if not superseded.

Ada went farther in suggesting what the Analytical Engine could be made to achieve than did Menabrea, but she rephrased Babbage's words of assurance for the men of Prony's first section, characteristically making a metaphysical virtue of the machine's requirements:

The Analytical Engine has no pretensions whatever to *originate* anything. It can do whatever we *know how to order it* to perform. It can *follow* analysis; but it has no power of *anticipating* any analytical relations or truths. . . . but it is likely to exert an *indirect* and reciprocal influence on science itself in another manner. For, in so distributing and combining the truths and the formulae of analysis, that they may become most easily and rapidly amenable to the mechanical combinations of the engine, the relations and the nature of many subjects in that science are necessarily thrown into new lights, and more profoundly investigated. This is a decidedly indirect, and a somewhat *speculative*, consequence of such an invention.[22]

In what might be the earliest of Ada's letters to Babbage concerning her preparation of the Notes, she wrote,

I want to know also something more about how you manage the *imaginary* quantities; because as they are *nonsense* when supposed to be numbers; & as your results are wholly *numerical*, & your engine is a strictly *numerical* engine, I do not see my way there.

Likewise *what* is the nature of the views you allude to (page 96, 97 of your Bridgewater), "& I was well aware that the mechanical generalisations &c, &c—which would lead &c &c." I particularly want to know to *what* that is wholly new, & valuable, you can allude, as being likely to be developped by the engine.[23]

Clearly, Ada's dependence on Babbage as sole authority on his machines did not prevent her from making her usual challenging "first queries" on unfamiliar subjects, which had so impressed Professor De Morgan; but it must have contributed to her being fobbed off with inadequate answers. Unfortunately, Babbage's answer to this letter has not survived.

The question about imaginary numbers is a curious one. Although mathematicians had for long been reluctant to accept them as inherently as "real" as the more conventional kind, it was well recognized that they could be used in computations as vector quantities and hence were perfectly susceptible to numerical methods, although a bit more complicated to manipulate. During her studies with De Morgan, as we have seen, a good deal of attention had been devoted to this topic, which was of special interest to him. What is really interesting about this query, however, is that, having been struck by the "strictly numerical" nature of the planned calculator, she was somehow induced to change her mind.

The statement in the *Ninth Bridgewater Treatise* to which Ada referred ran, "and I was well aware that the mechanical generalisations I had organised contained within them much more than I had leisure to study, and some things which will probably remain unproductive to a far distant day."[24] Ada devoted much space in her Notes to suggesting the implications of the "mechanical generalisations" of the Analytical

Engine. Many of these derived, actually or metaphorically, from the physical separation between the operating part of the engine and the numerical storage, on the one hand, and between the punched-card instructions for operations and the orders for numerical storage and transfer, on the other. These separations had both cognitive and cosmological significance. About the first she said,

It were much to be desired that when mathematical processes pass through the human brain instead of through the medium of inanimate mechanism, it were equally a necessity of things that the reasonings connected with *operations* should hold the same just place as a clear and well-defined branch of the subject of analysis, a fundamental but yet independent ingredient in the science, which they must do in studying the engine.[25]

Now, Babbage and his Cambridge friends had been of the view that "analysis," whose core was the differential and integral calculus, was, in principle, completely of a piece with the finite algebra; hence they criticized the theory of limits in their translation of Lacroix, noting that they preferred Lagrange's use of infinite series as a basis of the infinitesimal calculus. (This is quite contrary to modern views and also at variance with the position of Ada's teacher De Morgan, whose work began some ten years after Babbage had left the field of pure mathematics.) The "clear and well-defined reasonings" to be applied to operations, alluded to in the passage just quoted, refers to the rules for manipulating symbols. The development of the infinitesimal calculus—which took place before and independently of its establishment on a rigorous basis—arose from the analogy between the rule for summing exponents when multiplying numbers and those for the reiteration of other types of functional operation, such as differentiation. In applying this analogy, the method of "separation of symbols" consisted in permitting the symbols of function or operation to be detached from the symbols (or numbers) on which they operated, so that they might be handled independently.[26] For example:

$$x^a \cdot x^b = x^{a+b} \, ;$$

by analogy,

$$f^a \cdot f^b(x) = f^{a+b} \, ;$$

or in particular,

$$\frac{d^a}{dy^a} \cdot \frac{d^b x}{dy^b} = \frac{d^{a+b} x}{dy^{a+b}} \, .$$

Babbage's own work on the theory of functions (also called the calculus of operations) had been heavily dependent upon such analogies.[27] Thus it is interesting (and revealing) to find Ada continuing her polemic on the superiority of the Analytical Engine in these terms:

The calculus of operations is likewise in itself a topic of so much interest, and has of late years been so much more written on and thought on than formerly, that any bearing which that engine, from its mode of constitution, may possess upon the illustration of this branch of mathematical science should not be overlooked. Whether the inventor of this engine had any such views in his mind while working out the invention, or whether he may subsequently have regarded it under this phase, we do not know; but it is one that forcibly occurred to ourselves. . . . We cannot forbear suggesting one practical result. . . : we allude to the attainment of those combinations into which *imaginary quantities* enter. This is a branch of its processes into which we have not had the opportunity of inquiring.[28]

But Ada had had, as we have seen, every opportunity of inquiring, and I will return to the significance of her prevarication on this point. Meanwhile she herself passed from the practical to the speculative to the cosmological; the same separation of action and object was reflected in the design of the universe:

[W]hen it is remembered that this science [mathematics] constitutes the language through which alone we can adequately express the great facts of the natural world, . . . those who thus think on mathematical truth as the instrument through which the weak mind of man can most effectively read his Creator's works, will regard with especial interest all that can tend to facilitate the translation of its principles into explicit practical forms.[29]

The view that mathematical truths were a direct revelation of God's way of thinking had been on the wane for well over a century; but given the rapid technological and social changes that were taking place in the nineteenth century, almost any scientific discovery could be scrutinized for its implications for religious belief. In particular, there was much interest in demonstrating that scientific activity could lead to a strengthening rather than a weakening of faith. The Bridgewater Treatises were written with this aim in mind. But while the officially selected authors focused on some aspect of the physical or biological universe as a miracle of divine design, Babbage had chosen to consider the design of miracles.

Among his illustrations of the ways in which a really clever God could perform miracles, Babbage pointed to the possibility of adjusting his Difference Engine in the manner that had so impressed Lady Byron; it could be made to calculate a million terms of some regular progression, then produce an aberrant term as the million-and-first. After that it could return to the original series, continue with the new one, or proceed to yet another. In like manner the Analytical Engine could change its operations in accord with a previously planned change of programming. How much more estimable, he asked, must we consider a God who created the universee to behave in this manner, with miracles programmed in from the start, than one who must continually be adjusting and intervening in the workings of His mechanism? Babbage recommended his *Ninth Bridgewater Treatise* to the attention of Queen (then Princess) Victoria, as a work favorable to religion.

Was he serious? To judge by his autobiography, he was; and Ada and Mary Somerville, to judge by their correspondence, seem to have considered him so from the beginning. Even before Ada had read the treatise, she wrote to Mrs. Somerville,

I am longing to see Mr. B's book. From Mama's accounts of it . . . it is a pity it was written in such haste & is so fragmentary and undevelopped in its character. It seems to resemble one of those curious *multum in parvo* algebraical expressions of which you know infinitely more than I do, which under a few symbols involve & indicate to the initiated quantities endless in their complication & variety of mutual relations. But *what* a pity that such a mind has not in some degree filled up the crude outlines for the benefit of those who could not! I fear the work will be underrated, and the circumstances you mention of the extreme haste fully accounts for this, though it in *fact* enhances its merit & indicates the more what it might be. —However, I am criticizing what I have not read. I think when I *have* read it . . . I shall probably give my opinion to Mr. B. himself. Would this be presumptuous do you think?[30]

Babbage himself later presented a copy to her. Perhaps she saw in her work on the Notes an opportunity to fill in the outlines. Much of her comment on the significance of the Analytical Engine is on such an exalted metaphysical plane as to make hers actually the more mystical of the two accounts.

Most of the time Babbage devoted to his engines was necessarily taken up in working out to the minutest detail his plans for the arrangement and construction of the physical realizations of his grand visions. These were so complex that he had invented a system of mechanical notation or symbols to keep track of and coordinate all the moving and stationary parts at every moment during the cycles of calculation. In the area of physical mechanism his ideas were so concrete and practical that he has plausibly argued that the by-products of his industry, in the form of improved tools, engineering practices, and the training given to the machinists and workmen he employed, in themselves more than justified the investment by the government.

In contrast to this daily involvement, the Menabrea-Lovelace memoir dealt little with physical and mechanical details, and Babbage's discussions with its authors gave him scope to expand upon his visions. With authors so young and so impressed by his genius, in both cases the temptation to suggest functions or capabilities of the engine that were not quite worked out must have been great. Alone on a new frontier, Babbage was ever optimistic of success and ever underestimating the time necessary for completion of his plans. He had, for example, assured both his interpreters that the process of division, like the other arithmetic operations, could be executed by his machine in a straightforward manner. Yet in a letter to Ada dated 30 June 1843 (while she was at work on the Notes), he indicated he was still not satisfied with this operation: "I am still working at some most entangled notations of Division but see my way through them at the expense of heavy labour, from which I shall not shrink as long as my head can bear it."[31]

In addition to their relative freedom from consideration of the limitations imposed by mechanism, and Babbage's own tendency to enthusiasm and optimism, another circumstance must have contributed to the exuberance of the claims made for the engine by both authors, but especially by Ada. This is the pretense, adopted by both, of not having consulted the inventor about certain of the conjectures they made. In both cases, though the disclaimers were made for certain specific points, they seem by implication to cover the entire memoir. And while they may have been technically true when first written, they need not, as we have seen, have been true by the time of publication. Ada's assertion, for example, that she had not consulted Babbage regarding his plans for the handling of imaginary quantities was certainly false. Babbage seems to have encouraged these claims and disclaimers, which had the effect of freeing both author and inventor from responsibility for the statements made. Concerning her first Note, in which her unfathered disclaimers and conjectures first appear, he wrote her, "I am very reluctant to return the admirable and philosophic view of the Anal. Engine contained in Note A. Pray do not alter it."

That Babbage did not hesitate to correct even small misrepresentations of his ideas is demonstrated in the same letter, for he goes on to say, "There is still one trifling misapprehension about the variable cards."[32]

The conjecture that Menabrea attached to his disclaimer of consultation was on the manner of handling the signs of the numbers in the process of multiplication. Ada commended him on his penetration and, characteristically, took her conjectures farther. Why, she speculated, if the machine could be made automatically to combine the plus and minus signs of the pairs of numbers multiplied together, could it not be arranged to deal appropriately with any other algebraic symbols that might accompany numerical coefficients? Why could not symbols themselves be operated on? It should be easy to make the engine do algebra.

The suggestion was a plausible though imaginative extension, not only of the sign feature but also of the metaphysical hierarchy between the Difference and the Analytical Engine. The development of modern algebra, as Ada might have learned from De Morgan and Babbage (both of whom were among its pioneers), was based on the freeing of algebraic symbols from the presumption that they could represent only numbers and on the subsequent elaboration of rules for manipulating symbols in the abstract.

The idea of a machine that could transcend number, as the Analytical Engine had transcended addition and been generalized to other operations, had been in Babbage's thoughts for some years. In a letter to Mary Somerville written 12 July 1836, he spoke of having "a kind of vision of a possible developing machine."[33] This was only twelve days after he had taken the decision to adopt punched cards as input to the Analytical Engine, and two days after he had mused in his notebook,

This day I had for the first time a general but very indistinct conception of the possibility of making an engine work out *algebraic* developments— I mean without any reference to the *value* of the letters. My notion is that as the cards (Jacquards) of the calc. engine direct a series of operations and then recommence with the first, so it might be possible to cause the same cards to punch others equivalent to any given number of repetitions. But these hole[s] might perhaps be small pieces of formulae previously made by the first cards and possibly some mode might be found for arranging such detached parts according to the powers of nine numbers and of collecting similar ones [the entry breaks off here][34]

What he was groping for here was some means of bypassing or replacing the columns of numbers that were ordinarily the objects to be operated on, so that he could operate on symbols instead. The crux of the difficulty lay in the very separation between operation and number over which the Notes expressed such pride. In generalizing from the Difference Engine, Babbage had added a control system with a very different physical and conceptual basis from the number columns, which he retained. The control was essentially a yes-or-no affair (binary), while the columns were decimal. Now he seemed to wish to operate on the binary-coded cards, which could represent numbers, symbols, instructions, or locations; but, as Ada's account conceded, no results could be produced unless the system operated on the columns of decimally divided disks, with the seemingly intractable constant interval between successive divisions. It was this fixed relationship, not the inscription of digits on the edges of the disks (which were placed there for the benefit of observers only), that made the Analytical Engine "numerical."

There was another sense in which the engine was "numerical." Babbage sometimes applied his knowledge of number theory (as well as algebra) to simplify the mechanism or shorten the projected time required to execute certain types of operations. The representation of the positive or negative signs of the numbers as corresponding to odd or even digits on the top wheel of each number column, which Ada pointed to as an instance of the symbolic powers of the engine, was actually one such "arithmetical artifice." If a positive algebraic sign is represented by an even digit and a negative sign by an odd digit, then the sign of the result of multiplying (or dividing) two numbers is represented by the result of adding their sign-digits: the sum of two even or two odd digits is an even number; the sum of an even and an odd digit is an odd number. Correspondingly, the sign of the product of two positive or two negative numbers is positive, and the sign of the product of a positive and a negative number is negative. But to have generalized beyond a few such specific instances would have required a recognition of the correspondence between number and symbol of a kind that Babbage's plans and the argument of the memoir show he was then moving away from. (In fact, even the handling of the signs of numbers during addition and subtraction was a good deal less elegant.)

In his notebook in December 1837 Babbage returned to the problem of symbolic operation, but with less clarity than in the passage I have quoted, and he finally concluded that it would be "better to construct a new engine for such purposes."[35] But although he continued to evolve new plans and modify old ones for the remaining thirty-five years of his life, he never produced plans for a "development engine." In 1843 the solutions to his difficulties lay in mathematical (if not technical) developments between four and ninety years in the future.

There is nothing to indicate that Ada was even aware of the problems involved in making a machine do algebra, a subject over which she several times waved her hand (or wand, for she had taken to calling herself a fairy). While Menabrea cautiously interpreted the purported algebraic capability as that of calculating the coefficients of power or functional series, Ada made more ambitious claims in several of her Notes, unrestrained by Babbage. In Note A she observed airily,

It seems to us obvious, however, that where operations are so independent in their mode of acting it must be easy by means of a few simple provisions and additions in arranging the mechanism, to bring out a *double* set of results, viz. — 1st, the *numerical magnitudes* which are the results of operations performed on *numerical data.* . . . 2ndly, the *symbolical results* to be attached to those numerical results, which symbolical results are not less the necessary and logical consequences of operations performed upon *symbolical data*, than are numerical results when the data are numerical.[36]

It might be supposed that the method of "attaching" the symbolic results would be for some human programmer to have worked them out and arranged for them to be printed with the corresponding numerical results, but in Note E the claim is repeated with a variation that makes it clear that Ada had symbolic processing by machine in mind.

The engine can arrange and combine its numerical quantities exactly as if they were *letters* or any other *general* symbols; and in fact it might bring out its results in algebraical *notation*, were provisions made accordingly. It might develope three sets of results simultaneously, viz. *symbolic* results (as already alluded to in Notes A. and B.); *numerical* results (its chief and primary object); and *algebraical* results in *literal* notation. This latter, however, has not been deemed a necessary or desirable addition to its powers.[37]

The distinction made here between "symbolic results" and "algebraical results in literal notation" might imply that the former referred simply to the printing of formulae previously worked out by human "analysts" and "attached" to the numerical results. When we turn back to Note B to confirm this supposition, however, we find confusion compounded. Now Ada makes the claim for the algebraic capability of the engine with the aid of a diagram, reproduced here, in which circles appear

$$V_1 \quad V_2 \quad V_3 \quad V_4 \quad \&c.$$

Diagram of columns of disks in the Analytical Engine.

at the tops of the representations of the columns of stored numbers. The circles represent the top disks on which the signs of the numbers stored below are coded by means of odd and even digits. About these Ada says,

[E]ach circle at the top is intended to contain the algebraic sign plus or minus . . . according as the number represented on the column below is positive or negative. In a similar manner any other purely *symbolical* results of algebraical processes might be made to appear in these circles. In Note A. the practicality of developing *symbolical* with no less ease than *numerical* results has been touched on.[38]

Touched on but not explained. In the passage just quoted, moreover, Ada seems to be succumbing to a curious confusion between the planned engine and the representation on her diagram; although any-thing at all might be written with equal ease in circles on a diagram, it would be far more difficult to represent sines than signs on the top disks of a metal column. This kind of confusion is particularly tempting when the mechanism exists only in the form of drawings; Menabrea, too, occasionally fell victim. But it was a trap that proved especially difficult for Ada to avoid. Her turn of mind, revealed in much of her writing, was essentially intuitive and mystical. Against these pro-pensities, encouraged by her mother, Babbage fitfully fought hard. Still, it was the formal beauty and the surprising, seemingly magical results of mathematical reasoning and processes that entranced her. For the subject itself she had little natural talent; its techniques, despite hard work, continued to elude her; its symbols remained the doings of "mathematical sprites & fairies." Because of this she was never able to turn her probing questions and picturesque conceits into in-genious and fruitful answers. She could speculate that some wonderful accomplishment might be possible, but without a firm understanding of her subject matter she was not only unable to suggest how it might be achieved but indeed unable to see that such explanation was necessary.

Ada was well aware of her technical deficiencies and was invariably timid and tentative with Babbage in these matters. A paragraph in a letter of 14 August 1843 illustrates this; it also reveals that Babbage's supervision of the Notes was so close that only one footnote in the published version had been inserted without his previous knowledge:

I have ventured inserting into one passage of Note G a small Foot-Note, which I am not sure is *quite tenable*. I say in it that the engine is remarkably well adapted to include the whole *Calculus of Finite Differences* & I allude to the computation of the *Bernoullian Numbers by means of the Differences of Nothing*, as a beautiful example of its processes. I hope it *is* correctly the case.[39]

(The "Differences of Nothing," or zero, according to De Morgan's *Differential and Integral Calculus*, which was Ada's textbook on the subject, refer to the successive differences—first differences, second differences, and so forth—of the first terms of the series 0^n, 1^n, 2^n, . . . , where n is a positive whole integer. For $n = 1$, the successive differences are 1, 0, 0, 0 . . . ; for $n = 2$, they are 1, 6, 6, 0. . . . In her footnote, and in her letter to Babbage, Ada is—deliberately?—vague about how she proposes to use these numbers.)

Ada's letters also effectively give the lie to Babbage's autobiographical statement that he had worked out the example of the Bernoulli numbers "to save Lady Lovelace the trouble"—an unlikely assertion, since a number of letters proclaim the time and trouble she was taking over this "illustration." She announced her decision to include it in a letter (incorrectly) dated 10 July in Babbage's hand:

I am working *very* hard for you, like the Devil in fact; (which perhaps I *am*). I *think* you will be pleased. I have made what appear to me some very important extensions and improvements. . . . It appears to me that I am working up the Notes with much success; & that even if the book be delayed in its publication, a week or two in consequence, it will be worth Mr. Taylor's while to wait.* I will have it *well* and *fully* done; or not at all.

I want to put in something about Bernoulli's numbers, in one of my Notes, as an example of how an implicit function may be worked

Taylor's Scientific Memoirs came out in the form of complete volumes, and Ada's paper was the final one in volume 3.

out by the engine, without having been worked out by human head & hands first. Give me the necessary data & formulae.[40]

This letter must have been written some time earlier than the date on it indicates, because by 5 July it is clear that she was deep in the study of the subject:

I am doggedly attacking and sifting to the very bottom, all the ways of deducing the Bernoulli Numbers. In the manner I am grappling with this subject; & *connecting* it with others, I shall be some days upon it. I shall then take in succession the *other* subjects that have been suggested to me during my late labours, & treat them similarly.
"*Labor ipse voluptas*" is in *very* deed my motto![41]

As was too often the case, however, the *voluptas* ended before the labor:

I am in much dismay at having got into so amazing a quagmire and botheration with these *Numbers*, that I cannot possibly get the thing done today. . . . at this moment I am in a charming state of confusion.[42]

The letter, signed "Yours, Puzzle-pate," reassures him that her confusion is "of a very *bubble* nature." But the usually admiring and supportive William was beginning to show impatience at the interminable "quagmire" of corrections and revisions:

Lord L– is so vexed too at everything not being done that I am half beside myself. He is pressing me in several ways just now, most unfortunately and amongst it all, I really *shall* be a long time, & shall lose my head for everything. . . . I really *cannot* believe it to be incorrect, for nothing can exceed the care with which I have gone over it. But the fact is I am plagued out of my life, just now.[43]

There is no sign in the surviving correspondence on either side that she ever sent his calculations back to him for correction. In the example of the Bernoulli numbers, a draft of which she had already sent him before her comment on how "doggedly" she was attacking the subject, most of her time seems to have been spent in preparing the chart or "Table & diagram" that was to accompany the discussion. The general form of this was similar to Babbage's other programs, but Ada added embellishments and improvements:

I have been hard at work all day, intending to send you the diagram & all, quite complete. Think of my horror then at just discovering that the Table & diagram (over which I have been spending infinite patience & pains) are seriously *wrong*, in one or two points. I have done them however in a beautiful manner, much improved upon our *first* edition of a Table & diagram.[44]

The "Table & diagram" received such lavish care and attention that eventually the vexed and vexing William was called in to assist:

I have worked incessantly, & most successfully, all day. You will admire the *Table & diagram* extremely. They have been made out with extreme care, & all the *indices* most minutely & scrupulously attended to. Lord L– is at this moment kindly *inking it all over* for me. I had to do it in pencil.[45]

The experience of working with a man of Babbage's caliber, on terms that friendship and her rank made appear ambiguously like equality, was heady and inspiring. It was also exhausting and nerve-wracking, pursued in the face of nagging illnesses. It fed both her growing confidence in her own ability and an irritability with her colleague's seemingly careless ways. Her satisfaction with the way the Notes were taking shape inspired ideas about the proper division of labor between them, ideas that were quite in line with those she had expressed to her prospective governess. When he presumed to delete one of her paragraphs and merely sent it back so that she could "see the change," she responded with some asperity.

I am much annoyed at your having altered my Note. You know I am always willing to make *any* required alterations myself, but that I cannot endure another person to meddle with my sentences. If I disapprove therefore, I hope I may be able to alter it in the *revise*, supposing you have sent away the proof & notes.[46]

He apologized at once, but she began to scold him for mixing up and mislaying her papers:

I wish you were as accurate, & as much to be relied on, as I am myself. You might often *save* me much trouble, if you were; whereas you in reality *add* to my trouble not unfrequently; and there is at any rate always the anxiety of *doubting* if you will not get me into a scrape; even when you don't. By the way, I hope you do not take upon yourself to alter my corrections. I *must* beg you not. They all have some very sufficient reason. And you have made a pretty mess & confusion in one or two places (which I will show you sometime), where you have ventured in my *M.S.'s*, to *insert* or *alter* a phrase or word; & have utterly muddled the sense. . . . I fear you will think this is a very *cross* letter. Never mind. I am a good little thing, after all.[47]

At one point Babbage lost the final Note, the one with the elaborate "Table & diagram." It was found again shortly afterward, when the printer reminded Babbage that he had taken it home for reconsideration. Ada remonstrated:

I do not think you possess half *my* forethought, & power of seeing all *possible* contingencies (*probable* & *improbable*, just alike). . . . How *very* careless of you to forget that Note, & how much *waiting on* & *service* you owe me, to compensate.[48]

She was to dwell long and hard on this contretemps.

Her forte once again was in the role of administrator, executive, general:

No one knows what almost *awful* energy & power lie yet undevelopped in that *wiry* little system of mine. I say *awful*, because you may imagine what it *might* be under certain circumstances. Lord L– sometimes says "What a *General* you would make!" Fancy me in times of social & political trouble, (had *worldly* power, rule & ambition been my line, which it now never could be).

A *desperate* spirit truly; & with a degree of deep & fathomless *prudence*, which is strangely at variance with the *daring* & the *enterprise* of the character, a union that would have given me unlimited sway & success, in all probability. *My* kingdom however is not to be a *temporal* one, thank Heaven! . . . it is perhaps well for the world that my line & ambition is ever *spiritual*; & that I have not taken it into my head, or lived in times & circumstances calculated to put it into my head, to deal with the sword, poison & intrigue, in place of x, y, & z.[49]

Ada was living through a decade of social unrest that her contemporaries dubbed "the hungry forties"; she managed to remain blissfully unaware of it. What was happening to her was a reenactment of the grandiose plans and illusions first released during her studies with De Morgan. With Babbage, however, she apparently felt comfortable enough to reveal thoughts that she had previously confided only to her mother and, more tentatively, to Sophia De Morgan. Unlike Lady Byron, Babbage apparently attempted to bring her back to reality and moderate some of her more manic transports by demurring at the mystical elements they contained, for in the same letter she responded to his cavils:

"Why does my friend prefer *imaginary* roots for our friendship?" Just because she happens to have some of that very imagination which *you* would deny her to possess; & therefore she enjoys a little *play* & *scope* for it now & then. Besides this, I deny the *Fairyism* to be entirely *imaginary*; (& it is to the *fairy* similies [*sic*] that I suppose you allude).

That *brain* of mine is something more than merely *mortal*; as time will show; (if only my breathing & some other et-ceteras do not make too rapid a progress *towards* instead of *from* mortality). Before ten years are out, the Devil's in it if I haven't sucked out some of the life-blood from the mysteries of this universe, in a way that no purely mortal lips or brains could do.

The letter was signed defiantly "Yours [*sic*] *Fairy* for ever."

She continued to insist on being his fairy, until he was forced to accept the conceit, not realizing that the combination of fairyism with deep and fathomless prudence qualified her to have the decisive voice in their collaboration, both in high-level strategy and in routine details. Relieved of the burden of decision making and publicizing, she thought, he would be free to concentrate all his genius on his drafting table and workshop. She convinced herself that hers were the more systematic habits and superior forethought—a belief, to judge by Lady Byron's and Lord Lovelace's complaints of her want of order, as quixotic as her belief in her mathematical ability. Her confidence in her work, and thence in herself, was rising immoderately. When Lovelace's and Babbage's praises seemed inadequate, she did not hesitate to add her own:

I cannot refrain from expressing amazement at my own child. The *pithy* & *vigorous* nature of the style seem to me to be most striking; and there is at times a *half-satirical* & *humorous dryness*, which would I suspect make me a most formidable *reviewer*. I am quite thunderstruck at the *power* of the writing. It is especially unlike a *woman's* style surely; but neither can I compare it with any *man's* exactly.

To say the truth, I am rather *amazed* at them [the Notes]; & cannot help being struck quite *malgré moi*, with the really masterly nature of the style, & its Superiority to that of the Memoir itself. I have made Lord L— laugh much, by the dryness with which I remarked, "Well, I am very much satisfied with this first child of mine. He is an uncommonly fine baby, & will grow to be a *man* of the first magnitude & power."[50]

From admiration of the child, it is but a short step to adulation of the parent who could produce such a prodigy:

I am in good spirits; for I hope another year will make me *really* something of an *Analyst.* The more I study, the more irresistible do I feel my genius for it to be. I do *not* believe that my father was (or ever could have been) such a *Poet* as *I shall* be an *Analyst,* (& Metaphysician); for with me the two go together indissolubly.[51]

Curiously, despite such declarations, she continued to defer to Babbage on all technical and scientific points, no matter how trivial. "I wonder," she ventured timidly in one instance, " if you will like my further *addition* to the upper indices. I half fear *not.* But I can cancel it, if you disapprove."[52] And sometimes she even managed to combine technical uncertainty with tactical self-congratulation:

I have (I think very judiciously & warily) touched on the only departures from *perfect* identity which *could* exist during the repetitions . . . & yet I have not *committed* myself by saying if these departures would require to be met by the introduction of one or more *new* cards, or not, but have simply indicated that as the variations follow a regular rule, they would be easily provided for. I think I have done it admirably & diplomatically. (*Here* comes in the *intrigante* & the *politician!*)[53]

In this instance, apparently, Babbage's explanations enabled her in the end to "commit" herself by incorporating the "departures" within the scheme of cycles within cycles of operations, thus clarifying her passage in a manner more like a scientist and less like a politician.

It is one of the overwrought ironies of the situation that Babbage, in his polemical *Reflections on the Decline of Science* more than a decade earlier, had devoted a lengthy and scathing section to various methods of fraud and misrepresentation in science, which of course implied that he himself was a model of full, clear, and unbiased disclosure in his own proceedings.[54] Now there was one point on which Babbage must have found Ada's use of the disingenuous disclaimer less pleasing and less useful than her exaggerated assertions about his engine's powers. The "philosophical" Note A had also included a reference to his difficulties with the Government: "Respecting the circumstances which have interfered with the actual completion of either invention, we offer no opinion; and are in fact not possessed of the data for doing so, had we the inclination."[55] This declaration was Ada's way of distancing herself from Babbage's disputes over funding and responsibility, in order to establish a separate and disinterested position for herself and her future writings. As she later explained her stance to her mother, "I declared at once to Babbage, that no power should induce me to lend myself to any of his quarrels, or to become in any way his *organ.*"[56] No longer did she consider herself "subservient to some of *your* purposes & plans." When he read this passage, however, Babbage immediately set about to remedy any absence of opinion on the part of the readers of the memoir by composing his own statement of the "circumstances," to be included (anonymously) with Ada's publication. He was anxious that it should not seem to emanate from himself.

Ada was well aware of Babbage's intentions. She and her husband examined his draft with care, and her comments indicate that she was somewhat gingerly about the whole affair: "I keep back *your* note; wishing to consider it a little more. I think it unobjectionable, as far as I have *yet* considered it. Pray take care that the printing is so managed as to separate distinctly the *translator's* notes, from either *your* note, or one there is of Menabrea's own." And, after consideration, "I hope you will attend carefully to my criticisms about the Preface. I think them of consequence. If Lord L– suggests any further ones, you shall hear."[57]

Babbage had taken over Wheatstone's role of delivering the manuscript to the printer and publisher, while Ada, for the most part, remained at the Lovelaces' country estate. But even as she was beavering away over last-minute alterations and revisions, correcting proof, and fuming over the printer's carelessness, Babbage, Wheatstone, and another eminent crony—Sir Charles Lyell, the geologist—held hurried consultations with Taylor's coeditor, William Francis, about the insertion of this "preface." Francis was anxious to get the third volume of *Taylor's Memoirs*, which Ada's paper was to complete, published before the approaching meetings of the British Association in early September. Taylor, the editor and publisher, would be out of the country and out of contact until after that date. The inclusion in *Taylor's* of such a statement as Babbage's had no precedent and would be likely to set one that the editors would find awkward; Francis was most reluctant to approve it without Taylor's assent. We owe to Lyell an amusing glimpse of these illustrious scientists' negotiations over Ada's "first child." Lyell, who had weak eyesight, was forced to attend one of the meetings alone, having been unable to find Wheatstone at their agreed rendezvous, and later sent a report to Babbage:

I have spent two hours in negociating with Mr. Francis & in seeking in vain after twelve o'clock for Wheatstone at K.C. [King's Cross?]. I began by explaining to Mr. Francis the proposed alterations of which he highly approved but having heard me out he said finally that he would not print it . . . on his responsibility. . . . On Francis proposing

that you should wait fourteen days for T[aylor]'s return I remarked
that before that he must be out [published], to which he assented but
said that they could promise in six weeks after to come out with
another part of the translation.

I then asked, "Do you think that Mr. Taylor will then consent to
insert Mr. B's article?"—"I do not—but we could exert ourselves to
get out the translation with the notes in the present No. if Mr. G.
[Gyde, the printer] would omit the preface." We then entered upon
the pros & cons of the Phil. Mag. & ended as yesterday with objecting
to the translatress making her debut there. On my saying that the
notes by Lady L. must go to the Phil. Trans. as a separate paper, F.
expressed great regret at the loss of it to them & their work & on
my dwelling on the great interest the public would take in the Preface
& on my asking him whether he could not devise some way of their
profiting by both, he suggested that the Preface should make a separate
notice signed by you, addressed to the Public in which you say that
as a translation was about to appear you wished to explain &c. He
could then stitch this in to every No. of the translations & although
it could not form part of the work everybody seeing your name could
bind it up with the rest. He remarked that a friend of his seeing the
Preface on his table, said that he could swear to it's being all Mr.
Babbage's therefore why should he not put his name?

I left him, asking if he could promise to get out the whole (preface
& all) before the Cork meeting [of the British Association] & without
waiting for Taylor, provided you assented that the Preface should
appear as your own or as Wheatstone's. He said he would promise
but no time must be lost. As these are better terms than I expected
after the beginning of the conference I hope you will agree & put
your name manfully.[58]

The *Philosophical Magazine* was another publication of the Taylor
brothers, but the *Philosophical Transactions* was not. The former contained
short articles and notes of a technical and scientific nature, notices of
meetings, and so forth. Many of the articles were responses to previous
publications, especially those that had appeared in *Taylor's Memoirs*.
Most were in the form of letters to the editor, only a page or two
long, and they were generally signed. The objection to Ada's making
her "debut" in its pages might have been on the grounds of its lack
of anonymity, always a sensitive concern for Ada, and, in connection
with her translation, much worried over. (However, her article as it
appeared in *Taylor's* was irregular in this respect there also.) Alternatively, it might have been on the basis of the incongruity between
her paper and the usual contents, or because the *Philosophical Magazine*
was less prestigious than *Taylor's*. In any case, Lyell's letter shows that

all parties considered Ada to be at the threshold of a promising career in scientific writing and translating.

It is not difficult to surmise what happened next, despite the only partially surviving correspondence. Babbage was less than satisfied with Lyell's diplomacy; he did not want his preface separated from Ada's paper and published "manfully" under his own name. Perhaps Wheatstone refused to become the adoptive father. At last Babbage informed Ada what was afoot and asked her to withdraw her paper and publish it elsewhere, probably in the *Philosophical Transactions*. He was astonished and dismayed to receive an indignant refusal.

For months now, Ada's project had made heavy demands on his time and attention, diverted from work on the evolving plans for his beloved engine. He had been willing enough to comply with her requests, however imperiously made, for papers, books, meetings, explanations. Yet despite the stream of commands, the summonses to her townhouse when his presence was required, the orders to take charge of this or that detail or to run errands to the printer's, the reproofs given as if to a subordinate, it had not occurred to him that she considered the paper her own property or that her first object was not the pursuit of his interests.

Now the deadline was approaching, and clearly conciliation was in order. A new plan was devised, but once more the editor would be approached before Ada was told what it was. On 5 August Babbage wrote her, "I waited with some anxiety for a communication from you yesterday. Today I saw Wheatstone and proposed to him a plan which will fulfill *all* your conditions and some of mine—He approves of it and thinks it will be adopted. If it is I shall write by Monday Post."[59] For all his brilliance, he could not understand what it was to be immured in the country while important decisions regarding her work were taken by free-roving men in the city. A further blast must have ensued, for by 8 August he was forced into a completely defensive position:

I have nothing to add at present except that you did me an injustice in supposing I wished you to break any engagement with the Editor. I wished you to ask him to allow you to withdraw from it. Had the Editor been in England I believe he would at my request have inserted my defense [i.e., the preface] or forborne to have printed the paper— As it stands I have done all I can at present do to defend myself and having failed in the most important part shall make the best I can of the rest.[60]

Under the circumstances, this kind of extenuation could hardly have been expected to soothe her wounded sense of her prerogatives, and his casuistry only inflamed her further. As it turned out, too, his confidence that the editor would have complied with his wishes proved mistaken: on his return Taylor sent him a note of adamant retrospective refusal.

Ada's version of the quarrel was related to her mother in two letters, the first written on the same day as Babbage's rather truculent excuses, the second a week later and obviously referring to that irritating missive.

I have been harassed & pressed in a most perplexing manner by the conduct of Mr. Babbage. We are in fact *at issue*; & I am sorry to have to come to the conclusion that he is one of the most *impracticable, selfish, intemperate*, persons one can have to do with. I do not anticipate an absolute *alienation* between us; but there must ever be a degree of coolness & reserve I fancy in the future. I have had in W[illiam]'s absence, to act quite unadvised in the matter; but I am happy to find that W– & Wheatstone entirely approve my conduct & views. I declared at once to Babbage, that no power should induce me to lend myself to his quarrels, or to become in any way his *organ*; & that I should myself communicate in a direct manner with the editors on the subject, as I did not choose to commit a dishonourable breach of engagement, even to promote *his* advantage (if it *were* to his advantage, which I doubted).

He was *Furious*; *I* imperturbable & unmoved. He will never forgive me. I had tried to conciliate, & gently to *advise* & *suggest*, until I found that it was necessary to be very determined & explicit.

I am uncertain as yet how the Babbage business will end. He has written *unkindly* to me. For many reasons however, I still desire to *work* upon his subjects & affairs, if I can do so with any reasonable prospect of *peace*. I have written to him therefore, very explicitly, stating my own *conditions*, without which I positively refuse to take any further part *in conjunction* with *him*, upon any subject whatever. He has so strong an idea, I suspect, of the *advantages* of having my pen as his servant, that he will probably yield; tho' I demand very strong concessions.

If he *does* consent to what I propose, I shall probably be enabled to keep him out of much hot water; & to bring his engine to a consummation, (which all I have seen of his habits the last three months, makes me scarcely anticipate it ever *will* be, unless someone really exercises a strong coercive influence over him). He is beyond measure *careless* & *desultory* at times. — I shall be willing to be his Whipper-in during the next 3 years; if I see fair prospect of success. Much of this is W's suggestion; (altho' W– thinks B's conduct to me has recently been *very blameable*).[61]

Ada's sense of injury receives justification from an unexpected source, the memoir, written many decades later, of Cornelia Crosse, stepmother of John Crosse, of whom we will hear more.

Early in the Sixties Miss Kinglake and I went one evening to take tea with Mr. Babbage. He had promised to show us some interesting papers respecting Lady Lovelace's mathematical studies, and by arrangement there were no other guests.[62]

Maddeningly, there is no further mention of the interesting papers. Instead there is a description of Babbage's house, perhaps overinfluenced by the reading of Gothic novels:

It was large and rambling for a London house, having several spacious sitting rooms, all of which, with the exception of the drawing-room, were crammed with books, papers, and apparatus in apparent confusion, but the philosopher knew where to put his hand on everything. He received us in his unused drawing-room. . . . [T]he brilliant receptions Babbage used to give in the Forties [were held] in this same dreary, ghost-haunted room where we sat.

Then, suddenly:

His grievance was ever present; even the subject of Lady Lovelace, his friend and pupil in science, was not touched upon without reference to an angry dispute with Wheatstone and other of Lady Lovelace's friends, who objected to his making a publication of hers a medium for his own griefs. He told us the whole story, but the conviction remained with me that Mr. Babbage was wrong.*

Among Ada's "other friends," Lady Byron, whose enthusiasm for Babbage had cooled considerably since her first exposure to the "sublime views" opened by his Difference Engine, heard of the quarrel "with more pain than surprise." In due course she would refer to the story melodramatically as "Babbage's attempt to commit murder — I might almost call it Suicide considering your value to him."[64]

Ada's assertion that it had been her husband who suggested that she become the governor of Babbage's professional papers, habits, and affairs is startling, in view of Lovelace's repeated complaints about

---

*It is difficult to know what to make of some of Mrs. Crosse's statements. For example, "Babbage was very fond of talking of Byron's daughter: to him she was always 'Ada' for he had carried her in his arms as a child." If Mrs. Crosse was not embroidering her recollections, Babbage was embroidering his.[63]

her own disorganization of books and papers and her forgetfulness with respect to rented and borrowed equipment. But she was quite correct in stating that her conditions were very severe. They constituted a demand for a turnabout in the treatment she had recently received from him, embedded in a letter of astonishing length and presumption, in which she undertook to clear the air with her old friend the day before her second letter to her mother.

My dear Babbage
You would have heard from me several days ago, but for the *hot* work that has been going on between me and the printers. This is now all happily concluded. . . .
    You say you did not wish me to "break my engagement, but merely to ask to be released from it." My dear friend, if the engagement was such that I had no right to break it *without leave*, I had still *less right* to appeal to the *courtesy* of parties, in order to obtain an apparent sanction & excuse for doing that which their *justice* & *sense of their own rights* could not have conceded. . . . You will deny & dispute this. . . . Remember however . . . that your question becomes not whether Lady L– ought to oblige two parties, *you & the editors*, who both tho' on different grounds wish to dispose of my publication thro' another channel . . . but whether Lady L– ought tacitly to lend herself to certain possible or probable unworthy motives entertained by the editors. Now to this the reply is perfectly plain in the opinion of all parties accustomed to fair & honourable dealings. . . . My engagement was *unconditional*, & had no reference to the *motives* of the parties with whom I contracted it.

Having demonstrated that truth and justice were on her side in their dispute, she went on to justify herself in all possible other ways: she was actually protecting his own interests, though he did not see it that way and though his interests were completely incidental to her undertaking with the editor. After considerable rambling she turned abruptly to her conditions:

I must now come to a practical question respecting the future. *Your* affairs have been & are, deeply occupying both myself & Lord Lovelace. Our thoughts as well as our conversation have been earnest upon them. And the result is that I have plans for you, which I do not think fit at present to communicate to you, but which I shall either develop or else throw my energies, my time & pen into the service of some other department of truth & science, according to the reply I receive from you to what I am now going to state. . . . I give to *you* the *first* choice & offer of my services & my intellect. Do not lightly reject them. I say this entirely for *your own* sake, believe me.

My channels for developping & training my scientific & literary powers, are various, & some of them very attractive. But I wish my old friend to have the *refusal.*—

Firstly: I want to know whether if I continue to work *on* & *about* your own great subject, you will undertake to abide wholly by the judgment of myself (or of any persons whom you may *now* please to name as referees, whenever we may differ), on all *practical* matters relating to *whatever can involve relations with any fellow-creature or fellow creatures.*

Secondly: can you undertake to give your mind *wholly* & *individually* [sic], as a primary object that no engagement is to interfere with, to the consideration of all those matters in which I shall at times require your intellectual *assistance* & *supervision* & can you promise not to *slur* & *hurry* things over; or to mislay, & allow confusion & mistakes to enter into documents, &c?

Thirdly: If I am able to lay before you in the course of a year or two, explicit & honorable propositions for *executing your engine* (such as are approved by persons whom you may *now* name to be referred to for their approbation), could there be any chance of your allowing myself & such parties to conduct the business for you; your own *undivided* energies being devoted to the execution of the work; & all other matters being arranged for you on terms which your *own* friends should approve.

In these terms, reminiscent of a marriage settlement, complete with trustees, did Ada propose to become Babbage's guardian, should he agree to bring his engine as dowry into their alliance. It was also a vision of reliving and prolonging the intense involvement and exaltation of the past months, but without the petty mishaps and annoyances that had caused so much anxiety. Characteristically, it was enclosed in a high-flown moral and religious framework:

Our motives, & ways of viewing things, are very widely apart; & it may be an anxious question for you to decide how far the advantages & expediencey of enlisting a mind of my particular class in your service, can overbalance the annoyance to you of that divergency on perhaps many occasions. My own uncompromising principle is to endeavour to love *truth* & *God before fame* & *glory* or *even just appreciation*; & to believe generously & unwaveringly in the *good* of human nature, (however dormant & latent it may often seem).

Yours is to love truth & God (yes, deeply & constantly); but to love *fame, glory, honours, yet more.*

Not that she herself was without great ambition:

I wish to add my mite towards *expounding* & *interpreting* the Almighty, & his laws & works, for the most effective use of mankind; and certainly,

I should feel it no small *glory* if I were enabled to be one of his most noted prophets (using this word in my own peculiar sense) in this world. . . .

At the same time, I am not sure that 30 years hence, I may put so much value as *this*, upon human fame. Every year adds to the unlimited nature of my trust & hope in the Creator, & *decreases* my value for my relations with mankind *excepting as his minister*; & in *this* point of view those relations become yearly more interesting to me. Thro' my present relations with *man*, I am doubtless to become fit for relations of another order hereafter; perhaps *directly* with the great Power himself. . . .

Such as my principles are, & the conditions (founded on them) on which alone you may command my services, I have now stated them; to just such extent as I think is absolutely necessary for any comfortable understanding & cooperation between us in a course of systematized & continued intellectual labour. It is now for *you* to decide.

She was offering him a chance to work with a prophet and minister of God. Then, suddenly, she dropped back to earth out of the rarefied moral atmosphere in which she had all but disappeared: "Will you come *here* for some days on Monday. I hope so. Lord L— is very anxious to see and converse with you; & was vexed that the Rail called him away on Tuesday, before he had heard from yourself your own views about the recent affair." And then, still imperiously but somewhat incongruously for God's most noted prophet, "I sadly want your Calculus of Functions. So *Pray* get it for me. I cannot understand the *Examples*." Exhausted and depleted, she ended her marathon letter on a subdued note: "This letter is sadly blotted & corrected. Never mind that however. I wonder if you will choose to retain the lady-fairy in your service or not."[65]

The wandering, confused, and delusional elements of this document are obvious, though curiously intermittent. It is possible that William had indeed suggested to Ada that she should clarify the conditions of her future collaboration with Babbage before embarking on the long-term effort she proposed; but it is unlikely that she showed him this letter before posting it. To Lady Byron he only remarked laconically that "the long letter to Babbage" had been sent off.

What Babbage's reactions were can only be imagined; the old courtier left conflicting evidence. We do not know how much he knew, at this point or later, of her medical or personal problems, of which her letter contained a number of hints:

My dear friend, if you knew what *sad* & *direful* experience I have had, in ways of which you cannot be aware, you would feel that *some* weight *is* due my feelings about God & man. As it is, you will only smile & say, "poor little thing; she knows nothing of life or its wickedness!"

He had passed over, apparently without comment, a succession of grandiose plans and self-congratulation in the weeks past, caviling only at fairyism. But now he lost no time. At the top of this letter, in Babbage's hand, appear the words, "Tuesday 15 Saw AAL this morning and refused all the conditions."*

Volume 3 of *Taylor's Scientific Memoirs*, containing Menabrea's "Sketch of the Analytical Engine Invented by Charles Babbage, Esq.," translated and with notes by "A. A. L.," appeared in print a few days later. Babbage's preface came out in the *Philosophical Magazine* the following month. The quarrel between Ada and Babbage was patched up, though on somewhat ambiguous terms. Their relationship from this point was marked by unfailing patience, chivalry, and flattery on his part—but no repetition of the intense commitment and exchange of the previous months. Babbage was now over fifty; he would pursue the development of his plans and the recognition of his work and himself in his own fashion. But for Ada, the realization of her soaring ambitions, the fulfillment of her genius, the justification of her sufferings, the redemption of her father's sins—all of which were to validate her release from domestic restriction—became at once more pressing and more perplexing than before.

*The meeting cannot have taken place on the 15th, for there is no mention of it in Ada's letter to her mother of that date, written late in the day and indicating only suspense. Possibly it occurred on the day following.

4

In Time I Will Do All, I Dare Say

In refusing Ada's conditions, Babbage proved himself more diplomatic than he is usually credited with being. She was left unaware that she had been refused; the letters they exchanged during the ensuing weeks, concerned with tidying up a few minor points that had arisen from her work on the Notes, indicate that she was proceeding as if their friendship and understanding had been fully reestablished.

On 18 August Babbage wrote to announce that he would bring a set of drawings with him the following Monday; they would finally have time to examine "the mechanical part." On 9 September he announced his determination to visit her at Ashley Combe bearing books about the three-body problem, which had aroused her curiosity when she had occasion to mention it in one of the Notes. This letter and the one he wrote three days later were flatteringly signed "Farewell my dear and much admired interpreter" and "Ever my fair Interpretress, Your faithful slave." He had learned that the only real conditions between them were to be unstinting admiration and adulation on his part, for in between these two letters Ada had written,

Your letter is *charming*, and Lord L— & I have smiled over it most *approbatively*. You must forgive me for showing it to him. It contains such *simple, honest, unfeigned* admiration for myself, that I could not resist giving *him* the pleasure of seeing it. . . . I send you De Morgan's *kind & approving* letter about my article. I never expected that *he* would view my crude young composition so favourably.

You understand that I send you his letter in strict *confidence*. He might perhaps not like you to see his remarks about the relative *times* of the invention of the two engines. I am going to inform him of my grounds for feeling satisfied of the literal correctness of my statement on that point. I cannot say how much his letter pleased me.

You are a brave man to give yourself wholly up to Fairy-Guidance!—
I advise you to allow yourself to be unresistingly bewitched, neck &
crop, out & out, whole seas over, &c, &c, &c, by that curious little
being![1]

In fact, as we have seen, De Morgan's opinion of the "pretty tract"
was one of qualified enthusiasm, but both Babbage and Ada chose to
take any reservations he might have simply as a compliment to her
abilities. So Babbage responded, "It is gratifying to me also; for you
know I had arrived at the same conclusion. You should have written
an original paper. The postponement of that will however only render
it more perfect."[2]

Once Babbage had confessed his conversion to fairy guidance, a
project had to be found that, at least for a time, would enable the
excitable fairy to sustain the illusion of continued collaboration. This
is the simplest and most plausible explanation of the occasional references that appear in their correspondence between 1844 and 1849
to a "book" that is being passed back and forth between them. It is,
moreover, an explanation that fits the scanty evidence available better
than the tortuous alternative hypothesis that has been offered: that
of a gambling scheme or conspiracy. The references cease before there
is any sign that Ada had become involved in betting, and there is no
evidence that Babbage ever did.

The first occurrences of "the book" especially seem to lend themselves to the suggestion that they refer to a coauthored work in progress
and that the drawings, papers, and, eventually, "book" she demanded
in her letters all relate to the postponed "original paper," a more
detailed and complete elaboration of the practical possibilities and
philosophical implications of the Analytical Engine. The book itself is
not mentioned until over a year after the completion of the Menabrea
memoir. In October 1844 Babbage was visited by Sir David Brewster,
his cofounder, along with Herschel, of the British Association; he apparently asked Ada for the book so that he could show it to his guest.
Perhaps he thought Sir David might be of help in publishing the work;
he was an influential Edinburgh publisher and would later be the
author of a work entitled *Martyrs to Science*, among whom he included
Galileo, Tycho Brahe, and Johannes Kepler.

Ada replied that she was unable to spare it that day, since she was
referring to it constantly. She suggested that Sir David call upon her
the following morning.[3] At this point Ada still considered herself the
"High Priestess of Babbage's Engine," as she termed it, though her
plans were soon to change quite drastically. It is perfectly possible,

indeed, that later underlined references to "the book" refer to quite a different work (or works); Babbage's son thought that an 1849 mention referred to what later became *The Exposition of 1851.*

In the year between the memoir's publication and the first mention of "the book," both Babbage and Ada were active in promoting the memoir, distributing copies wherever they thought they might do some good. In his letter of acknowledgment to Babbage, Menabrea (who had been given a copy of the translation by Babbage's son in Rome) referred to the author of the Notes as a "savante" named "Lady Lovely."[4]

Ada lost no opportunity to dazzle and mystify acquaintances who were in no position to judge the true extent of her mathematical knowledge and achievements. The necessity of anonymous publication proved only a minor inconvenience, and one that could often be turned to attention-getting advantage. The question of how to signal her authorship without signing her name had been mooted in several of her letters to Babbage during the month of July. This was a matter (like John Stuart Mill's dedication of his *Political Economy* to Harriet Taylor) in which her husband's views were of more moment than her own. At one point in the deliberations, the decision was held up by "Rent-day," the ritual by which Lord Lovelace and other landlords collected their dues from their tenants. On those days he saw no one else. That over, she was finally able to announce that he had decided that she should append her initials to each of the Notes and to the translation; she went on to say, "It is not my wish to *proclaim* who has written it; at the same time that I rather wish to append anything that may tend hereafter to *individualize & identify* it, with other productions of the said A. A. L."[5]

Reluctant to proclaim or not, both she and William freely gave out copies to a number of unlikely as well as likely recipients. Soon the thanks came flowing back, registering varying proportions of gratitude, admiration, and bewilderment. The elderly playwright Joanna Baillie thanked her for her "learned tract"; the art historian Anna Jameson was reduced to "babbling," according to Lady Byron; the actor John Kemble confessed himself overawed at the "high and practical order" of the intelligence she had revealed: "And if I expressed some very natural surprise at meeting a young lady who was familiar with a work (Comte's) that I have found few men inclined to grapple with, you may believe that the last communication with which you favoured me has not tended to diminish the feeling of respect with which so very odd a conversation in so very odd a place inspired me."[6]

Ada, who made something of a specialty of holding odd conversations in odd places, had encountered Kemble in a railway carriage and had followed up the advantage she gained there by sending him a copy of her publication to show him what young ladies were capable of. On hearing the story, Lady Byron chortled with satisfaction, "The impression produced is the more amusing because he has, I find, been in the habit of indulging in gross abuse of literary & learned ladies." He was the editor of a review "in which those coarse attacks on distinguished women have appeared." She reminded Ada that De Morgan's approval was of more value than Kemble's, and added, "The 'Mother of Ada' may perhaps be as good a passport to posterity (if I am to have one) as the 'wife of Byron'—you are right in believing that Science itself depends mainly on 'religious & moral training' for its *highest* success."[7]

Both mother and husband were quite willing to concede to Ada intellectual supremacy, provided the former could maintain her authority over matters religious and moral, and the latter his control of their joint wealth. From time to time they would pass their scientific questions to her, and the contrast between them was striking and characteristic. Lady Byron wished to know what happens to light when it is absorbed by a dark, rough material. Unlike water, it cannot be retrieved. Does light perish when cut off from its source? William, more concretely, wished to know whether a square inscribed on the side of a hexagon had any relation to the sides of the triangles that determine the hexagon. (Unfortunately, her answers to these questions were not preserved.)

And now Lord Lovelace, thinking of his present position rather than his passport to posterity, eagerly explained to all his acquaintance the reason for his wife's eccentricities. She was not mad, just working.

William especially conceives that it places *me* in a much *juster & truer* position & light, than anything else can. And he tells me that it has already placed *him* in a far more agreeable position in this county. Besides the many *other* motives which concur to urge me on to perseverance & success in a studious & literary concern, I must name how very important an addition it is to their weight that I see W— looks to it as what is to place *him* (even more perhaps than myself) in the most advantageous & natural position, from its various indirect effects. Oh dear! How mercilessly he carried off my proofs & revises to some of his friends who came here.[8]

In October, while Ada was visiting her mother, he took the opportunity to enlighten another visitor, his own cousin, Lord Ebrington:

I told him a little the reasons why you went so little into any society—health—& pursuits which latter I said were of such a nature to exclude the world in its ordinary way. . . . The conversation insensibly led me to mention what you had been occupied about and it ended by my telling him the analytical engine had been the subject and I gave him a copy of the translation.

He also explained what Babbage most wished cleared up in the minds of the influential: the financial arrangements between the inventor and the government. Then William went on,

My dearest I hope I have not done amiss in letting so much be known about you—I did not parade the information—but I did not wish either to be mysterious with him about you. . . . I think too you will admit that it is desirable that some of our and his less discreet but very worthy relatives should know a little the reason why they can know & understand you so little & that it is well to take such opportunities as may occur for allowing such information to ooze out.[9]

He need not have worried. Ada was not at all shy about presenting herself in her new role as scientific writer. Of course she sent a copy to Mary Somerville, which was acknowledged from Italy with praise as gracious as it was now distant and vague. She congratulated Ada on her proficiency "in the highest branches of mathematics, & the clearness with which you have illustrated a very difficult subject."[10] Mr. Babbage must be gratified. Then she went on to caution Ada against overwork. Mutual friends had reported that she looked thin and peaked. Was she pregnant, by any chance?

Ada saw herself on the threshold of possibilities almost unlimited in scope and promise. The problem was how to choose, or at least in what order. Even before she had completed her "hot work" with the printer, she was arranging to clear away such obstacles as her children might in the future present, by asking a cousin to engage a tutor for them. She explained that they could not expect to see much of her because "I am now a completely *professional* person, to speak plainly; & am engaged in studies & in literary & scientific avocations, which render me quite unable (were I even fitted by *nature*, which I am *not*), to associate much personally with my children, or to exercise a favorable influence over them by attempting to do so."[11]

She went on to say that she had in mind a tutor who could take complete responsibility and manage the children on his own, but who would nevertheless welcome interventions by herself or Lord Lovelace "at any moment, or on any occasions." (Indeed, she enjoyed making out long and elaborate schedules of lessons for her tutors to follow.) The desirable tutor, she thought, would have strong religious convictions, "but unitarian or rationalist," and a "*logical & well balanced* turn of mind rather than highly speculative or imaginative."

Of course she promised her cousin a copy of the Menabrea translation "with copious Notes of my own." It had been a troublesome undertaking, but she was pleased "to have got launched, in however *humble & dry* a form." She was planning to supplement her occasionally inconvenient flesh-and-blood children with "a *large* family of brothers & sisters" of this her first publication. This letter was written as her quarrel with Babbage was approaching its climax, and she included in it a glimpse of some of her alternative plans to promoting the Analytical Engine. Her cousin was going to Germany; would he bring her information on "*microscopial* structure & changes in the *brain & nervous matter* & also in the *blood?*"

Even before the quarrel she had never intended her work with Babbage to absorb all her time and energy, at least not for more than the two or three years that continually seemed to be all it would take to complete the current engine. Now she was even less certain that she wished to concentrate indefinitely on a task that had begun to seem too mundane and circumscribed for her rising intellectual and spiritual aspirations. To William she confided,

I am very much *afraid* as yet of exciting the powers I *know I have over others*, & the *evidence* of which I have certainly been *most unwilling to admit*, in fact for a long time considered quite fanciful & absurd. . . . I therefore carefully refrain from all attempts *intentionally* to exercise unusual powers. . . . I had better continue to be simply the High Priestess of Babbage's Engine & serve my apprenticeship faithfully therein, before I fancy myself worthy to approach a step higher towards being the High Priestess of God Almighty Himself. . . . By the way, I am particularly *ill* at present.[12]

The obvious way forward in her scientific apprenticeship seemed to lie in writing more reviews. During a lull in her work with the Menabrea memoir, she had at one point envisaged a slashing attack on Whewell, "when I have knowledge enough. . . . Do you not pity him?" At the same time she considered a review of "Ohm's little work," which she almost actually began.[13] The "little work" was probably "The Galvanic Circuit Investigated Mathematically," which had appeared in *Taylor's* in 1841, translated by the same William Francis who had proved so obdurate about Babbage's preface. It was through this translation that many English men of science first became aware of Ohm's contributions to the theory of electric circuits, although his famous law—that the current in a circuit is proportional to the voltage across it—had been announced in 1827. Both Whewell and Ohm sank without a trace in Ada's plans, but her interest in electricity continued. Electricity seemed to be everywhere, and it was credited, through such concepts as "animal magnetism," with influencing many occult phenomena and a good deal of human behavior.

Wheatstone continued to ply her with papers for translation. No sooner had she laid the memoir to rest at the printer's than he reminded her of a translation of "Seebeck's memoir in the transactions of the Berlin Academy," which she had too long neglected. If she felt she could not undertake it, the book had better be'returned to Mr. Taylor. Seebeck was the discoverer of the principle of the thermocouple, another electrical phenomenon, though not an occult one. She did attempt a review of this work but failed to complete it. Nonetheless, a year later Wheatstone had far grander plans for her, as she explained to William:

He has given me much important information & still more important *advice*. He is anxious I should take such a position as may enable me to influence *Prince Albert*, who is (he knows) a very clever young man. . . . The Prince's whole & sole desire is to be at the head of a *scientific* circle in England, & he has expressed his utter mortification at the opposition & cold water which have been thrown on all his desires in this respect.

Wheatstone says none but some *woman* can put him in the right way, & open the *door* to him towards all he desires; & that a *woman* can say that which any *man* would get into a scrape by doing. Wheatstone does not wish me to think of doing anything immediate. By no means. But he says it would occur in the *natural course* of things that *if* I can take a certain standing in the course of the next few years, the Prince would on some occasion speak to me about *Science*; and that in that case, if I happily seize the moment, I may do for science an *inestimable benefit*; for all that the Prince wants is a sensible *adviser & suggester*, to indicate to him the *channels* for his exercising a scientific influence. He is very clever they say, but in a *slow* way; not a *brilliant* man.[14]

It was all very curious; within the same week not only Wheatstone but a new scientific friend of hers, John Crosse, had given her the very same counsel that she should try to become Prince Albert's scientific adviser. It is strange indeed that Ada's advisers, who were men of the world, should have outstripped their protégée in naïveté. The prince was at this time a very young man, and in a very delicate position. A foreigner in a xenophobic court, he was still in the process of establishing his hold over an emotional and self-important young woman, accustomed to the most chivalrous, flattering, and subtle manipulation by her ministers. The combination of such a situation with Ada's propensity for unconventional opinions, candor, and startling confidences is delicious to contemplate. Ada herself had some reservations about this plan with respect to the Queen. "I don't like some traits I hear, I must say; not from *Tories* but from liberal people. I fear she is selfish in certain ways, & about her *husband* too."

Unfortunately for biography, nothing came of the scheme to loose Ada in the path of the Prince, as 'twere by accident. Yet it is rather pathetic that scientific reformers and activists should have felt reduced to grasping at such straws. Nor was this the only such plan to waylay the powerful. Babbage's friends and supporters were given to discussing the benefits that might be derived from chance encounters with the Duke of Wellington, and before Ada's Notes were even completed, Babbage himself was planning to send a copy to the Prince. But even when Albert was nearing the zenith of his influence, his favor was insufficient to secure for the inventor the position on the committee for the Exhibition of 1851 that he so longed for and so richly deserved. The Calculator had added too many enemies.

In addition to becoming a high priestess of science, Ada wanted to make important discoveries of her own. At the beginning of 1841, she had made a floundering entry in her journal on the subject of imagination. It teaches us, she said, that "Death is Birth." After chastising her father's dark vision as excessively morbid, she wandered a bit in circles:

Imagination is the *Discovering* Faculty, pre-eminently. . . . It is that which feels & discovers what *is*, the REAL which we see not, which exists not for our *senses*. . . . Mathematical science shows what *is*. It is the language of unseen relations between things. . . . Imagination too shows what *is*. . . . Hence she is or should be especially cultivated by the truly Scientific, those who wish to enter into the worlds around us![15]

Finding this venture into Coleridge's territory seemed to lead nowhere, she desisted; and by the time she was working with Babbage, and no doubt under his influence, she had decided to be more methodical about the "Discovering Faculty." In July 1843 she reported to him,

I intend to incorporate with one department of my labours a complete reduction to a system, of the principles & methods of *discovery*, elucidating the same with examples. I am already noting down a list of discoveries hitherto made, in order myself to examine into their *history, origin, & progress.* One first & main point, *whenever & wherever* I introduce the subject, will be to *define & classify* all that is to be legitimately included under the term *discovery*.

Here will be a fine field for my *clear, logical, & accurate* mind, to work its powers; & to develop its *metaphysical* genius, which is not the least among its qualifications & characteristics.[16]

Babbage was himself the veteran of many attempts to reduce discoveries and inventions to a system, and this was written while she was in the full flush of her eagerness to be associated with him. By the fall of 1843 she was no longer so sure, and on his side the disengagement was even more rapid and complete. There is no mention of their "book" in any of his correspondence with other people. The only surviving mention of it in a letter to her—years later, it is true— shows its lack of priority: he had an engagement, he told her, that would prevent him from attending to her wishes concerning it.[17] The "book" remained a convenient pretext around which stimulating conversations and visits could be fashioned; meanwhile she began to consider that a more experimental approach and direct experience might be in order for the kinds of discoveries she now wished to make. The person she turned to as her new teacher, collaborator, and assistant was none other than Michael Faraday.

Of all the congratulations and acknowledgments that followed the distribution of the Menabrea memoir, perhaps none is so surprising at the present time as that sent to Babbage by Faraday, the most revered physical scientist of his day. "I think I may thank you for the Translation," he wrote, "& though I cannot understand your great work yet I can well comprehend by its effect on those who do understand it how great a work it is."[18]

That Faraday knew no mathematics is particularly surprising in view of the wide-ranging interests and expertise of so many of his mathematically-inclined friends and contemporaries: Babbage himself, John Herschel, Augustus De Morgan, and Charles Wheatstone, to name but a few. But Faraday's case was a peculiar one. He had never been to a university, obtaining his entire scientific education from private reading, public lectures, and on-the-job training. In his case, unlike those of Mary Somerville, George Boole, and Babbage, his autodidactic activities had not led him into mathematics. In fact, he took an almost perverse pleasure in the view that he could make scientific discoveries as well, or better, without any formulae, mathematical or chemical. Once he even suggested to James Clerk-Maxwell, who worked out the mathematics of Faraday's brilliant and empirically arrived-at electromagnetic theory, that mathematicians should publish their results in duplicate, the second version being a translation of the "hieroglyphics" of the first. He boasted that the only mathematical operation he had ever performed was turning the handle of the Difference Engine, and he was the last important physical scientist to be able to take such a position.

Faraday came from an impoverished family. He was the third of ten children of a hard-working blacksmith, whose health had failed under the onslaughts of misfortune and family responsibility. Michael was apprenticed to a bookbinder at the age of thirteen; before then his schooling had been of the most rudimentary kind. However, his employer encouraged him to read the books he handled, at least after working hours. From the *Encyclopaedia Britannica* he learned the elements of what was then known about electricity, and from *Conversations on Chemistry*, by Mrs. Jane Marcet, he absorbed his first notions of chemistry. Fascinated by "facts," he checked Mrs. Marcet's assertions by means of such experiments as he could perform with the meager resources at his command and his own ingenuity. He was indelibly impressed when he succeeded in confirming many of her statements. The ability to conduct research on a pittance, thus acquired, was to prove of lifelong value.

Conversations on Chemistry was the first of Mrs. Marcet's many educational books. It was originally published in 1806, with the sort of subtitle then thought appropriate for a female-authored work: "Intended more especially for the Female Sex." Her most popular work was *Conversations on Political Economy*, published in 1816, the inspiration for Harriet Martineau's *Illustrations of Political Economy*; another of her books was *Conversations on Natural Philosophy*, an introduction to the physical sciences for young children. Mrs. Marcet too knew no mathematics. Commenting on her role as a science popularizer, Faraday's biographer remarked in 1965,

Mrs. Marcet wrote for an audience newly created by Humphry Davy. . . . Davy captivated the young ladies of the higher ranks of English Society who flocked to the Royal Institution. So successful was he, indeed, that he served to subvert the original purpose for which the Royal Institution had been founded in 1799. Benjamin Thompson, Count Rumford, had originally envisioned it as a center for the dissemination of practical knowledge to the artisan class, but financial difficulties and Davy's genius soon turned it into a centre for chemical research and popular scientific lectures. . . . It was to this audience that Mrs. Marcet addressed her *Conversations*, so that Davy's followers could understand as well as admire him.[19]

It is a common and still persistent view that the enthusiastic attendance of women at a scientific lecture during that period was due mainly to the glamour of the lecturer. No consideration is given to the paucity of opportunities then available for women to learn about science — any more than to the possibility that men too might profit from simple scientific texts. It is also notable how often the interests (in both senses) of two educationally deprived groups, upper-class women and working-class men, were set in opposition, with the instruction of the one deemed to subvert that of the other. If Faraday was considered a legitimate beneficiary of the facilities for informal instruction, Ada and Mary Somerville, who were eager attendants at lectures, demonstrations, and exhibitions, were somehow suspect.

In due course Faraday progressed from reading Mrs. Marcet and performing simple experiments to attending the public lectures he could afford. There he took copious and meticulous notes, which he bound himself. To be able to illustrate his notes, he took lessons in perspective drawing, dusting his teacher's room and blacking his boots as payment. Then one of the bookshop's customers, who was a member of the Royal Institution, gave him tickets to four of Davy's lectures. There he took his usual careful notes, fair-copied them in copperplate calligraphy, and forwarded them to Davy in support of a request for employment. Eventually he got his wish and was hired as Davy's assistant. Taken by his employer on a Continental tour, he was forced to act as Davy's valet as well, but he did meet a number of continental men of science. Once returned, he settled down to surpass his patron in a career of fundamental discoveries in chemistry and electricity.

Despite the recognition and honor that eventually came his way, Faraday remained as sharply aware of his class origins as Mary Somerville was of her sex. Both disarmed the resentment of the gentlemen with whom they had to deal by resolutely modest and deferential behavior. Both kept scrapbooks of letters received from "celebrities," among whom Ada was counted, and Faraday kept portraits as well. It was Babbage who in 1840 relayed his request to Ada for a picture, and she obliged with one of the engravings from a set of hasty sketches by Châlon that had been executed several years previously. Ada had been even less pleased with the dashing and exotic look this artist had given her than she had been with Mrs. Carpenter's wide jaw. She complained to her mother that his dresses were in very bad taste, stiff, and extreme in fashion. But Faraday could find no flaw in the offering. He thanked her effusively, and their friendship developed.

Beginning in 1841, when her lessons with De Morgan were almost as enthralling as her collaboration with Babbage was to be, Ada began to reveal hints of an occult purpose. To Mrs. De Morgan she wrote, "I hope before I die to throw light on *some* of the dark things of the world."[20] The dark things of the world included, besides the flaws in her father's character, her premonitions of her own preternatural powers, the management of her treacherous physical and emotional state, and the ultimate design of the universe by a deity in whom she only fitfully (but then extravagantly) believed. Then, too, she had to contend with her granite-willed mother's dogmatic certainty about all matters medical, spiritual, and phrenological.

Emboldened by her seeming progress in mathematics, she applied to the Greigs—Woronzow was now married to a fellow Scot—for help in some of her more arcane enquiries. She was interested in the action of poisons, she said. They obliged by sending her a copy of Robert Christison's 1829 *Treatise on Poisons*. Christison was a professor at the University of Edinburgh and a world authority on the subject. He had a special interest in the toxic action of opium, which was just then becoming an object of growing medical unease after centuries of acceptance by both professionals and the public. The book, however, was not what Ada had in mind. It had too many details and not enough general principles: "just a few specific instances by way of illustration, is what I want . . . an epitome of the *deductions* to be made by philosophical & learned minds, from all the facts known about poisons in conjunction with organized life."[21]

This was a tall order; with her mother she was even more sweeping in her requests for reading matter, now that she was "under the dominion of electricity":

I must manage to collect everything I can relating to the Nervous System. . . . I mean to go and see the Animal Magnetism. . . . Dr. Elliotson is to show me by and bye *his* present most striking case. I hear that one theory of the *mode of action* of Mesmerism is that it is an *abstraction of the Electricity* of the body, electricity being the bond of union between the *mind & muscular action*. I only heard this crudely stated just as I now repeat it.[22]

In her next letter she asked Lady Byron, then in Paris, to find out anything she could "curious, mysterious, marvellous, electrical, etc."

Lady Byron was very much interested in mesmerism, or "electro-biology" as it was also called; but unlike Ada, once convinced she never wavered in her faith in its more occult manifestations.

Mesmerism took its name from Franz Anton Mesmer, a Vienna-trained German physician who popularized hypnosis as a medical procedure and evolved a theory, later very much elaborated by others, to explain its effects. Proceeding from an interest in astrology, which posited an influence by the stars on human behavior, he at first considered that the stars' force was electrical in nature and later that it was magnetic. Hence he began to stroke the diseased bodies of his patients with a large magnet to cure them. The first news of his success was published in 1766. Then, after observing a priest attempting to exorcise a demon and able to obtain the same effects without a magnet, Mesmer too discarded the magnet. The force, he realized, must reside in himself. He theorized that the universe was permeated with a rarefied cosmic fluid, complete with tides, that had a special affinity for the human nervous system. His term for the effects produced by this fluid was "animal magnetism"; there were to be several changes of nomenclature before the modern term "hypnosis" became quite general.

Mesmer moved to Paris about a decade before the French Revolution. In the ferment of ideas then current, he soon attracted many clients, despite the condemnation of the medical faculty of the University of Paris. His technique was quite theatrical. His patients sat around a kind of vat of water, sometimes with their feet immersed, and, holding hands in a circle, conjured up an image of an electric battery and circuit. Mesmer, robed in black, flitted among them, making "passes," or cryptic hand gestures, at one or another. The more susceptible often fell into nervous fits or fainted.

Finally, a government commission, including the chemist Antoine Lavoisier and the American ambassador Benjamin Franklin, was set up to investigate Mesmer's practice. They too condemned it, and the movement went into a temporary decline. Popular interest revived in the nineteenth century, and in 1831 a committee of the Academy of Medicine found that "magnetism" had merit as a therapeutic technique. From the first, some observers had attributed the effects to psychological causes but this view was felt to depreciate its value.

Compared with the Continent, England was rather slow to take up mesmerism. Many respectable phrenologists feared it was tainted with quackery. Despite the suggestions that the effects were due to the expectations and suggestibility of the patients rather than to a cosmic fluid or an occult force within the mesmerizer, most practitioners and patients continued to associate it with such phenomena as sleepwalking, "tableturning," astrology, clairvoyance, and the other accouterments of mediums and spiritualists. Since then, of course, the positions have been reversed: phrenology is forgotten, or condemned as pseudo-science, whereas hypnosis has attained a certain standing, especially as an adjunct of psychoanalysis, which now plays a role similar to that once occupied by phrenology.

Mesmerism was finally given respectability in England, temporarily at least, by the advocacy of Dr. John Elliotson, a London University professor of medicine and physician at the University College Hospital. He was one of the first to exploit the value of clinical teaching and to foster the use of the stethoscope. He had also been the first president of the London Phrenological Society, and after he became interested in mesmerism he insisted that the two "sciences" be linked.

The technique of mesmerizing, or hypnotizing, then as now, involved inducing eyestrain by asking the subject to fixate upon some small object while simultaneously relaxing the rest of the body. The mes-merizer passed his hands slowly and rhythmically before her face, intoning monotonously. All this induced a sleepy state in which the subject was relatively open to suggestion. The mesmerizer attached more importance to the "passes," however; they were the means of imparting the "magnetic fluid" to the patient. The fits and fainting that often followed fluid passage were held to be part of the cure, though inevitably they were endowed with sexual significance by hostile observers. Mesmerizers were predominantly male, and subjects pre-dominantly female.

Since the mechanism by which the mesmeric effect was produced was so little understood, positive instances of mesmeric power pro-liferated; it was claimed by various investigators that all sorts of ma-terials and objects could produce a variety of effects at greater and greater remove. One German investigator, Baron Karl von Reichen-bach, wrote learnedly of his experiments with "sensitives," people who could see emanations from human fingers, from magnets, and from "magnetized water." Magnetic emanations could be seen, by those with eyes to see, emerging from graves; they could darken daguerreotype plates. Above all, however, mesmerism became as-sociated in the public mind with Victorian fascinations with energy, will power, and barely controlled sexuality. It was frequently alleged that mesmerizers used their powers to seduce their patients, who were always suspected of repressed sexual leanings.

Unfortunately, Elliotson's investigative techniques included public demonstrations of the mesmeric trance in its more melodramatic man-ifestations; he thought the presence of an audience made the trials more scientific and objective. He failed to realize, however, that the repeated use of the same pair of well-trained, lower-class sisters as subjects (his "most striking case") could damage his credibility, especially in the eyes of the editor of the influential medical journal *Lancet*, which was then used to retail scandalous allegations against him. He also came in conflict with the medical committee of the University College Hospital, and was forced to resign and continue his investigations at home—where Ada presumably went to see him. Elliotson founded both a journal and a hospital devoted to mesmerism, but eventually he gave up his belief in clairvoyance.

In her Notes to the Menabrea memoir Ada had at one point cautioned her readers, "In considering any new subject, there is frequently a tendency, first, to *overrate* what we find to be already interesting or remarkable; and, secondly, by a sort of natural reaction, to *undervalue* the true state of the case, when we do discover that our notions have surpassed those that were really tenable."[23] It was a tendency to which she herself was particularly prone. She was surrounded by a number of strong proponents, as well as opponents, of the faith in mesmerism. Of the former, her mother, who was curiously oblivious to the improprieties attributed to it, was of course the foremost. By the autumn of 1844 Ada's interest and credence had reached such a pitch that, as she informed Lady Byron, her advancing studies of the nervous system left her "more persuaded that I was mesmerized [in 1841], & that all my ill-health had its *foundations* in that." Although she thought the mesmerism had affected her stomach instead of the "sensations & mental system," yet it gave rise to "at least 3 years of sufferings & of unnatural feelings mental & bodily." If her mother approved, she planned to consult her friend Miss Martineau, who was then busily spreading the news of a recent mesmeric cure that she herself had experienced. "I want to draw her powerful understanding to the subject," said Ada, once more in the role of executive, "more *systematically & forcibly*, than perhaps would occur unless she is a little *suggested to*."[24]

In spite of a prodigious output of writing, fancywork, good works, and well-publicized trips (including an exhausting sojourn in America), Harriet Martineau was, during some periods of her life, one of the famous Victorian invalids and a celebrated medical case. Like Ada, she was wont to blame her "nervous system" for her woes. As an infant she had been fostered out and had almost starved when her poverty-stricken wet nurse was reluctant to reveal that she had lost her milk. After this was discovered, her mother compensated by forced feeding, and she developed the usual delicate stomach. She was stricken with deafness in adolescence; in addition, she claimed she had always lacked a sense of smell and hence of taste. Her late thirties brought a complete collapse of health, which was diagnosed by her physician brother-in-law as a prolapsed and tumorous uterus. Surgery was considered too risky, so he treated her with opium and an iodine tonic; but she remained largely confined to her sofa and in pain for six years.

No longer able to write enough to support herself, she was offered a government pension, which she refused, believing that acceptance would compromise her principled objections to the tax laws and her political independence generally. For this she was much criticized, many of her acquaintance considering her refusal to be evidence of insanity. Nevertheless, her friends got up a testimonial fund to supply her needed income. Lady Byron, while of the opinion that her reduced circumstances were freely chosen—and therefore not to be pitied— was moved by her inability to continue her former charitable work and sent her £100 a year, delicately made out to a third party, in order to remedy this loss.

In the spring of 1844 Miss Martineau's condition began suddenly to improve. Just at this point her doctor suggested that she try the services of an itinerant mesmerizer. After he junketed on, she and her maid continued the treatments as best they could until a suitable substitute—a female practitioner this time—was found. Soon Miss Martineau was able to dispense with her opiates and tonics, and by the time Ada wrote to her she was walking several miles a day. Convinced that mesmerism had cured her, she was now eager to develop the clairvoyance that was said to accompany it. Instead, her maid, who was undergoing mesmerism to improve her eyesight, suddenly revealed herself as a "somnambule" and also accurately predicted the safe return of the crew of a wrecked ship. Swept by enthusiasm, Miss Martineau published the whole account as a series of "Letters on Mesmerism" in the *Athenaeum* late in 1844. Ada must have seen some of the letters before publication.

A storm of criticism followed, as well as approval; it was pointed out that the news of the crew's rescue could have reached the ears of the "clairvoyant" before she made her prediction, but Miss Martineau was unshaken, choosing instead to break with those friends and relations who found her new faith an embarrassment. She never did recant, but when her symptoms returned ten years later—eventually killing her after decades of slow growth—she reverted to opiates instead of mesmeric treatment.

At the same time as she determined to write to Miss Martineau, Ada also wrote to Faraday, proposing to become his student and to repeat all his experiments under his supervision, doubtless with the intention of eventually investigating the mechanism of mesmerism (which in fact Faraday later did). His reply was delayed, as he explained, by a breakdown in his own health. Now he was recovering and had "the difficult pleasure of writing to you." She had made her "high object" and her "mind and powers . . . fully manifest" to him, and he would rejoice to help her; but "nature is against you."

You have all the confidence of unbaulked health of youth both in body & mind; I . . . feel the decay of powers, and am constrained to a continual process of lessening my intentions and curtailing my pursuits. Many a fair discovery stands before me in thought which I once intended, and even now desire, to work out; but I lose all hope respecting them. . . . Understand me in this:—I am not saying that my *mind* is wearing out; but those physico-mental faculties by which the mind and body are kept in conjunction and work together, and especially the memory, fail me. . . . It is this which . . . has tended to withdraw me from the communion & pursuits of men of science my contemporaries . . . and which, in conjunction with its effects, makes me say, most unwillingly, that I dare not undertake what you propose, to go with you through even my own experiments.[25]

He then confessed to her that he often had attacks of giddiness and reeling of the head, for which his "medical friend" prescribed rest and a sojourn at the seaside.

Beginning in 1831, Faraday had indeed progressively curtailed his social activities and contacts. After 1834 he refused all dinner invitations, and after 1838, all visitors at the Royal Institution—where he lived as well as worked—for three days every week. By 1841 the giddiness, headaches, memory loss, and confusion were so severe as to preclude all research activities for the next three years, although after a lengthy visit to Switzerland that year he recovered enough to resume his lecturing schedule. At the time of his correspondence with Ada he was returning to his electrical experiments.

Faraday's condition has been called a "nervous breakdown," or rather a succession of breakdowns, supposedly due to the usual causes: overwork, stress, worry, guilt, anxiety, and childhood experiences—in short, psychological causes. It is clear from his correspondence, however, that his symptoms were more consistent with those of chronic chemical poisoning; they tended to clear during his therapeutic trips to the fresh air of the sea or the mountains—in any case, away from his laboratory. From very early in his career he made efforts to contrive methods for improving his memory, which nevertheless sustained gradual, cumulative, and irreversible damage. It is all the more strange that he never saw the possibilities of mathematical notation for this purpose, unlike Babbage, who was also concerned with maximizing his mental capacity.

In his laboratory Faraday produced, handled, and isolated a number of highly toxic and dangerous substances, as had his predecessor Humphry Davy. Davy had even made a practice of sniffing and tasting many of the new chemicals he discovered. Not surprisingly, his health declined rapidly and he died fairly young. During Faraday's first years at the Royal Institution he was involved in several violent explosions while handling such substances as nitrogen trichloride and muriatic acid. More than once he found himself picking glass out of his skin and eyes. Less spectacular but more insidious were the effects of such compounds as hydrofluoric acid, nitrous oxide, and ether. He also worked with several heavy metals. The first electric motor, which Faraday invented in 1821, involved a basin of liquid mercury, known to penetrate even the intact human skin with poisonous effect. The same is true of the aniline based on benzine, another of Faraday's discoveries. Later in the nineteenth century aniline dyes were used, first by German manufacturers, to produce brilliantly colored cloth, much in demand for fashionable clothing, that sometimes produced sickness and even death in the wearer. Dizziness, confusion, difficulty in concentration, and memory loss are among the symptoms of aniline poisoning.

An interesting point about Faraday's refusal to work with Ada is that the reason he gave was different from his usual justification for denying such requests. He did not make a practice of accepting pupil-assistants at all, even when they were referred and recommended by prominent men of science. As he wrote in reply to one such inquiry, he always prepared and carried out his own experiments, thinking to himself in silence as he did so. The assistant who set up his lecture demonstrations was supposed to be conscientious and obedient but to display no initiative. This may have been an outcome of the jealousy and bitterness with which Faraday's own relationship with Davy had ended or it may have been connected with his increasing difficulties in concentration and recollection—which he tried for some time to conceal from other scientists, if not from Ada. Yet Babbage, who was not so handicapped, also preferred employees to students.

Nevertheless, it is fascinating to speculate on what might have happened had he granted her request. The female experimental scientist was an even greater rarity than the female scientific writer. Mary Somerville had tried a few simple experiments, but she gave up and regretted the whole enterprise when one of her results was later disconfirmed. Only Caroline Herschel had had a successful and honored laboratory career, and she had been very much her brother's research assistant. Yet Faraday's example and training could have shown Ada what was involved in translating vast and general speculations into concrete testable propositions, and how to proceed step by step from observation to hypothesis to experiment. He might have enabled her to apply the painstaking intensity of which she was sometimes capable to investigations that did not require the symbolic manipulations she found so difficult. Of course, too, she might have found that there was more to chemistry and electricity (and to the investigation of mesmerism) than she really wished to know; and even the science-dazzled William might have raised objections when she began blowing up beakers and test tubes. But the eventual consequences of not working with Faraday were in some ways even more explosive.

His expression "of unbaulked health" must have struck Ada as particularly ironic, and indeed reveals how little he really knew of her. But his refusal, like Babbage's was couched in terms that in no way discouraged her ambitions or her search for a teacher and collaborator to help her realize them. Quite the contrary, as she reported to her mother:

Do you know it is to me quite delightful to have a frame so susceptible that it is an *experimental laboratory* always about me, & inseparable from me. I walk about, not in a Snail-Shell, but in a *Molecular Laboratory*. This is a new view to take of one's physical frame; & amply compensates me for all sufferings, had they been even greater.

By the bye, Faraday expresses himself in absolute amazement at what he (I think most happily & beautifully) designates the "elasticity of my intellect." Even from the little correspondence we have lately had, he seems quite *strangely* impressed with this characteristic; & says he feels himself a "mere tortoise" in comparison. As far as regards *himself*, he is so humble-minded, that I cannot take *his* estimate of his powers. As regards *me*, I see the *fact*, that he is (justly or not) in great astonishment. It is evidently his impression that I am the *rising star* of Science. . . . I may be the Deborah, the Elijah of Science.[26]

Curiously, the phrases Ada quotes from her correspondence with Faraday do not appear in the letters from him that survive; in fact, it is clear from his long letter that it was she who provided the description of her intellect, to which he merely assented. Yet this was enough for her to impute to him the impression that she was the rising star of science, and she now announced that her dedication to science, with

Faraday's flattering encouragement, justified her in declaring herself no longer the property of mother or husband.

Lady Byron's response to all this was remarkably calm. At another point she remarked on Ada's ravings, "I hope the Self-esteem will not conduct its Owner to a Madhouse—it looks rather like it," but now she only declared herself amused. She was "glad you are in communication with Faraday—his tone of mind seems to me so truly scientific—more so than that of another friend of yours." Meaning, of course, Babbage.[27]

Ada had not neglected to query Faraday on his moral and religious as well as his scientific beliefs. His answer, he said, would disappoint her. He did not agree that high mental powers ensured high moral sense. Nor did he see his religious beliefs as necessarily connected with his science. "There is no philosophy in my religion," he said; "I do not think it at all necessary to tie the study of the natural sciences and religion together."[28] Faraday belonged to a small fundamentalist Christian sect, now extinct. Unlike Ada and Lady Byron, and unlike the exponents of natural theology whose views are represented by the Bridgewater Treatises, he felt that faith was primary and needed no justification by science; he had no difficulty in separating the two.

His materialism in his professional activities was eventually to lose him Lady Byron's approval. Although by then so concerned with Ada's loss of faith that he discussed it with a mutual friend, he agreed to visit her during her final illness to give her his opinion of the mesmerical treatments Lady Byron was attempting to foist on her. "It happens," Ada wrote her mother, "to be [a subject] he has seriously considered."[29] Lady Byron was furious at his skepticism, which became public the year after Ada's death, when he reported in letters to the *Times* and the *Athenaeum* that according to his investigations, "table turning" (psychokinesis) was due to inadvertent motions of the muscles of the hands placed upon the table's surface during seances. His views pained many, including Elizabeth Barrett Browning, and drew down upon himself the withering logic of Lady Byron, who wrote to her friend and fellow spiritualist Sophia De Morgan,

Will not the Professor bring his logic to bear on Faraday's illogical letter in the Athenaeum? It is before me—"The influence of Expectation" assumes as a fact what certainly was *not* the fact in some of the Experiments—For instance when a child of 3 years old produced very strong effects—"A *quasi* involuntary action"—This is a new class of action to be added to the Voluntary & the Involuntary. There are assumptions without end. . . .

After all, what has Faraday proved, granting his proofs? 1^{ly} That the Tables do not *turn* the hands but the hands the Tables 2^{ly} That the Force which affects this in the hands, is not Volition, nor the direct intention to produce motion, but a self-deceptive Wish—A curious Metaphysical question at least—& the Philosopher had better leave it to that School & not be so angry.[30]

At the top of this letter Mrs. De Morgan noted, "My husband wrote a strong critique on Faraday in the Athenaeum." Mrs. De Morgan presumed heavily upon her husband's position of *ignoramus* regarding occult phenomena; after his death she took the widow's advantage of claiming in her memoirs that he at last became a convert.

A few days after Ada declared herself the Elijah of Science to her mother, she wrote Woronzow Greig to discuss her plans in somewhat more detail: "I am so much occupied at this moment, in preparations & arrangements for *writing* [the "book"?] & also in recovering with a *very unsteady* progress from an attack of my but too common *Gastritis*, that I cannot say to you all I could desire." After thanking him for collecting books and materials for her, she continued,

My own scientific plans become more & more definite. But this is quite *in confidence* to you.

I have my hopes, & very distinct ones too, of one day getting *cerebral* phenomena such that I can put them into mathematical equations; in short a *law*, or *laws*, for the mutual actions of the molecules of brain; (equivalent to the *law of gravitation* for the *planetary & sidereal* world). I am proceeding on a track quite peculiar & my own, I believe. There are many & great difficulties, but at present I see no reason to think them insurmountable.

In case Grieg had any doubts about this, she reassured him, "L– knows *all* my plans & views, & seems to think them not absurd, and that if not *feasible* in themselves they may lead to what *is*, in the course of the investigations." The great difficulties she foresaw were experimental, not mathematical:

The grand difficulty is in the *practical experiments*. In order to get the exact phenomena I require, I must be a most skilful *practical manipulator* in experimental tests; & that in materials difficult to deal with; viz: the brain, blood & nerves of animals. In *time*, I will do all, I dare say. And if not, why it don't signify, & I shall have amused *myself* at least. It appears to me that none of the physiologists have yet got on the right tack;—I can't think *why*.

Have you heard about Miss Martineau & the Mesmerism? There *can* be *no* doubt of the facts, I am persuaded. I have seen her letters, some of them.

All this bears on *my* subject—It does not appear to me that *cerebral* matter need be more unmanageable to the mathematicians than *sidereal & planetary* matter & movements; if they would but inspect it from the *right point of view*.

I hope to bequeath to the generations a *Calculus* of the *Nervous System*.[31]

In other words, she was planning to subject biological tissues to electric and magnetic fields—to investigate and even quantify mesmeric influences on the mind and body. She was hoping to reduce mesmerism to mathematical formulae!

Greig was in no position to evaluate Ada's plans to explain mesmerism scientifically. Babbage and Wheatstone, who might have done so, were both extremely skeptical about the more occult aspects. Yet there is no evidence that Ada consulted Babbage about her project, unless it somehow related to the subject matter of "the book." The two notes she sent him just at this period were those relating to the book and to the visit of Sir David Brewster, who was sympathetic to mesmerism and was sometimes cited by Lady Byron as an authority. Yet in spite of her fascination and near conviction in the autumn of 1844, and without having performed her planned experiments, by the end of the following January she was writing to Greig's wife, Agnes, "I have heard of the new Mesmerism, which I think a new *humbug.* You will think me very crabbed I am afraid. All these yearly new marvels provoke me with the *folly* of people who really ought to know better. I cannot believe in these *oppositions* to all experience, science & philosophy."[32] What had happened to change her mind?

With Faraday unable to oblige her, Ada turned, within a few days of her letter to Greig, to Andrew Crosse, another gentleman scientist and a specialist in experiments with the voltaic battery. Like Babbage, he financed his scientific activities out of his patrimony. Again like Babbage, he was an embittered man, having been wronged by the Royal Society. In 1837 he had observed some insects apparently growing in a metallic solution, one in which he had been attempting to grow crystals under electrolytic influence and in which he had believed no life could be sustained. Although he later protested that he had merely described his observations and had offered no opinion regarding the insects' origins when he reported the phenomenon to the Royal Society, he was derided as having made claims of spontaneous generation. Hurt and angry, he retired to his family estate in Somerset to pursue his work in isolation.

After his first wife died in 1850, he married that Cornelia Crosse whose memoirs were quoted in the previous chapter. Mrs. Crosse inherited her husband's letters from Ada; in 1869—a year of renewed interest in Byroniana—she produced an anonymous article in which extracts from some of these letters were published. The article repeats some of the memoir's phrases and is identical with it in style.

The Crosse estate, called Fyne Court, was situated not far from the Lovelace "hermitage" at Ashley Combe. Ada had been acquainted with Crosse for some years (in 1842 she had corresponded with him about possibility of her repeating his experiments on the production of electrical insects) when she decided to become his houseguest in order to learn more about the techniques of experimenting with electricity. Despite a statement in Mrs. Crosse's article to the effect that Ada's "visits" took place in 1841 and 1842, it is clear from her reports to William in November 1844 that this was her first sojourn at Fyne Court. Internal evidence in Mrs. Crosse's published extracts also places it at the same period. In one letter to Mr. Crosse, which seems by comparison with her letters to Greig and Lady Byron to have been written in preparation for her visit, Ada enclosed some papers and commented,

I am anxious that *we should try the experiments* mentioned; and you may require a little preparation possibly for the purpose. (One of these experiments was on sound, produced in a bar of iron by electro-magnetism)* . . . The letter in the large handwriting is an account of an experiment with the muscles of frogs, which I hope we may manage; but I should think it required delicate manipulation. . . . I am anxious to consult you about the most convenient and manageable and portable forms for obtaining constantly acting batteries; not great intensity, but continual and uninterrupted action. Some of my own views make it necessary for me to use electricity as my prime-minister, in order to test certain points experimentally as to the nature and *putting together (con-sti-tu-tion)* of the molecules of matter. . . . By eventually bringing *high analysis* to bear on my experimental studies I hope one day to do much.

Another letter confirms her arrangements for the visit and continues,

---

*The sentence in parentheses seems to be an interpolation by Mrs. Crosse. The elisions in the first of the extracts are reproduced as they appeared in the published article; those in the succeeding extracts are my own.

Perhaps you have already felt from the tone of my letter, that I am more than ever now the bride of science. Religion to me is science, and science is religion. . . . The intellectual, the moral, the religious seem to me all naturally bound up and interlinked together in one great harmonious whole.

Faraday's opinion notwithstanding. Having established her theological position, she now felt compelled to forewarn her host about the unpredictability of her bodily state.

I think I may as well give you a hint that I am subject at times to dreadful physical sufferings. If such should come over me at Broomfield, I may have to keep my room for a time. In that case all I require is to be *let alone.* . . . With all my wiry power and strength, I am prone at times to bodily sufferings, connected chiefly with the digestive organs, of no common degree or kind.[33]

She added that she did not regret her "sufferings and peculiarities"; they were a form of instruction and discipline—as we have seen, she liked to think of her "susceptible frame" as a "Molecular Laboratory."

Ada arrived at Fyne Court, in the village of Broomfield, on 22 November. William remained behind, and the following morning she sent him an account of her journey and her first hours in the Crosse household.

Finding no *symptom* of either breakfast—or of human beings, I have sat down to write to *you*, in my *shawl & bon* in their *very* cold Sitting Room.
I was down at nine, having been told there [would] be breakfast. We all sat up reading & talking philosophy till *one o'clock* last night. I suppose *that* is the secret of the dawdle this morning. My head is very *muzzy* this morning, from that cause I think, & I shall take the liberty *this* evening of calling my hosts to a recollection of *Time.* The droll thing was we were discussing the metaphysics of *Time & Space*; & in so doing we forgot real Time & Space. . . .
Our journey was agreeable. In the course of it I was able to give Crosse a complete outline & vista of all my scientific plans & ideas. He seemed to think there is much merit in them & remarked "I see no mere enthusiasm in all this. It is all so *quietly reasoned*, so soundly based."

Indeed. In those heady days, when the nature of electrical action was barely beginning to be revealed, perhaps anything seemed possible, at least to the man who had discovered *Acarus crossii* improbably living in an electrolytic solution. Or perhaps this was another case of Ada's embroidered reporting, in which she took polite assent for conviction.

Ada went on to describe a member of the family, recently returned from Germany, with whom she had apparently not previously been acquainted:

I find my visit is of *more* importance to me than I had anticipated. The eldest son John is giving me information as to scientific doings in Germany which it is of the utmost importance I should be in possession of, & he will undertake, if I choose, to be my *organ* to any extent he can in procuring me every means of keeping *au courant* as to *German* mathematics & natural philosophy. He is a most *extensively* read, & also a very clever young man. I can get from & by means of him, what I could from *no* one else. He is to be 6 months at Berlin next year, & he will occupy himself in *catering* for me. — John Crosse says that in Germany books of *dry science sell* even better than light works; — that for instance such a one as *Wilkinson's* would *pay richly*, had it been written in German. This is a hint which I shall endeavour to get more information upon; for my own future purposes. I believe that many of my subjects would be read to an extent, in *that* country, which they never would here, (besides *paying* well). . . . It would be a proud thing to write German scientific treatises. I know very nearly enough of the language.[34]

The next day she wrote again. She was obviously enjoying herself and finding her visit stimulating beyond expectations; but, thanks to the disorder of the establishment and her host, she was not making much progress in the laboratory lore for which she had come.

This is certainly a most extraordinary domicile to visit at. It appears to me to be the most *unorganized* domestic system I ever saw. . . . All this is just so much of a novel field for observation & analysis of the human atom, and as the interests of my avocations will doubtless oblige me at times to be more or less a guest in this dwelling, I am obliged to study how to fall into the circumstances for the time being, in the most easy & agreeable manner.

There is at least *one* very unusual & agreeable circumstance. I am treated without *let* or *hindrance*, & left to do exactly as I like. . . . This is just what I want. I am a nobody here. . . .

The oldest son is a *frank & cordial* person; & we take to each other I think. He has a mine of reading & references in him to me quite invaluable. *Me* he evidently regards more as he would a *young man* than as a fine lady. I do not mean to say that he is wanting in any point of manners due to a *woman*. Far from it, but the *lady* & the woman are quite merged in his simple consciousness of the *intellect & pursuits.* He addresses *them,* not *Lady Lovelace.* . . .

There is in Crosse the most *utter* lack of *system* even in his Science. At least so it strikes *me.* I may be mistaken. Perhaps I don't see *enough,* as yet, to discover his *system.* . . . I have quite a difficulty to get him to show me what I want. *Nothing* is ever *ready.* All chaos & chance.[35]

Her recurrent "gastritis" had been causing some discomfort, but she assured William that it "didn't signify." Mrs. Crosse, the first Mrs. Crosse, was at that time an invalid, a circumstance that might help to explain the chaos, if the causal connection was not the other way around. From the next day's installment it was clear that Crosse's system was undiscoverable; Crosse's son, however, continued to be all that was charming, valuable, and engrossing.

Dearest mate,
All well, but not much *time.* I could stay here a *fortnight* with advantage. I have made the *whole family* laugh heartily with my witty fun about the *chaotic* nature of the establishment & proceedings. Old Crosse delights in my *quizzing* him. The *playful Bird,* you know!! . . .

Young Crosse is an excellent *mathematician* as well as meta*physician* &c; & he *works* my brain famously for he opposes everything I advance, *intentionally*; but with perfect good humour. This is very useful & good for me. He will be an addition to my catalogue of *useful & intellectual* friends, in many respects. . . .

There is *no order.* Everyone straggles down whenever he pleases. . . . The post &c ALL *by chance,* & at all sorts of hours. I never saw the like.[36]

After leaving the turbulent Crosse household, Ada continued her witty fun by letter, but now a note of nostalgia entered. She had managed to adjust to the disorder rather well, after all:

My dear Mr. Crosse,—I found my gold pencil this morning in the pocket of the gown I wore on Tuesday evening. I believe I had put it there to *prevent* losing it, as I went up to bed that night. My journey was very wretched, so late, so cold, so dreary. I could not help lending my cloak to a lady who was my companion, and who seemed to me more delicate and in need of it than myself. This did not, however, add to my own physical comfort. Many times after it became dusk did I think of your hospitable "chaos," and wish myself back, and imagine to myself if you were all sitting down to dinner, and if you missed me at all or not. In short, I had in my own brain a very comical chaos composed of what I had left behind, and a thousand hetero-geneous ideas, all of them but half alive and stagnant through physical cold. . . . My gold pin does not come forth—but it is not a thing of much consequence. If a stray gold pin, however, does develop itself, don't fancy it is an *electrical* production, but send it to me. My kind recollections to the various heterogeneous atoms (organic and inorganic) of your chaotic mass.[37]

Especially to John. Ada made no secret in her letters to William that she did not like all the Crosses equally: she found the second son, Robert, and Miss Crosse rather disagreeable.

The physical discomfort that Ada had borne so cheerfully at Fyne Court can be appreciated only in the light of her daughter's complaints, years later, about the miseries of the Lovelace estate. For all his interest in architecture and building projects, the ascetic earl had little concern for the creature comforts of his household. In addition, Ada was at this time without the services of a lady's maid—a traveling aid for the genteel woman only slightly less necessary than a portmanteau. Ada had dismissed her maid shortly before this journey, explaining to her mother that her servant would "come between [me] & God." Lady Byron remonstrated, but William was able to reassure her on practical grounds. He had let the experiment go forward on a trial basis and had found to his surprise that Ada was capable not only of taking care of her own clothes but even of ministering to herself when she was sick. Only her books and papers remained in wild disorder—though he hoped, from her letters from Broomfield, that her stay was having a salutary effect there too, "by representing forcibly the want of system pushed to its fullest extent."[38]

Back in London, Ada wrote again to William, revealing that she had lost much of her enthusiasm for laboratory procedures as practiced by "old Crosse" and was apparently abandoning her plans for experimental research; but she was full of a long visit Wheatstone had paid her that morning, and of the advice he had given her, so uncannily identical to that she had received from the frank and cordial young Crosse. "My dear Crow," she had begun, "I feel rather disappointed at not hearing from you this morning. I wonder why you did not write me a line." But then she went on to announce that Wheatstone had been with her for five hours and had given her the important suggestion that she try to become Prince Albert's scientific adviser, which had been John Crosse's idea as well.

Oddly enough too, Wheatstone has spoken to me of the great importance of studying the *German* philosophy & science. Altogether, my visit to Broomfield seems to have been very opportune, & to have laid the foundation for much that is wanted for me just at this epoch in my progress. I am glad however that I stuck to my original day for leaving it. I should not have derived *much* more by *staying there* a few days longer at the present time.

As far as the *German* studies are concerned, *they* cannot be carried on while on a *visit*; and it is *at home* only that I should make any real progress as such. If young Crosse stays with *us*, some hours can be daily given for the time-being, to that purpose; but when one is visiting a *family* one must be *diluted* a little amongst *all* the members of it, of course; besides which the *electrical experiments* are a main engrosser of one's attention at Broomfield. Add to this that the *irregularity* of the family habits are a most serious drawback to my *health*, & to anything like *study* there. Wheatstone burst out laughing at the very *idea* of my having *stayed* there any time & yet Wheatstone is no man of *luxuries* certainly. He says it is the strangest & most *uncomfortable* house he ever stayed in.

Having got started on the subject, she could not resist going on:

There are certainly inconceivable *oddities* there. For instance: The *Water-closet* can only be got at *thro' the Drawing Room*; & of course it is perfectly evident the errand one is going on, since the exit leads nowhere else. I don't mind that sort of thing in the least, when it is inevitable, & I take everything coolly & as a matter of course. That is the only way. Sometimes they *lock up* the Water closet, & then one has to make a hue & cry after the key.

Although she might have boggled a bit at its location, Ada seems to have taken it for granted that a large manor house should have only one flush toilet in it. Nevertheless, it is interesting that she delayed her description of the sanitary arrangements at Fyne Court—if ever a place belied its name—until after she had left. Had she been afraid that William might order her away sooner or swoop down to rescue her if he knew the indignity in which she was living? The opening of her letter, too, suggests some uneasiness over his possible reaction to the very good time she had been having. It is similar to the anxiety she betrayed when her mother failed to write. Now she went to great lengths to justify the change on which she had decided:

I soon determined that I must try & get what I want with *young* Crosse at *my* house.

Wheatstone has given me *some* very striking counsels. I did not think the little man had such *depth* in him. I can't *write* it all to you, in even small part, but I know you will agree fully with him, when you do hear it.

The plan for my writing on Babbage's subject clearly *won't* do. Don't be vexed at this; a subject is fixed on instead, so it will make no difference, & I can as easily do the one as the other.[39]

Ten years earlier Ada had attempted to elope with her young tutor to the house of his relations; now she was planning to install her new tutor in her own home. Her constant references to John Crosse as "young" might have served to reassure William; in fact, John Crosse, born in 1810, was then in his middle thirties and about six years older than Ada. John Crosse had been educated at Exeter College, Oxford, where he took a second in classics in 1833. He went on to Lincoln's Inn to study law but left within a year. At the time he made Ada's acquaintance he seems to have been somewhat at loose ends.

From hindsight Ada's condescending treatment of such eminent men as Babbage and Wheatstone seems shocking—but that is the function of hindsight. Not having the benefit of the judgment of history, or of mesmeric clairvoyance, as the years went by Ada's friendship with Babbage was increasingly characterized by an affectionate high-handedness, and less by respect and admiration, on her part. Occasionally she would compare her esteem for him, not generally to his advantage, with that for her doctors, Gamlen and Locock.

In any case, now that her plans to write "on Babbage's subject" were to be shelved in favor of German studies, she plunged into her usual preparations. The very next day she cautioned Greig not to expect her "to shine forth as a discoverer *for the present.*" The following week she dispatched a more positive request. Assuring him she had "become as much tied to a *profession* as *you* are," she continued,

I think you could do me a service, unless you see in it anything *objectionable.* Pray consider well over it. —

When Lovelace became a member of the Royal Society several years ago, it was entirely on *my* account. But the inconvenience of *my* not being able to go there to look at the particular papers &c which I want, continually renders the advantages I might derive quite nugatory. . . . Could you ask the secretary if I might go in now & then (of a *morning* of course) to hunt out the things I require, being *cense* [supposed] to do so in *L's* name & *for him,* tho' it would *really* be for *myself?*—As you know the Secretary, you can judge if he is a discreet man, who would not *talk* about the thing or make it *notorious;* one who in short could understand *why & how* I want to get entrée to their library in a quiet & unobtrusive manner. . . . Perhaps early in the mornings.[40]

Although Grieg's mother, Mrs. Somerville, had been honored by the Royal Society, which had placed her bust in the Hall, she herself was not admitted there. Ironically, too, the admission of aristocratic amateurs, like Lord Lovelace, had been one of the abuses cited by Babbage in his *Reflections on the Decline of Science in England*, a work that had made him so many enemies in the scientific and political establishment as to doom his hopes of suitable recognition for himself.

There is no evidence that Ada was ever permitted to slip quietly in and out of the Royal Society library; but with the encouragement of Wheatstone and John Crosse she drafted, probably soon afterwards, the beginning of an essay on the molecular structure of matter. This might have been the other subject that had been fixed on. It is only a fragment, and written in quite general terms. In it she speculated that although to the naked eye the sets of nerves relating to different perceptual systems appear quite similar, invisible differences in molecular motion might account for differences in function:

Here lies a deep mystery as yet. A Newton for the *Molecular* universe is a crying want; but the nature of the subject renders this desideratum of improbable fulfillment. Such a discovery, (if possible at all), could only be made thro' very *indirect* methods; —& would demand a mind that should unite habits of *matter of fact* reasoning & observation, with the *highest imagination*; a union unlikely in itself.[41]

It was, of course, on just this unlikely union in her own mind that Ada piqued herself.

More substantial, and certainly done in consultation with John Crosse, was a review of an "Abstract of 'Researches on Magnetism and on certain allied subjects,' including a supposed new Imponderable by Baron von Reichenbach." This was probably composed in 1846, shortly after the appearance of the "Abstract," which had been translated and abridged from the German by William Gregory, a professor of chemistry at Edinburgh University. Dr. Gregory—who had clearly had access to the library of the Royal Society—was much impressed with the scientific learning evidenced by von Reichenbach's experiments and was quite willing to accept his findings. Ada's review was cautious, but it gives us an idea of what she might have had in mind for her proposed electrical experiments. It began, "This is an important pamphlet, both for the strictly scientific and for the intelligent public. It points out definite and feasible tracks for experiment on some of the more occult forces of nature." She noted that many of the experiments described might be performed by amateurs:

[W]e point this out, not because we should attach the same degree of
value to the experiments of amateurs as to those of the philosophical
& experienced manipulator with Nature's secrets. But very *numerous*
observations & experiments will be desirable in order completely to
test the reality of the facts brought forward by Baron von Reichenbach.
If amateurs, of either sex, would amuse their idle hours with experimenting on this subject, & would keep an accurate journal of their
daily observations, we should in a few years have a mass of registered
facts to compare with the observation of the Scientific.[42]

In view of her assertions of "professional" status to Greig and her
cousin, and of the encouragement of Wheatstone and Crosse, it is not
clear whether Ada placed herself in the ranks of the amateurs or of
the "scientific." In fact, she was writing at a time when the latter were
only beginning to distinguish themselves as a self-conscious group
from the educated upper classes, thanks in part to the *Decline of Science*
controversy and the founding of the British Association for the Advancement of Science in 1831, which initiated the process of dividing
Science into sciences. Except for a few professors of natural philosophy
and such special cases as Faraday, scarcely anyone actually earned
his living in the pursuit of science. Recounting his injustices in a statement to the prime minister written in December 1846, Babbage listed
the employment he had applied for and been refused, and the honors
and rewards that had come to other men of science. It was clear from
his list that the more lucrative positions were sinecures awarded to
clergymen.

In the passage here reproduced from the Reichenbach review, the
words "of either sex" were crossed out in pencil. Possibly Ada performed this operation herself; but other corrections and alterations—
of a purely stylistic nature—are definitely in another hand, very like
that of John Crosse. Though the result is hardly an improvement in
her prose—indeed, the changes sometimes "muddled the sense"—
she was now permitting another person to "meddle with my sentences,"
a liberty she had strictly forbidden to Babbage. In the margin of one
page there is a penciled note in Ada's hand: "Here follows the extract
of occurrence in *Pfeffel's Garden & c.* To be given when T. C. returns
me the rest." The inference is inescapable that young Crosse had
achieved the intimacy of a nickname.

Ada began a descriptive summary of some of the "facts brought
forward" by Reichenbach with the following: "It appears that certain
persons of highly sensitive Nervous organization, see luminous emanations from powerful Magnets, in complete darkness." Such magnetic
emanations had also been known to darken daguerreotype plates—

a fortunate circumstance in Ada's view, since it permitted experiments free from subjective bias. At this point she admitted a distrust of "mesmeric" phenomena and even of certain overenthusiastic scientists.

Some crystals, her review continued, are also luminous in the dark to the sensitive, who furthermore can feel warm or cool currents issuing from them. Perhaps this "Magneto-Crystalline" force, which could be conducted through wires and other substances, might also be tested on daguerreotype plates. "We believe," ventured Ada, "that it is as yet quite unsuspected how important a part Photography is to play in the advance of human knowledge." One experiment—which she considered to have been arranged for the least possible subjectivity and bias—utilized human subjects at opposite ends of a wire. It was found that when one grasped the near end, the other observed currents and emanations.

Ada then touched on the possible influence of terrestrial magnetism; here, however, she explicitly condemned any belief in mesmerism: "We associate it with a disgusting tissue of human imposture & weakness." Reichenbach, it seems, had instead suggested ways of placing the investigation of the connection between terrestrial magnetism and human physiology on a proper footing. But of greatest interest to her was the study of the reaction of nervous tissue to electrical influence:

It would be well to experiment with the prepared animal Nerve (see Mattenici's "Traité des Phenomènes Electro-Physiologiques" 1844,—in order to discover if any apparent effects, luminous or dynamical, are produced on this sensitive instrument while it is undergoing *electrical* influences,—or when brought into Juxtaposition with Magnets, Crystals &c. . . . [T]his hint is carelessly & hastily thrown out, & would require nice & close consideration as to *how* it might be followed up. . . . We have not ourselves fully examined the subject.

The vague prophecies and "hastily thrown out" suggestions, which are then modestly hedged as not fully thought through—although she had been thinking along these lines for years—are reminiscent of similar expressions in the Menabrea Notes.

The review went on to hint at another paper to follow, which would compare the work of Reichenbach with Faraday's recent research "on a somewhat similar subject." Finally, she looked toward a unified view of nature, including "what unsensed forces surround & influence us." The reference to Faraday's "recent researches" confirms the period of the writing of this review and does indeed indicate an interesting comparison with Reichenbach's science. On 9 November 1845 Ada shot Babbage an urgent query: "Has not Faraday made a *most important* experimental advance? Pray write to me directly."[43] He had. In September 1845 Faraday demonstrated that when a ray of polarized light is passed through a certain kind of "heavy glass" (containing borosilicate of lead) between the poles of a powerful magnet, the plane of polarization of the light is twisted. This discovery is still considered to be among his most important, since it established the interaction between an external magnetic field and the oscillating electromagnetic field of a light ray. (The role of the "heavy glass," which has a high refractive index, is simply to slow down the speed of travel of the light ray between the magnetic poles enough for the rotation of the plane of polarization to be detectable.)

There is no sign of the projected essay on Faraday among Ada's papers, and no indication in her correspondence that the Reichenbach review was ever submitted for publication. Since Ada's correspondence with John Crosse was almost all destroyed, it is impossible to know why not, but one guess is that she was forestalled by the appearance of a review of the same work by James Braid, a Manchester surgeon whose interest in mesmerism led him to establish what was essentially the modern view of "hypnosis," as he named it. His pamphlet, *The power of the Mind over the Body*, published in 1846, was the first English critique of Reichenbach. In it he attacked the "facts" that Reichenbach reported, offering alternative explanations of his observations, subjecting his methods to a withering scrutiny, and reporting the results of his own attempts to repeat Reichenbach's experiments—he accomplished the sort of criticism that was precluded by Ada's insecurity and caution in science.

Braid rejected effects that could be demonstrated only on a small number of highly selected, highly suggestible subjects. The "magnetized water," he suggested, could possibly have been detected by the absorption of the smell of the magnetizer breathing over it! The daguerreotype experiment he had repeated without results. The luminous emanations sometimes seen over graves, he suggested, might be due to decaying vegetable matter. Neither he nor Ada, however, mentioned that at the time there were indeed a number of human beings who glowed in the dark. These were mostly women who worked in the match factories and developed a form of phosphorus poisoning known as "phossy jaw," which ended in a fatal crumbling of the bone.

Neither the Reichenbach review nor the comparison with Faraday's work (if ever attempted) succeeded in enlarging the corpus of "the productions of the said A. A. L." That achievement finally arose in a most unexpected quarter. In a letter to her mother, undated but from internal evidence possibly written as early as 1840, Ada had remarked, "W wants to set me about an Essay on Planting."[44] In 1848 her husband got his wish.

Early in August of that year, in a letter to Babbage, William mentioned that he was engaged in writing a review of a book on the effects of climate on the growth of crops, to be published in the *Agricultural Society Journal*. "I do not suppose it will be above 15 pages—and being for the leather-gaiter-and-top-boot mind, it must not be or appear too learned."[45] Fatal words. By 20 August he himself was deeper into the complexities of the subject than he cared to be:

I am occupying myself most disagreeably with a short review of a part of Gasparin's [book on] agriculture. He is a man of scientific attainment but not accurate—and you know enough of the French savant to understand how the precision of their language & neatness of figure may often successfully disguise a good deal of blunders and arrogant error. It is to some extent so in his book . . . & while drawing attention to many believable facts he has collected together I wish to indicate where I think his reasoning is erroneous.[46]

The error that most disturbed him, he told Babbage, was Gasparin's failure to take account of differences in the number of daylight hours, as well as the heat intensity as measured by the temperature, in calculating the contribution of sunlight to the total amount of "caloric" or heat equivalent necessary to bring a crop from germination to maturity. In his calculations to demonstrate his point, however, Lovelace used the Fahrenheit temperature as a measure of "heat," forgetting that the zero point of both the Fahrenheit scale and the centigrade scale (which Gasparin used) are quite arbitrarily set: neither of them can be used as an absolute measure of heat, nor can ratios between temperatures be meaningful. (That is, a temperature of 40 degrees, in either scale, is not twice as hot as 20 degrees.) In addition, temperatures taken on the two scales do not in general have the same ratios, though the physical effects produced by the temperatures are presumably the same no matter what scale is used. Lovelace asked Babbage for his opinion and, in closing, mentioned that Ada was again well and climbing the hills around the estate.

Babbage's reply to this letter has not survived, but the problem of the temperature scale would arise again. In his reference to Ada's health and activities William made no mention of the fact that she too had become involved with his review. Shortly thereafter she herself sent Babbage a note, in connection with a visit he was planning to make to their retreat at Ashley Combe, and hinted at what she was up to. "Could you bring down *Herschel's Astronomy?* We have it at *Horsley*, but . . . it would be rather useful just now."[47]

Babbage's stay with the Lovelaces stretched out over some weeks, to judge by his correspondence. On 24 September Mary Somerville, temporarily back in England, dispatched a request for consultation on a new edition she was preparing:

I want your advice as to what portion of statistics I should add to the next edition and would be truly obliged to you to mention the particular subjects you think would be useful and suitable to the work. The progress of every kind has been so great especially in this country during the last twenty years that though I have mentioned many things concisely I hardly think I have said enough, but the subject is so extensive that the selection is difficult & there I want your help & know that you are both able & willing to give it—We do not leave England for ten days or a fortnight so perhaps we may see you but at all events write—

I hope you are laying in a stock of health by idleness & country air to enable you to resume your labour when you come to town with more comfort—With kind remembrance to Lord and Lady Lovelace.[48]

The decorous dignity of Mrs. Somerville's expressions of regard and requests for assistance are in amusing contrast to Ada's often imperious or impetuous urgency. Soon after Babbage returned to London—brown as a berry, it is to be hoped—he was pursued by a dramatic account of events surrounding his departure:

The skies are weeping unceasingly over your departure. The morg you went, it *set in* wet;—& it has scarcely intermitted for 10 minutes since.—

You must have had a very wretched journey. You cannot think how we miss you.—Even the *dogs*, & the brace of Thrushes (Sprite & Harry), look as if there was something wanting.—

My chief reason for writing so soon is to mention that Lovelace has been really quite *unhappy* because he was unfortunately *just* too late to see you on Thursdy Morng. He rushed after you to the Lodge & saw you *driving on. He* shouted, *Mrs. Court* tried to *run after* the two *Pegasi*(!), the men on the *lower terrace* (seeing there was something wrong), *all* yelled & shouted.—

But in vain. Neither you nor the beasts would hear;—tho' I really wonder that the *latter* did not run away again, thro' the *fracas.*—Lovelace is afraid you must think he neglected you.—

How well we managed to effect [an excursion to] Dunster & that beautiful tour round the magnificent *Grabbish, just* before hopeless wet set in!— . . . Sometimes I think *en passant* of all the *games,* & of notations for them. If any *good* idea should accidentally strike me, I will take care to mention it to you. But this is not likely. . . . I am very anxious to hear from you.[49]

Ten days later Lovelace wrote himself to add his regrets over Babbage's departure. It was only then that Ada's role in his review finally appeared in the correspondence.

The proofs of the climatic review were dispatched yesterday—one went to Herschel with whom Lady L. had exchanged some correspondence. She has appended two notes—one on the mathematical development of T. (temperature) another suggesting a nebulometer by means of a roll of iodized paper unwound by clock work past a slit in a box the rest of which is darkened. This would give the sunshine in black—the cloudy minutes light, for every day. . . . The great lawn much misses the pleasure of your company.[50]

The "nebulometer" was a device described by Sir John Herschel to Ada in answer to a need Lovelace had expressed for some way of measuring the heat from sunlight as distinct from "atmospheric heat."

It is odd that Lovelace should have needed to inform Babbage of Ada's contribution at that late point, and after they had all been together for a month—but not nearly as strange as the correspondence that Ada had in fact had with Sir John. On 10 August he had written her,

Dear Lady Lovelace:
I have rec'd and read with much interest . . . Lord L's *very able* review of Gasparin. In reference to the several points in your two notes I think—first as concerns my observation of the Lunar influence in clearing the sky, I cannot do better than [illegible word] for you and Lord Lovelace peruse such notes as I have made. They are irregular and loose no doubt, but I think they speak a *very positive* language (extending as they do to 80 full moons) which however to be fully felt the circumstances of each must be attended to. It is very difficult to reduce the sort of impression of such an induction to a numerical statement of n failing cases and n' successful and n'' doubtful. What appear doubtful at first sight are in reality often decidedly corroborative ones—I have not the smallest objection to your mentioning my notions on this subject. My own impression is that the fact is an established one and there really is a *very considerably* greater probability than the

average . . . that a night when the moon is "Round to the Eye" will be an eminently clear night.[51]

The techniques of statistical inference from a set of empirical observations began to be worked out only at the end of the century; before then not even the foremost astronomer of his time could arrive at anything but a subjective judgment as to whether a given set of observations indicated that a full moon was more likely to be accompanied by a clear sky than a crescent moon. Here, Sir John, with notes on 80 lunar cycles, was forced to express his confidence by underlining his words rather than by presenting numerical probabilities drawn from his data. (And, as it happens he was wrong.)

The foundation of the methods now applied to empirical evidence lay in mathematical theories of probability, which were among the interests of both Babbage and De Morgan; but neither had worked out this type of application. In his *Decline of Science* Babbage had looked on the inevitable scatter of values obtained in scientific measurements as only a sign of inaccuracy to be got rid of, if possible. Later his *Ninth Bridgewater Treatise* had contained arguments on the likelihood of miracles being reported by independent witnesses, which De Morgan gently chided. "It is true that the misfortunes of our friends are not displeasing to us," he began. However, "it puts me in charity with anyone to find he can make a slip in that — — —theory of probabilities. . . . The fact is, that evidence cannot be of the nature of an hypothesis from which the chances of an event are to be derived, but of that of an event observed, from which the chance of an hypothesis is to be found."[52] (Babbage duly amended the second edition.)

This was no further than the Reverend Thomas Bayes, an eighteenth-century mathematical divine, had got, but De Morgan must have enjoyed the chance to prick the know-it-all Babbage. His correction administered, he concluded puckishly, "Loose talk is often true if $a + b = c$, where it is entirely false as to $a^r + b^r = c^r$." This last expression is a reference to the still unproved "Fermat's Last Theorem," of which Fermat, in the seventeenth century, claimed he had discovered a marvelous proof, but the margin of the letter he was writing was too narrow to contain it. Fermat died without finding a wide enough strip of paper, and only now are computer-aided techniques enabling mathematicians to approach a solution. Mathematics can be very amusing, and Ada occasionally tried her hand at mathematical allusions, but there are no letters in her correspondence that begin to approach this level of sophistication.

De Morgan's and Babbage's German contemporary Gauss, in his work on the relative proportions of chance events, laid the basis for statistical methods that would be developed only much later. Another contemporary, and one within Ada's social circle—the Belgian Adolphe Quêtelet—had actually noticed, without being able to describe mathematically, that there were associations between certain measurable characteristics of organisms, such as height and weight. It was not until the very end of the nineteenth century that Karl Pearson, like De Morgan a professor of mathematics at University College, London, published a series of papers describing such now commonplace tests of statistical association as the chi-square and the correlation coefficient that bears his name. And it was only in the twentieth century that R. A. Fisher applied statistical methods to the design of experiments to be used in agricultural studies. Thus, although mathematicians of the nineteenth century were possessed of many abstruse concepts in the "purer" branches of algebra and analysis, what are at present the most ordinary and everyday applications of statistics, used—and misused—by countless students and workers with no mathematical pretensions at all, are of surprisingly recent vintage.

Recommending some works of astronomy to Ada, with the comment that there was nothing special about the moon as distinct from other heavenly bodies (presumably as far as determining the weather went), Sir John went on to discuss the problems of using temperature as a measure of the effects of heat on plant growth. He considered "Quêtelet's notion about the *vis viva* [energy of motion] (t^2) as a measure of the effect of temperature rather fanciful." He could not understand "how a sum of thermometric degrees or of their squares" could be taken as a measure of the quantity of heat. He did not elaborate his objections, but in fact there were several sources of possible confusion in his discussion.

The basic problem he was discussing was that of finding the total quantity of energy in the form of heat necessary to bring a crop from germination to harvest. Now, the absolute temperature of a substance is a function of the random motions of that substance's molecules, which is its heat energy. Consequently, the (absolute) temperature of the air can be considered a measure of the average heat energy of its molecules. So the total amount of energy available for transfer to the plants over the agricultural season is some function of the average energy during that period, multiplied by the time. Part of Sir John's bewilderment seems to have sprung from a formula in which the temperatures were averaged over one-day periods, so that the times did not appear in the resulting expressions.

But the real problems with the formulae proposed had to do with the temperature scales used. Clearly the values of the sums, whether squared or multiplied by time or not, depended on where the zero point of the temperature was placed. (They also depend on the size of the degree unit, but this only multiplies everything by a constant factor and does not affect the ratios in the kinds of calculations that Lovelace wrote to Babbage about.) For both the Fahrenheit and centigrade scales the zero points were rather arbitrarily chosen. The Fahrenheit zero was placed at the coldest point to which a mixture of ice and salt could be brought. The centigrade zero, like that of a scale previously devised by Réaumur, was set at the freezing point of pure water. In none of these scales, then, does the zero point indicate a total absence of heat.

That point, the point of zero heat, or complete absence of random motions of the molecules of matter, is taken as the zero point of the Kelvin scale, or absolute temperature scale, which was first described in 1848, the year this correspondence took place—although the point of zero heat had been identified as long before as 1800. It is only on such a scale as Lord Kelvin's that temperature readings have a direct energy significance. On all others it is temperature differences that are meaningful. For the purposes of developing an index of the rate of plant growth, neither absolute zero nor the freezing point is really useful, but rather some temperature below which plants do not grow. This point, then and now, is taken as 42 degrees Fahrenheit (6 degrees centigrade) and must be subtracted from each time-averaged temperature reading before the latter can be multiplied by the time period and summed.

In the same letter Sir John described an invention of his for measuring and cumulating "quantity of caloric"; he had, however, realized too late that the invention could be made useful only by the addition of an internal thermometer, to provide the comparison against which the ambient temperature deviations of the atmosphere could be compared. A second letter from Sir John, dated 14 August, contained further discussion of instruments for measuring heat and light. It is curious that Lovelace mentioned none of Ada's exchanges with Sir John when he plaintively asked Babbage some days later "whether in England we have any register of *solar* as distinct from atmospheric heat? It appears to me an undeniable but most evasive element, something like an eel when grasped. Lady L. is quite healthy again & when weather permits ascends the hills."[53]

A third letter from Sir John was dated 16 September (during Babbage's visit to the Lovelaces) and went once more into the question

of "the lunar influence on weather." But without appropriate mathematical and statistical models, attempts to set the influence of the
climate on plant growth on a scientific, that is, measurable, basis were
bound to be confused.

The Gasparin review, when it appeared in the *Journal of the Royal
Agricultural Society* in December, thus had the benefit of consultation
and comment from two of the best scientific minds of the period; it
had stretched their expertise to the limit and beyond. Ada had once
more become, albeit peripherally, involved in a subject whose current
development is dated from the middle of the present century, with
almost all the theories and observations of previous centuries forgotten
or ignored. Yet some of the issues raised by the Lovelaces in the
review are still of interest and still under discussion.[54]

The review cites a great many observations, calculations, and speculations, with recommendations that more should be garnered and
refined. But there is also confusion and uncertainty admitted as to
how all the information collected should be combined and interpreted.
All the sample calculations were now presented in centigrade, though
the problem of the location of the zero point was still ignored; so it
does not appear that Babbage counseled Lord Lovelace on this point.

Lovelace made a careful distinction between "atmospheric heat"
and "solar heat" but took the latter as half the difference between
thermometer readings in the sun and in the shade; he explained that
he had arrived at this formula by trial and error, attempting to make
the results of his computations agree with those of Gasparin. But he
criticized Gasparin for neglecting the effect of light, as compared to
heat energy, in ripening plants.

The latter part of the review deals with the effect of moisture in
general, and rain in particular. Here the possible effect of the moon
in clearing the night sky figures large, and a number of observations
are presented but no definite conclusions drawn. Instead of relying
on the authority of Sir John Herschel, however, Lovelace surprisingly
falls back on folklore: "The precise nature of the moon's influence
has never been exactly substantiated, though it has been more or less
believed in old as well as modern times, and in distant countries not
deriving their traditions from each other."[55] In other words, a number
of independent observers have believed in the moon's influence. This
reasoning harks back to Babbage's arguments on the likelihood of
miracles based on the testimony of independent witnesses. If Babbage
saw this section of the review, he apparently did not object to the
argument based on "loose talk," called in by Lovelace to support
Herschel's subjective impressions.

In the footnotes Ada furnished, her severe and didactic "Notes" style might easily have been recognized even if she had not appended her initials to them. Leaving the uncertainties of lunar influences to her mate, she concentrated on questions of mathematical formulae and instrumentation. In her first note she applies "the notation of the calculus of finite differences" to the sum of the temperatures, a formula "originated with Réaumur." Then she does the same to the sum of the squares of the temperature, the formula favored by Quêtelet and considered fanciful by Sir John. She observes, however, that this latter formula, unlike the former, involves the deviations from the average, and interprets it to mean "that departures from the *mean* are more favourable for developing vegetation than uniform temperatures: a most interesting deduction should it be verified by facts." Here, in spite of the reservations expressed by Sir John, Ada (following Quêtelet) had hit upon a point that indeed has since been partially confirmed (although once more the problem of the zero point of the temperature scale was overlooked). More recent work has suggested that maximum temperature readings (whether daily or weekly) are a more sensitive index of the "intensity" of growth than the length of the growing season. These maxima would appear in the deviation terms of the Quêtelet formula.

Space and, presumably, the limitations of the intended readership precluded a discussion of the subject as lengthy and detailed as its mathematical complexity warranted, but Ada could not resist throwing in another formula before regretfully abandoning the topic: "we will only slightly allude to a neat equation of Pouillet's in which *thermometrical elevation* [temperature rise] caused by the direct solar action . . . is made a function of . . . the *atmospheric thickness* traversed by the solar ray." She then presented the formula and concluded her note: "These reasonings and computations seem to contain the elements of correctness, and the equation probably expresses the true physical relations. . . . Any notice of these further conclusions or disussion of their value and exactness, is however here inappropriate. A. A. L." The formula would have to stand as an unexplained magical incantation.

The second note contained the discussion of instruments for measuring heat and light, profiting from her correspondence with Sir John. She called them to Gasparin's attention, since he "seems to write unaware of the means which photography has offered toward the easy and delicate appreciation of degrees of *nebulosity*."

So, despite the cloudiness of the state of knowledge of the subjects treated, the Lovelaces were well pleased with their joint production. Ada wrote her mother that it was the best work William had done.

"Lovelace is anxious I should tell you how highly I think of his digest," she said graciously; "His literary powers improve rapidly."[56] Lady Byron responded with a rare flash of humor: she produced a satirical skit on the collaboration that perceptively catches his practical and her speculative, theoretical turn of mind.

Scientific scene between Lord L & Lady L who meet to determine the problem of the weight of a certain mass of matter, standing on a certain rock —
A. I have now taken all the angles and measured the sides of this figure, and have estimated the density by that of the Geological stratum. . . . I find the weight of the whole to be $a + a/2$.
L. But a fragment of which I have ascertained the specific gravity, is nearly $2a$. . . .
A. You could not have attended to the conditions — it was damp perhaps.
L. No, my dear — see it is quite dry. . . .
A. This is very strange (Organ of Wonder bgins to act). There must be an extraordinary state of the Atmosphere. I feel as if my body could not sustain the pressure — Possibly some planetary influence. . . . (speculates 10 minutes). William, do you think we shall be able to lift our feet from the ground soon? — I declare I already find it difficult —
L. Well, my avis mirabilis, I shall try (looking very comical) for I am going to fetch some Gunpowder & blast this rock on which we stand — Just go a little to one side. . . . (An explosion takes place & Lady L's Parasol is broken) Just as I thought — I observed a *ferruginous* stain in the superincumbent mass, and *here* you see is *Lodestone* — so, although the theory you formed is a very clever one, I think we shall be able to walk to the top of Dunberry still.[57]

When the momentary excitement had passed, however, it was clear that this was neither the "original paper" nor the intellectual collaboration she sought. Agriculture was even more "humble and dry" a subject to write on than Babbage's engine, and nothing could fend off for long the spasmodic fluctuation of her ambitions, nor the inexorable return of illness and depression:

Last night I was suddenly taken UNWELL, & *very much* indeed so. . . . but as Dr. L— observes, it is a *great misfortune*, & what will tend to prevent my success in life with everything I may undertake. . . . I do not see how, with my particular constitution, I can ever do any good in the world. It seems to be a *physical impossibility* for me to carry on *anything* CONTINUOUSLY. The objects most liked at one time, may at any moment be *hated*. For instance I detest now both my Harp & my Studies. And despite the utmost exertion on my part, I am forced to own the utter fruitlessness of all hopes of such CONTINUOUS attention to any subject whatever, as could ensure any great ultimate success. So it is I fear. I am one of those genius's who will merely run to grass; owing to my unfortunate *physical* temperament. Pray don't be angry with me.[58]

5

My Dearest Mate

Given her wide-ranging and intense interests in science, her debilitating illnesses, and the social and family duties that—welcome or unwelcome—often intruded upon her time, it is startling to realize that at the same period in which she was preparing herself to be Babbage's "High Priestess," Ada was also considering a career in music. This was, if anything, a more daring and unusual course for an aristocratic lady to contemplate than that of a scientific writer, for it involved the possibility of appearing before an audience.

She had begun to take lessons on the harp before she was married; but afterward, and especially during her pregnancies, she attacked it with renewed vigor. Early in 1836 we find her playing duets with Mary Somerville's daughters and practicing until her fingers are sore. A few days after the accession of Queen Victoria, she admitted to Mrs. Somerville that she was slighting her mathematics in favor of music:

I should be devoting some hours to it now, but that I am at present a condemned slave to my *harp*, no easy task master either. . . . I play 4 & 5 hours generally, & *never* less than 3. I am not tired at the end of it . . . [but] I have suffered dreadfully with my fingers, from there having been a fortnight's intermission.[1]

William thoroughly approved; her drawing-room performances made Ada so much more intelligible and acceptable to the neighboring gentry of the county, as a lady of conventional accomplishments.

We had a neighbour or two to dine latterly—and last evg. the sweet sounds of the harp were heard too—to my (scarcely less than their) delight. I am very glad that there has been this little intercourse with the one or two neighbours we boast of for besides being as they must be struck with the grandeur & nobleness of her intellect—she has but to be natural to be as much loved as she is to be admired and wondered at.[2]

He had reason to be relieved. Her devotion to her instrument fluctuated wildly. In 1842 she had assured him that "as a matter of *reason* & *principle* more than of *mania*, I have made up my mind to make the utmost for *one year* of my *Harp*. . . . Just for that limited time & avoiding *mania*. . . . I am sure you will approve all this."[3] But the following year, as she traveled with her mother, he had occasion to write,

Dearest I hope you practise the harp. I fear your fickle-ness about it—you know what a treat it is to me—yet seldom am I permitted to enjoy it—you sometimes ask me if I think you will attain celebrity on it—you are displeased if I doubt it. Write me word then if you can that you have been practising & that I *shall* have some times the pleasure of hearing what I have so often heard the subject of much care, anxiety & complaint. Surely having bestowed so much upon it, it should not be without return.[4]

If music and mathematics sometimes competed for her attention, at other times she peevishly rejected both; and even her musical interest could take a form whose prospect her admiring husband found less delightful than evenings with the harp. Ada seems to have taken her first singing lessons at age 26, but it was not long before this new outlet threatened to overstep the drawing-room decorousness proper to a countess.

The fact is that my powers seem to be developping so wonderfully in the musical & dramatic line, that they say I may in 3 years equal Pasta, Adelaide Kemble, or anyone. (But pray don't hint a word of this to anyone of course.)[5]

Roupetch [her teacher] is astonished with my powers, both *musical* & *dramatic*, & says that I must sing *scenas* by & bye, as such a genius must not be lost to my friends & society.

In SCENA singing you know there is *real* ACTING, just as on the stage. For example I should sing a scena from Norma, (one of those between her & Pollio say), in the little library at Ockham, I & my Pollio being in ordinary dresses, but doing it exactly as it is done on the stage, & the audience sitting in the large library, we merely placing ourselves so as to be seen thro' the folding-doors. This would be to me the greatest of enjoyments, & will give *you* much to be proud & fond of. They all say that my dramatic genius is something wonderful;

& the more scope I have in prospect for it, the more settled, calm, & happy, does my mind become. The *style* of song that best suits me is that in which there is the expression of deep & stirring & generous sentiments, like those in Norma for instance, & also those in which there is *vigorous scorn & indignation*. Perhaps the latter you would not have imagined *would* be my line. But it is marvellous, they tell me, how powerfully I can express *scorn & fury*, & yet my action being all the while so tranquil & so removed from the stage *centre*. . . . A singing & acting Avis, would be delightful. Think how merrily & joyously evenings would go; & how delightful when we have company, to be able to *inspire* them a little, as I know *I* could, thro' song! It is a great & glorious gift to have, where as with me it results from deep & powerful & generous *feelings*. For there is a mysterious kind of mesmerism in *such* expression as *I* am likely to be able to give, which ennobles the hearts of those who LISTEN.[6]

Just who were "they" who so admired her operatic talents? Aside from her teacher, the only reference in these undated letters is to Lovelace's sister Hester, who was married early in 1843 but stayed with Ada in her London house much of the time before that. The letters, then, were probably written in 1842, after the first flush of her studies with De Morgan had subsided but before the translation of the Menabrea memoir and its exciting sequelae. During this period she suffered from frequent asthma-like attacks, which sometimes— like her other illnesses—passed off so suddenly that Lovelace had added "Phoenix" to his avian menagerie of names for her.

Clearly, Ada's musical aspirations during the period in which she entertained them were no more modest or limited in scope than her scientific ones. Norma is among the most exacting soprano roles in the operatic repertoire. It demands an unusual lung capacity as well as fiery dramatic talents. But William was not altogether convinced by her latest ambitions, and, in a series of pleading harangues, she added to her other arguments and inducements the seemingly unanswerable one of her health. Singing for several hours a day, she assured him, expanded the lungs and chest and alleviated her asthma. Moreover, the dramatic form was required for her mental health. The agreeable Dr. Locock was of the opinion that she needed outlets for her "hysterical" tendencies.

The dramatic line . . . is clearly the only one which directs my *Hysteria* from all its mischievous & irritating channels. . . . I perceive that the scientific & sedentary line are (at any rate for the *present*), quite pernicious & against the grain; but *were* I a perfect *Artiste*, my opinion is that I should *then* need the *scientific* very probably. There seems to be a vast mass of useless & irritating POWER OF EXPRESSION which longs to have full scope in *active* manifestation such as neither the ordinary active pursuits or duties of life, nor the *literary* line of expression, can give vent to.[7]

Dr. Locock's theory, which Ada reported here, is strongly reminiscent of the ancient belief that hysteria was a manifestation of the womb's migration to various unsanctioned locations in the body. (Hysteria and other medical theories, as they relate to Ada's multiple symptoms, are examined in the Appendix.) What was local and concentrated in others, he had told her, was diffused all over her system. "The results are very singular altogether, & make you require very peculiar & artificial excitements, as a matter of *safety* even for your life & happiness."[8] It was not to be the last compliant diagnostic service Dr. Locock rendered his patient.

In answer to William's rather philistine objections that acting and singing (lower-class occupations, for the most part) represented a comedown from her scientific aspirations, she retorted that music and drama as *she* meant to study them were neither superficial nor wholly unconnected with science. (A pity she did not elaborate this interesting point.) The struggle for his consent continued over several more lengthy, cajoling letters, almost impossible to order because of their changing arguments, themes, and moods. One, whose date—April 1842—is written in Lovelace's hand, ran:

My dearest Mate

I quite agree with the wisdom of your remarks; but I think I have perhaps led you slightly into an error as to the nature & intensity of my *dramatic* wishes & plans.

I consider the dramatic REPRESENTATION as only *one* ingredient (tho' an important & to me essential one), of far more enlarged views; & I would on no account make it my *mainstay*.

My genius in that line seems not unlikely to be so very prominent, that with very moderate cultivation I must even *excel*; but *I* never would look to excellence of mere *representation* being satisfactory to me as an ultimate goal, or exclusive object. . . . I consider that all my vocal & dramatic studies will be merely an *education* & *preparation* for that which is ultimately to be. . . . Hester says it is evident I no longer seek excitements *as* excitements; & that the Theatres are to me not places to which (as formerly) I must fly in order *not to sit at home*; but objects of lively interest for *study*. . . .

I will not however mistate [*sic*] to tell you my *impression*. I think that to which all is tending is the development of a very high order of *practical* genius; & that poetry, in conjunction with *musical composition*,

MUST be my destiny and *if* so, it will be poetry of an *unique* kind; — far more *philosophical* & higher in it's *nature* than aught the world has perhaps yet seen.[9]

In her buoyant moods, nothing seemed impossible:

Tell the Hen I am vexed at her thinking I can only take a mathematical astronomical view of the heavens, & that if *I* were to write verses on the Moon, the subject would be the *living* things of our Satellite, and that I should wonder whether it's surface so fair & so bright would open as perplexing a perspective of mixed weal & woe as our orb does . . . & whether the *shadows* which are dimly visible in her glittering countenance are truly emblematical of the spiritual state suited to her physical conditions. — Then I should go on the aspects our planet presents to the spirits of the Moon, &c. &c. &c., & in short I could compose a very sublime poem, but not a word therein of mathematics & the laws of motion. So the Hen has not quite such a kooky of a daughter as she supposes. . . . [A]nd tell the Hen I half think it is the growth of my Ideality & Causality that has made me ill, & that the excitement of writing all this has *swelled* one side of my face.[10]

This was neither the first nor the last time she would construct an optimistic theory in a desperate attempt not only to understand but also to justify and turn to account the physical and mental suffering with which she was bewilderingly visited.

I feel that there has been a frightful crisis in my existence during the past year; & Heaven knows what intense suffering & agony I have gone thro'; & how *mad* & how *reckless* & how *desperate* I have at times felt. Now all these dreadful things are passing away, & my great powers seem to be finding their legitimate vents; my health to be more promising than perhaps ever in my life; & my mind in all respects settling. . . . And I feel that every gentler & social feeling [is] developping more & more, as my imagination comes to find surer prospects of employment & vent.[11]

Dearest Mate I entirely agree with most, (& perhaps indeed with *all*) the suggestions in your kind delightful letter. . . . But pray do not feel disappointed or fancy me in a *combative* & *opposive* humour, when I say that *just now* I cannot bring myself to attend to Science at all, except in the incidental and more *active* way of occasionally going to a Lecture or going to see Scientific sights; & that for ABSTRACT Science I seem to be *wholly unfit* for the present. . . .

I have *no* idea of utterly or ultimately relinquishing Science; but years only can I fancy decide whether Science will be eventually merely my *subsidiary*, tho' profitable, amusement & relief from other pursuits;

or whether it will in the end be in any degree my *principal* & *grande* object. And my present impression is that it cannot be until my *Dramatic & Expressive* tendencies have had fullest scope & sway, that I shall even be able to judge fairly. . . . My plan & wish is therefore to devote myself with single & unremitting energies to the various branches in the Dramatic & Musical line; & merely *not* to FORGET Science & Scientific habits & connexions; but avoiding all cultivation of the latter which would interfere with attainment of perfection in the former.

I shall be able to account for & explain my slackened *scientific* energies & wishes, to Wheatstone & others, on the ground of health & medical injunctions; & I shall in no way hint at my change of present plans, except as being the mere temporary obedience to *medical orders*.

In this you will I know think me wise. I ought not to be committed *against* Science. It *may* after all turn out to be my eventual & most valuable object.

Now are we agreed? I imagine perfectly; & that you will perceive I am not under the influence of any violent mania, or hasty & ill-regulated conclusions. . . . I am at the same time so aware of the nature of *hysterical* constitutions like mine, (especially where there has been recently such very serious derangement of the most important functions), that I assure you I mistrust all present conclusions. . . . It is a great comfort to me that you, dear wise old Crow, do not feel disgusted with what I have indeed to own to you, from its *apparent* fickleness & versatility.[12]

But no plan, no goal, seemed capable of sustaining her for long:

As you say, Nothing can be in some respects less satisfactory than my condition. At the moment I am in a very wretched & uncomfortable state of mind altho' my *physical* condition is again pretty well in an average state. I cannot describe to you all the *misery* I am enduring mentally. . . . It is something quite malgré moi. I am wholly *uninterested* in anything & no efforts enable me to continue *any* of my pursuits at the present time. —

I feel that my aimless, objectless, useless life is *most deplorable*. The only *horror* I have is about ever again leaving Town. (Don't be angry with me for this pray, pray.)

Hester sees the whole. . . . I have the most painful & ardent desire & aspiring after *perfection* and *achievement* in SOMETHING, . . . & these in the want & struggle to vent & express much deep-hidden & un-developed *power* & feeling; & yet I can as yet find nothing that *fixes & forces* me in any way that could lead to definite & complete results. —

Without very powerful & CONTINUALLY-ACTING external stim-ulants, I shall be utterly wasted & lost to myself & my fellow-creatures. But how to provide them is the difficulty.

Her solution to the difficulty is surprising:

Money is the *rub*. If I had plenty of that, I think I could manage to turn & educate my powers to some account.

She closed this sad note with the parenthetical admonition, "Don't tell the Hen."[13]

It would be easy to form the impression from these plangent outbursts that Ada confided implicitly in her husband, much as she sometimes feared and evaded Lady Byron's complicity in her affairs. But in fact she kept many things, though not always the same things, back from him too; eventually she learned to manipulate and divide them against each other, shattering their close collaboration.

In the early 1840s she also began to tell Woronzow Greig—and eventually Babbage—things she could not have told either husband or mother. As Greig put it in his memoir,

I became very intimate with her—quite as much as it is possible for persons of different sexes to become consistently with honor. Her communications to me were most unreserved and to such an extent did she at times force upon me her confidences upon the most delicate and compromising matters that I often found it necessary to refuse to listen to her. She frequently made the most ghastly statements to me apparently for no earthly purpose but to commit herself as much as possible.[14]

Presumably the word "commit" is used here in the sense of "expose." But Greig's claim that Ada forced her confidences upon him seems belied by their correspondence, in which it is clear that he was breathlessly receptive to her developing revelations, protesting occasionally in scandalized fashion only for form's sake. The memoir itself set out to be as sensational as he dared make it; at one point he remarked, "Ada told me many anecdotes of her early days, but I omitted to take them down in writing at the time and have forgotten most of them. However, I recollect that they all exhibited more or less excentricity."

In the end there was much that Ada preferred to keep from Greig too and confided instead in Babbage, who may have been a less emotionally involved, or more worldly, audience; his autobiography reveals him as blunt and indelicate in both religious and sexual matters. Not so Greig. The probable date of composition of his memoir suggests it might have been called forth by the disclosures and scandals that surrounded Ada's untidy ending, and by his own emotional reaction to them. His notion of "excentricity," too, might be indicated by the

anecdotes with which he abruptly terminates his reminiscences. After
the birth of the Lovelaces' first child, he relates,

the only thing connected with his early childhood which I recollect is
that the father used to spin the child up (over a bed) horizontally nearly
to the ceiling of the nursery and catch him in his hands. About this
time [illegible word] the parents used to sleep without any night clothes
and every morning the nurse used to bring the child to them in bed
when they were in this state.[15]

Here the memoir breaks off. Perhaps even the writer was beginning
to suspect it revealed more about himself than about the subject with
which he, by this time, had become obsessed. Unfavorable though his
first impressions of Ada as a young girl had been, by the opening of
1841 he clearly found her confidences enthralling.

Still borne aloft by the prospects her lessons with De Morgan had
opened, she wrote to Greig on 15 January, "I am very much obliged
by your letter received yesterday morning, & anticipate with much
pleasure the rest of the series. —I have *so* much, on many subjects,
that I should like to tell you, and *so* little time to tell any of it, that I
am puzzled." She was, she informed him, getting on very well with
her mathematical studies and had determined that they would form
the basis of her future profession.

I am laying out a vast capital, which can bring no return for many
years; but a glorious speculation I believe it to be. . . . *You* are *right*:
I *ought* to do something; —to write something. But *not* at *present*. . . .
I will confess to you, (for *you* will not attribute it to a vain, empty self-
sufficient conceit), that I have on my mind most strongly the impression
that Heaven has alloted me some peculiar *intellectual-moral* mission to
perform. A *mission* of some kind or other, we *all* have to fulfill, but
whereas (if missions were classified) those of a hundred-thousand would
be all perhaps fitly placed under one head of a more obvious &
ordinary kind, there *are* missions for the *few*; . . . missions to make
better known to the many the laws & the Glory of God, and blessed
are those who fulfill faithfully such missions . . . for the glory of Him
who is so *darkly* known as yet in the world, & for the love of those
many whose greatest blessing it is (tho' they may yet appreciate it
not), to know him a little less imperfectly.
     Now to such a possible mission I will be true & faithful (to the best
of my ability) until the last pulse shall have closed for me the *present*
law of connection between the spiritual and the physical.[16]

It is a curious fact that (with the possible exception of Babbage) all of Ada's acquaintance seem to have taken her "intellectual-moral" pretensions absolutely seriously. Lord Lovelace and Lady Byron might sometimes have demurred at the extremity of her religious rhetoric, but on this occasion Greig was moved neither to laughter nor to embarrassment by her mathematical plans to enlighten the multitudes about the nature of divinity. In a style as prolix as her own, and even more florid, he expressed his approval at such length that a second letter was required to contain the overflow of his effusions, as well as the more practical matters he had forgotten to communicate.

His first response got off to a slow start as he recounted (not too truthfully, if we compare it with his later memoir) the history of the growth of his rapture:

Long before I had the pleasure of your acquaintance, I in common with the rest of your countrymen felt a peculiar interest for one who was cradled in celebrity, and I often used to speculate upon the character tastes habits pursuits and talents of Ada Byron. When I first met you at my mothers house, I rejoiced at having the opportunity of comparing the vision of my imagination with reality, and eagerly commenced the study of a mind and character which have since become the objects of my wonder and admiration—I hastily came to the conclusion that the portrait I had drawn did not possess one feature in common with the original.—Herein I was greatly mistaken, very many of my early ideas were correct, but a long time had to elapse before I got entirely rid of my false impressions.—Still, though you were not the being my fancy pictured, I distinguished in you from the first a peculiar development of extraordinary intellectual powers which combined with some striking traits of character appeared to me very remarkable at your early age as differing not only from persons of your own rank and condition but from the generality of mankind.

He went on to assure her that her marriage to his oldest and best friend had not diminished his interest. He urged her to proceed according to her announced intentions—only taking care not to overwork to the detriment of her health. He had long known she was ambitious, had begun to feel impatient for the results, had even begun to fear her talents were misdirected and, perhaps, to doubt her high aims: "Such was the state of my feelings when I arrived at Ockham a few weeks ago, and found you and Lord Lovelace involved in that most delicate and distressing affair . . . " (a reference to the news of Byron's incest and Medora Leigh's paternity that Lady Byron had dropped upon them shortly before).

The talent, judgment, deep feeling and total abandonment of self which you exhibited on that trying occasion . . . excited my astonishment and admiration to such a degree that I could no longer refrain from declaring to you my conviction that you were destined for something great, and that it was your duty to render the talents committed to your charge available for the good of mankind. My opinion thus frankly given you took in good part.

Greig was totally without humor, and it must be that he really thought Ada might have been offended by his panegyric; what she had done to bring forth this cloudburst of reverence was to write Medora a gracious letter "without blot or hesitation":

and you have since admitted me to a glimpse of your world of futurity by means of that letter which I *now begin* to answer. . . . 'Tis a noble object and worthy of you.—Most cordially do I agree with you in thinking that long and toilsome preparation must be made in secret.[17]

And on and on, with the inevitable "festina lente" thrown in for good measure.

In his burbling delight, however, there was one passage of her letter that he refused to take seriously; but it was to become perhaps the main topic of later exchanges. "I am now happier than ever in my life before," she had said; "I have never been happpy, even in the ordinary earthly sense of that term, until just lately." Naturally, he expostulated a bit over that. She had always had so much, ever since he had known her, especially since her marriage, "of every blessing this world affords." To say she had never been happy was surely wrong. Perhaps it was simply the contrast between her former doubts and the brilliant future that now awaited her.

A number of Ada's letters to Greig are dated in his hand and with dates that do not correspond to her own partial dating, making it difficult to place them with certainty. Nevertheless, within a year or two after the opening exchange between them—that is, by the end of 1842—we find her explaining to him the ebbing of her scientific ambitions and her decision to concentrate on music for the time being, although by then the peak of her interest in that too had passed. "As to *myself*," she began, "I think further explanation is necessary to you."

At this moment I can only say that I never dreamt of *excelling* in more than *two* things, viz: one *mental*, & *one* purely *executive* & *active* pursuit.

Say, for instance, my *harp & singing*; & whatever *mental* pursuit I might choose ultimately. I say harp & *singing* because I reckon them as *one*. The same principle precisely applies to both. In one I am already advanced three quarters of the way; & the other is not likely to be a matter of much difficulty owing to the accident of my having the most flexible voice *possible*.

It is not possible to form the kind of independent assessment of Ada's musical ability that, possessing some evidence, we can make of her mathematical and literary powers. But her accidental revelation that she had planned in detail the staging of her appearance as Norma before her vocal training had got well under way—indeed, it is doubtful that it ever went very far—should give us pause in accepting her claims. She continued,

I am not dropping the *thread* of Science & Mathematics, & this may probably still be my ultimate vocation. Altho' it is likely perhaps to have a formidable rival to its being other than just my pastime, should I take seriously with *"undivided mind"* to musical composition.

Time must show. To say the truth, I have less ambition than I had. And what I *really* care most about is now perhaps to establish in my mind those *principles & habits* that will best fit me for the next state. There is in my nervous system such utter want of *all* ballast & steadiness, that I cannot regard my life or powers as other than precarious.

And I am just the person to drop off some fine day when nobody knows anything about the matter or expects it.[18]

Her speculation was to be as wildly wrong as most life-and-death predictions are, but Greig must have taken alarm at her forebodings; for when she wrote again it was ostensibly to reassure him—and incidentally to hint at confidences to come, quite different from those of her intellectual-moral mission.

And now don't be *unhappy* about *me*. I am doing very well indeed;— as well as possible. And I have no notion or idea whatever of either taking myself *out of the world*, or living a useless invalid *in it*. So be easy.—You know I am a d——d ODD animal! And as my mother often says, she never has quite yet made up her mind if it is *Devil* or *Angel* that watches *peculiarly* over me; only that it IS one or the other, *without doubt!*—

(And for *my* part, I am quite indifferent *which*.)—

But if you knew one half the harum-scarum extraordinary things I do, you would certainly incline to the idea that I have a spell of *some* sort about me. I am positive that no other *she* Creature of *my years* COULD possibly *attempt* many of my everyday performances, with any impunity.

I think I must amuse you when we meet, by telling you some of them.[19]

A year later she told Greig, by post, of one of her performances. She had been attempting to interest the impresario Crivelli in her potential as a singer. As usual, when contacting anyone outside her close circle of relatives and suitably introduced acquaintances, she felt impelled to conceal her true identity.

My dear Greig,
I send you Signor Crivelli's reply, which is amusing enough; for it is very evident that Mrs. W. King is a personage of very great indifference to him.
Will you be kind enough to read what I have written, & then to put it into the Twopenny Post. I want to excite his *artistical* & *professional* interest.[20]

Almost immediately after this setback to her musical aspirations, science came once more to the fore, when Babbage proffered his assistance in appending notes to the Menabrea translation; and the direction of her ambition was diverted for some time thereafter.

Greig was by no means the only man with whom, in talk and in letters, Ada engaged in uninhibited confidences. Dr. James Phillips Kay was another. One of the many medical men whom Ada's mother knew socially, he had actually started out in banking, and he retained an interest in economics. After studying medicine at Edinburgh University, then perhaps the best medical school in the world, he had settled in Manchester, worked in a dispensary, and come to know something of the condition of the poor in the Lancashire factory districts. This experience qualified him for appointment as one of the Poor Law Commissioners in 1835. His convictions about the connection between the misery of the working class and their lack of educational opportunities led him to a concern with primary education. In 1839 he was appointed first secretary of the Privy Council committee charged with administering the newly established government grant for public education. He went on to found the first teacher training college in the country, and his influence on the structure of state schools continues to be felt to the present day.

None of his public interests appear in his correspondence with Ada or in her references to him. A letter from him written in October 1841 reveals that she had deputed him to obtain a book for her, just as she had asked Greig and his wife, earlier that year, to supply her with Christison's book of poisons. Possibly she felt that even the subjects of the books she read could become a topic of unwelcome gossip.

Dear Lady Lovelace
I sent for Francis' British Jews, but the people who publish it say it is out of print, and they are engaged in preparing another edition. This was four days ago.

I have bethought me of a new name for you which is chiefly prompted by your waywardness, beauty & intangibility—You always elude my grasp, and seem to delight in leading me into some bog while I am gazing on you half in admiration, half in wonder, somewhat in apprehension, and altogether in kindness.... Henceforth you are christened "*Will o' the Whisp.*" A delusive & beautiful light flickering with wayward course over every dangerous pitfall, deep morass and miry slough, and rather pleased when those who follow it with greatest constancy are thoroughly bogged to the neck. I can imagine this spirit of the wild . . . laughing at the sight of its own mother . . . and as to a poor friend on whatever services bent, taking the utmost pains to baffle his pursuit, and when he was knee deep in the mire telling him he had lost his road, by giving himself a great deal of unnecessary trouble, in point of fact making a much ado about nothing considering other peoples affairs. . . .

The sole inference that I draw, is that the course of these volatile spirits is uncertain, contradictory, vexatious, if it be made a matter of serious thought, and that therefore if no serious thought may be bestowed on its end, the beginning of the journey is equally delusive. I shall *cease* to try your course by ordinary means or to recommend anything dictated by *common sense* . . . strait is crooked, and wrong is right, words are wind, and professions "*windy respirations of forced breath*". . . . [T]o believe is too silly, to trust idiocy—there is no *pleasure* in this world, but only *duty*, which is its *opposite*—people say the nearest way is strait, this is false, for a crooked path is far more pleasant, & therefore easier & shorter. The common relationships of life are insufferable burthens—no tasks are pleasant but those which are self-imposed. Who ever has a right to her love is a *usurper*, and he only is to be preferred who has *least title*. There is a race of creatures, that have attempted to impose laws on their species, which are observed only by the *mean* and *earthy*. Spirituality of mind spurns every fetter that custom, or principle has sanctioned. The high-minded are a law *unto themselves*, and if they do *wrong*, that is *right*. To be anything but *indifferent-submissive* to the *penalty of existence*—paying a great *forfeit in living*, yet neither content to *die*, nor to be alive—this is the only philosophy of the *Whisp* genus.

Yet some of them are gentle, good in their hearts, beautiful, gifted, full of tenderness, bold, firm, and much to be loved, but all the while wayward, wandering, delusive, but worse than all deluded, & wasting their lives in a vain show.[21]

Above this letter are the words "A letter of Kay's from Battersea with the Character of AAL," in what looks very like Lady Byron's hand. In fact, Ada did have a hoard of unconventional opinions that she sprang on friends and dinner partners alike when the mood took her. While she might have delighted the irreverent and intellectually adventurous, such as Babbage, she must occasionally have disconcerted and discomforted others. In the former category, the roving novelist Eliot Warburton recalled in irony her sentiments on overpopulation when writing to announce the birth of his second son:

It would seem to be from a mere spirit of contradiction that I write to beg *your* ladyship congratulations on the birth of another crowder of the man-glutted globe. I remember when you so kindly brought yourself to congratulate me on the birth of my Devonian animalcule [his first son was born in Devonshire] that you confess[ed] you gasped for breath when you thought of another fellow sufferer being crammed into this black hole of a (Calcutta) world. I remember Walter Scott anticipated that we should have to eat one another. What high feeding we shall have once the shambles begin to fill. Well!! it is better to eat man in his lamb state than bedevilled by maturity.[22]

Less able to take in stride Ada's style of polite conversation was John Cam Hobhouse, Byron's old crony who had been so "exceedingly disappointed" in her at eighteen. Since then he had seen something of her from time to time, leaving a series of his impressions and observations in his diary. But although he had been an admirer of and apologist for Byron's social aberrations in his younger days, he had become more stuffy as he aged. At a dinner party in 1846 he was quite taken aback when she drew him into a discussion of death and its sequel.

I sat next to Lady Lovelace at dinner — poor thing she is looking very ill indeed and from what she told me I fear the worst consequences — she spoke to me very freely on subjects few men & no women venture to touch upon — e.g., she remarked that the common argument in favour of a future state derives from the pleasing hope the fond desire, the longing after immortality — [and] was evidently worth nothing for men entertained such hopes and expectations about things material which never happen.[23]

By 1849 he was able to report with some relief that "Lady Lovelace is not quite so eccentric in her manner as she was."[24] But in the early 1840s it was Kay who put a stop to Ada's daringly unconventional communications to him—more definitely perhaps than he intended.

When, early in 1842, he married the Shuttleworth heiress, Kay followed a custom common among the wealthy: he assumed his wife's name and arms. He also altered his premarital friendships. Nevertheless, at some point thereafter we find him performing the function of Ada's literary adviser that Wheatstone was performing for her scientific interests. She had written a book review, and Kay-Shuttleworth wrote to complain of her excessive praise of a simperingly conventional and moralistic Victorian romance.

Your review is too much of a panegyric. It is not sufficiently critical. The reader would imagine it to be written by a partial friend of the author.

I have tried but I never could read one of James's novels, and I was quite unprepared to believe that he could write a faultless novel. He seems to me a dull, undramatic dunce, who does the bookseller's work for the bookseller's *hire*, in a respectable fashion, by no means deserving the warm, and almost undiscriminating praise you lavish on "Morley Ernstein" . . . and as you are neither paid by Mr James, nor his friend, I think the tone of your review too favourable . . . because you describe scarcely any faults.[25]

In closing, he advised her to try her hand at another review and proposed to take both together to an editor. The review of *Morley Ernstein* that Ada wrote survives among her papers, but not that of the other novel, if it was ever written; neither, in any case, was ever published. In the case of *Morley Ernstein* and its author, George Payne Rainsford James, Kay-Shuttleworth's prejudices and his grudging school of criticism were well founded. Subtitled *The Tenants of the Heart*, the book was as hackneyed a treatment of the hoary Mephistophelian good-against-evil theme as has ever been produced. The young, noble, innocent eponymous hero successfully withstands the manifold temptations placed in his path by a Byronic villain—including the opportunity to seduce a virtuous young girl. He remains triumphantly faithful to his childhood sweetheart. Almost five hundred pages are required to bring Mr. James's preachy labors to their predictable conclusion.

Morley Ernstein was published in 1842. Into Ada's review, probably written within the year, has crept one of the rare social issues to engage her attention:

We feel gratified & hopeful when we see an author like Mr. James employing his best powers in the cause of mercy & enlightened justice; & adding his influence to that which we firmly believe & hope is steadily tho' slowly tending towards the entire purification of our penal code from the sanguinary & the damnatory which still remain it's disgrace. . . . We feel the greater pleasure too in these noble views of Mr. James because a philosophical poet of the present day for whom we have always entertained the highest respect, has grievously disappointed & pained us by throwing the weight of his genius, in a recent publication, into the opposite, the more barbarous, the less Christian scale.[26]

Her pained rebuke here was aimed at Wordsworth, who became increasingly conservative in his later years, opposing parlimentary reform, Catholic emancipation, and even abolition of the death penalty for various minor offenses. The only fault she was able to find with *Morley Ernstein* was the author's excessive use of coincidence to move the plot along.

Her review began with a nod at the Victorian view that novel reading should be frowned on as lacking in moral uplift and instruction. "We wish," she said, "that the term '*Novel*' were totally dissociated from a work such as this." Although she deplored many novels, she held it was possible for some to be edifying: "A novelist is sometimes a great preacher. We consider Mr. James to be pre-eminently of this highest class of novel-writers; and we consider him . . . a deep metaphysician; a *practical*, not a mere *theoretical* metaphysician." Coming from Ada, this was high praise indeed. But she did not stop there.

Mr. James' metaphysics are derived not from a-priori closet speculations . . . but simply from that which he has himself felt, & seen others feel. Contrary to our usual experience, . . . we were deeply interested in the very first chapter, in the very first page even. The distinction between the "spirit of the soul" & the "spirit of the flesh," never as far as our own knowledge extends, so forcefully & beautifully pointed out before, so brought home as it were to the very inmost heart & experience of the reader, at once engaged all our highest sympathies & interest.

What possessed Ada to write this high-flown cant, so at odds with the boldness and quirkiness of her usual opinions? She was clearly in one of her religious phases, but perhaps also impressed by her mother's moral posturing, which she occasionally attempted to imitate, as she had with Dr. King. Most curious, considering the evidence that she had been toying with Dr. Kay's sense of propriety, were her comments on the relations between the sexes (as depicted by James): "There is an accurate & delicate perception of the very different character of love in a man & a woman. A man, however deeply attached, may have episodes of passion & fascination for other women. . . . With a woman, this is less usual, perhaps indeed it *never* is, where once a real attachment exists in her mind." It was more an accurate and delicate perception of the sexual conventions of the Victorian novel. The solemn and pompous style of this review contrasts painfully with the gentle irony and humor of an adolescent effort of Ada's lying close by in the same box of papers. Assigned an essay on Scott's *The Heart of Midlothian*, the young Ada did not fail to point out that the book's heroine, Jeanie Dean, was very like her governess, Mrs. Dubourg, but she also expressed a good deal of sympathy for Jeanie's sister Effie and the latter's seducer. Nevertheless, she concluded, sagely passing on that classic advice of mother to daughter handed down through the ages, had Effie been a Jeanie, with "the steadiness to keep clear of all improper familiarity with George Staunton, there is no reason why she might not have been married to him." Of course, she conceded, "then we should have had no trial & no 'Heart of Midlothian,' and I should not have been late for dinner today." And Effie's prudence and prudishness would also have precluded the tastiest observation in all of Ada's juvenile oeuvre: "Jeanie's filial piety cannot be too much imitated by all who have parents."[27]

Soon after Ada had formed her acquaintance with John Crosse, so filled with exciting opportunities, Woronzow Greig wrote to warn her about certain rumors then in circulation. What prompted Greig's warning may possibly have been the first instance of malicious gossip about Ada, of one sort or another, being assiduously spread by Mrs. De Morgan. In any case, Ada replied to Greig indignantly:

I am trying to discover *the Traitor*. It is not because I think that particular instance of false report of much consequence to us, or that I am annoyed much at *that one special thing* having been promulgated. It is so absurd that it *could* not gain *much* currency. But it *is* of consequence to know *who* is the untrustworthy & malicious individual whom one ought on no account to harbour as a *spy* under our roof.—

And *if* ever I discover for certain who it is, I shall *tell him* the above, the very first time we meet by accident (if it be ten years hence);— doing so in *my* provokingly *cool* way, without *feeling* in *temper*; just fixing my great eyes on him, & saying my say in the most *gentle* way, taking care that *several* persons hear it.—No one could do this so well as *I* could, & it would be exceedingly amusing. My Mother quite chuckled over my notion. There is the advantage of being rather of a *cold-blooded* temperament. No insult, no impertinence, ever can *excite* me. So I can *venture* on what others must not trust themselves to attempt.[28]

The idea that Ada held from time to time about her cool, unruffled temperament must be classed with her other occasional fantasies about herself, such as that she was possessed of well-organized and systematic habits. This also appears to be another case, as when she relayed Faraday's praise or her altercation with Babbage to her mother, in which she has permitted her imagination to color her report. What Lady Byron actually told her regarding her practice of setting down her victims in front of witnesses was, "If you will but smile at your company during the Analyses, they will never find out they are taken to pieces."[29]

What the particular rumor on this occasion was, is perhaps hinted at by an entry Hobhouse made in his diary in 1847. He sorrowfully recorded that upon Lady Lovelace's absence from a dinner party due to illness, Lady Charlotte Berkeley, one of Ada's countless remote relations, "in a whisper, asked me if she was not mad."[30] Hobhouse himself did not believe it, but almost every fresh observation concerning his old friend's daughter seemed to come to him as a surprise.

Early in 1845, not long after their exchange of letters about the "false report," Ada began to reveal to Greig quite explicitly her growing dissatisfaction with her marriage.

Had you not touched on *one* terrible chord the other evening, you would have had no hint from me of my *life-less life*. That chord was simply to comfort in little kindly offices in matrimonial life, *especially in illness.*—It seemed to me so cruel & dreadful a *satire*, that much was revealed by me before I knew what I was about hardly. . . . *My* existence is one continuous and unbroken series of small disappointments—& has long been so. . . . It is not *his* fault that to *me* he is *nothing* whatsoever, but one who has given me a certain *social position.*— He does *all* that his perceptions can enable him to do. He is a good & just man. He is a *son* to me. . . . Unfortunately, *every* year adds to my utter *want* of pleasure in my children. They are to me irksome *duties*, & nothing more. Poor things! I am sorry for them. They will at least find me a *harmless* and *inoffensive* parent, if nothing more.[31]

If Lovelace was impatient or insensitive during Ada's frequent and often very trying illnesses, none of this is shown in his correspondence. Nor was it, apparently, in his behavior before Greig, who was so horrified by her assertion that Ada had to write back a few days later to explain herself more fully. By this time she had been able to construct one of her analyses, which had little to do with lack of sympathy during sickness or any other such common marital complaint; perhaps she hoped to confuse as much as reassure him into silencing his protests. "It is all in vain," she wrote; "I am a *Fairy*, you know. . . . I don't *want* a mortal husband, & he is a bit of an earthly clog [clay?] now & then."[32] However, she went on, she was Lovelace's fairy guardian for life, and had her own fairy resources with which to deal with him. The more she tried to reassure her correspondent in this curious fashion, the more he expostulated and the more she was induced to "commit herself."

That there was an undercurrent of flirtation in this extended confidential exchange is seen in another letter, in which she assures Greig that he is *"quite safe"* in her hands. But the expression of her domestic discontents was by no means confined to the gratifyingly shockable Greig. She could be even more vehement to her mother, who was no stranger to the violence and misery swirling beneath the placid surfaces of genteel households. And to her, Ada could be almost brutal at times about the children. Each, it seemed, had a turn at earning her special disfavor.

Her elder son, Byron, or Ockham—for a courtesy viscount could be referred to in this fashion even within the family—wished to be like a workman. He had, at the age of five, not the remotest conception of rank or of the dignity attached to titles.[33] As for her daughter Annabella, at the same age she had a "great pudding face, & eyes sunk & small with vile fat"; her insolence made her mother wish "to annihilate her."[34] In one of her (perhaps fortunately infrequent) interventions in her daughter's care, Ada, imitating her own mother, decided that Annabella was a "bilious child" who ate too much; after analyzing her phrenologically, she cured her with dose of calomel—mercurous chloride, a dangerous poison, then in common use as a laxative.[35] And Ralph, the baby whose birth had so signally failed to cure her menstrual complaint? He received relatively little of either praise or blame until he emerged from early childhood, and he lived with his grandmother from the age of nine.

Nevertheless, as they passed into later childhood, each was found to have some redeeming virtues. The day came when Ada found herself "charmed with my *metaphysical* child," the previously despicable Annabella.

If she will only be kind enough to be a metaphysician & a mathematician instead of a silly minikin dangling *miss* in leading strings I shall love her *mind* too much to care whether her *body* is male, female or neuter.

But really, all joking apart, I feel there is that in her which I shall delight to commune with as she comes to maturity (& which has nothing to do with her sex either way). William begins too to feel her superiority. He says he hopes she will marry a man whose position & circumstances may be such as to place his wife above the necessity of giving her intellect and energies to the mere daily affairs of life; that if A's mind is capacious & superior, it would be a shame that it were not free to follow out its own bent & studies. . . .

In short what *he* dreads in a girl, is evidently a *purposeless desultory* Miss, whose interests merely consist in flirting, embroidery, or perhaps (as a great God-send) the piano & miniature painting. What he would like is a *business-like* young lady; & yet she must not be a *busy body*, & she must be *feminine* & *elegant* in manner & appearance.[36]

Femininity and elegance were of great importance in a female child, no matter how young, and Lovelace fussed over Annabella's baby fat as much as he did over Ada's sudden (and temporary) weight gains. At least as important to the Earl was rank, as Ada reported to her mother:

How many aristocratic young ladies seem to *elope* & *choose for themselves* now-a-days. . . . Lovelace begins to make a wry face on the subject— & to talk much *aristocratic* principle—quite forgetting that people have *feelings* as well as *position*—He says that *in the end, position* is after all *the* grand thing to be considered in our rank of life![37]

Lovelace, though descended from a grocer and a salter on his father's side, was, through his mother, also descended from King Henry VII; and despite her declaration on behalf of "feelings," Ada too was acutely conscious of her rank. (Still, the subject of elopements of aristocratic young ladies with men of inferior social position was a delicate one for Ada to raise with either her mother or her husband.) Between the ages of seven and nine Annabella spent most of her time with Lady Byron, and when the grandmother complained of the girl's manners, Ada, from a distance, was able to advise serenely, "Tell Annabella that, as she likes to make *me* a *model*, People admire my gentle politeness to the *low,—shopkeepers* &c."[38]

A good deal of effort was put into separating the children as they grew older; when together, they tended to engage in "low nonsense." So, just as Annabella returned home to live with her parents, bad habits corrected, Ralph was dispatched in her place. As for Ockham, not even the formidable grandmaternal influence could subdue his sullen aspect and democratic opinions. His concerned parents decided when he was only eleven to give him to the navy, a transaction effected some two years later.

Keeping the children at home posed problems in addition to the trials to which they subjected parental patience, with their puerile impertinence and their demands for time and attention. The only approved alternative (approved by Lady Byron, that is) to educating them themselves was to consign this task to properly qualified governesses and tutors. But even the best recommended of these might not be proof against the dangers of exposing their charges to social or moral contamination. One governess, for example, committed the impropriety of introducing the children to a friend, who then presumed to send them gifts. Her own governess, Ada recalled sternly, had merely nodded at her sister when they passed in the street. In any case, educated hirelings could be awkward to have around. At one point Ada worried that there might be something wrong with her own and William's deportment: her son's tutor was too much at ease with them.

Not that finding suitable instructors in the first place was an easy task, for even religious training at the hands of the most conscientious parent was not without its pitfalls, as Lord Lovelace ruefully admitted to his mother-in-law:

[H]ere a difficulty comes in—the immaculate conception—I was embarrassed at *their* questions—I did not like to impress on them the miraculous account—nor did I like to speak with disrespect of the gospel or to allude to the possibilities that the writer was mistaken or the text have been interpolated with fables. This is the strong food fit for later years & more virile digestions. But if *I* feel these difficulties, how can *I not know* that others will be beat by them, & either teach that which I disapprove of, or excite a morbid curiosity to be gratified by the disappointed pupils at any cost. To be sure the immaculate conception may do in Spain . . . but it tends here to raise all sorts of inconvenient precocious surmises.[39]

To lessen the dangers to decorum, privacy, and sound theology, William and Ada did sometimes try to teach the children themselves. Ada reported with satisfaction her success in interesting Annabella in her lessons: "*I* look after *Algebra*, & her *Music*—She finds I explain *agreeably*, & give animation & zest to the subjects. But I take care to let her *find out* her great ignorance & imperfection."[40] There may have been an implied comparison here with her earlier report of William's attempts to instruct the children in French, a task he had taken over from their most recently sacked tutor: "I daresay they do not *learn* nearly as much; for I do not think W— has a particularly enlightened or skilful method of instruction; and besides that he has not above 20 minutes to give to each of the two. . . . I don't think he is over patient."[41]

The Lovelaces shared with most of their contemporaries the conviction that children from their earliest years owed their parents the most profound deference and unconditional obedience. At one point Ada even complained of Byron's having started a letter to his father with the words "I liked your letter," instead of Annabella's more suitably filial "Thank you for your letter." Moreover, he was given to slouching, and was morose and unresponsive; sometimes she even feared him. William tried in vain to make his heir stand up straight and fling back his shoulders: "He needs drill and a brace," said the Lord Lieutenant. But he did eventually admit to a sneaking sympathy and identification.

I am much concerned about the inertness of B's present temperament— nothing seems to excite him. At his age [12] I think I was to some extent in that way—from being bullied & abused by my mother with as I then supposed the assent of my father—So of course I voted myself a dolt—and remained in that belief for many years—but that is not the case with B. for everyone has been considerate with him.[42]

Both Lovelaces were also believers in and practitioners of corporal punishment, to which Lady Byron did not object in principle; but she was much more concerned to select the most effective means of control, with due regard to the character of each child. She observed them carefully and analyzed them endlessly, at one point remarking to Ada, "I think both you and L— exact manners from Byron which it has not been possible for him to acquire at his early years & under the circumst[ances] & if his attention be too much turned upon himself by reproof he will be shy & awkward for life."[43] Elsewhere she urged "moral influence" in place of slapping for Annabella, though later she recommended whipping to cure the girl of bedwetting. But above all, she cautioned against disciplinary methods that could backfire. "I hope the experiment of forcing A. to swallow will not be repeated," she wrote. "It is dangerous to put things down anyones throat during a struggle to resist (at least solids) . . . besides it is a foolish kind of coercion—& one that *might* be effectively resisted."[44]

Lady Byron's circumspection, and the elaborate instructions she laid down as to how she was to be spoken of to the children in her absence, bore fruit. Both of the younger children remained affectionately attached and loyal to their grandmother throughout life, blaming whatever defects they found in their upbringing on their parents. In the face of later revelations, Annabella would insist the Lady Byron "was really like a guardian angel, & I owe almost all the good I have learnt in some way or other to her."[45] More remarkable still, she also considered her grandmother very amusing.

Ralph, who was thought to be the inheritor of his mother's genius, also fixed the blame for his defects upon her, exculpating the woman who had chiefly reared him:

I should not have lamented any loss if I could have been entirely formed & instructed by her [Lady Byron]. . . . I certainly now think that some of the greatest mistakes made in my management were in accordance rather with my mother's desire than my grandmother's — I was supposed to possess such a vigorous genius as to need a very serious, solitary & depressing education — I had not as it turned out sufficient mental & moral vitality to thrive under such restrictions.[46]

As for the difficult elder son, with his bad posture, his contempt for titles, and his admiration of rebels, Ada was implacable about packing him off to sea at a tender age, insisting that he remain there, like it or not. She wrote to her cousin, who was one of his tutors, "I do not quite understand what you say as to his idea of *changing*. He surely does not imagine he is to go into the Navy just to amuse himself for a *year or two* only?"[47] After he was sent to sea, however, his father's judgment did soften a bit, enough for him to comment on a letter the thirteen-year-old sailor had written his brother, "On the whole I think it is satisfactory though the style of some of the jokes is neither elevated nor interesting." But boys, he conceded, will be boys.[48]

All three of the Lovelace children became united in their hostility to their father and resentment of his authoritarian demands. Yet nothing is more difficult to fathom than the character of the Earl. His surviving letters and writings present a very different picture from that reflected in the comments of those who were closest to him and presumably knew him best. His letters to Ada, for example, were invariably affectionate, if often paternal and admonitory:

Depend upon it that until you more successfully & religiously control your own careless propensities to disorder & delay — you will never be in a condition to improve yourself or satisfy others. You cannot expect good health — Your calculations should be based on indifferent bodily strength.[49]

Yet hints dropped by Ada to her mother indicate that he may not always have been so patient, may at times have resorted to abuse and blows with his exasperating wife as well as his obstreperous offspring. In an undated letter that from internal evidence seems to have been written about 1844, she remarked cryptically, "He is a good Crow, (tho' he *does* try to murder his thrush now & then). The attempts on my life & limbs are very droll (in retrospect)."[50]

This obscure comment might be passed off as metaphorical, or of only momentary significance; two years afterward, however, a much longer passage refers to a quarrel, this time witnessed by friends and neighbors, in which Ada fled from the scene in circumstances that made him believe she might not return. Her flight was apparently enough to shake him into renewed efforts at the "little kindly offices in matrimonial life" she craved.

The simple naïf joy of the Crow over my return is amusing & really touching. Just now I might do *anything*—absurd or ill-tempered. . . . I believe he had fully made up his mind that I should *not return* & he certainly has told everyone so. . . . I am rather struck with one thing;— L's very increased & very obvious *deference* for the thrush! . . .

The Corvus [Crow] has at times behaved so *very* ill, before the P[earce]s. . . . You will be amazed to hear that he seems to have made all kinds of blundering awkward apologies at the P's (at Minehead), *after I was gone*, for his strange conduct to me the evening before. P. himself understands him well—& sees that he has a *real affection* for me—tho' at times a very *bad manner* & a very *fidgety temper.*[51]

The Pearces at Minehead, like the Crosses at Broomfield, lived not far from the King estate at Ashley Combe. This same letter of Ada's suggests another reason, in addition to the children's own ungovern-ability, for Ockham's being sent to sea and the others' spending long periods away from home with tutors, governesses, friends, and relations: as they neared the end of childhood, they became sentient witnesses to the less decorous and aristocratic passions of their parents. In six months, Ada continued, they would decide about the entry of Byron (then aged ten) into the navy; he was becoming difficult to admonish. "Certainly when *thrushes* are compelled to *peck* at *full-grown Crows*, it often has a very great & permanent effect. . . . But in a boy of 11 years old, there is not the same *material* to act on."

Neither bad manner nor fidgety temper emerges from Lovelace's letters to Babbage any more than from those to Ada or Lady Byron. Rather, he seems sensitive, conscientious, and kindly, although proud and reserved. With Babbage he was not without a collegial, if exasperated, sense of humor over the difficulties with publishers they shared. That he was ambitious for Ada is clear, as is his disappointment when Queen Victoria somehow neglected to take her up as a kind of intellectual-moral adviser. "I hope," Lady Byron wrote him at the height of the season in 1850, "that she [Ada] will become able to take such a position relatively to the Queen, in the course of time, as the characters of both ought to ensure—this quite apart from any worldly objects."[52] William agreed but observed sadly that "it seems as if the links of the Court are accidently formed & yet firmly closed against admissions from without."[53]

By then it was Ada who was lecturing her husband on the importance of attending Court functions. "These sorts of *omissions*," she informed her mother, "are likely to cause our being neglected."[54] Lady Byron had often urged her son-in-law to take what she considered his proper place in the councils of government. At last she decided he was "too sensitive and not genial enough to be a Diplomatist"[55]—not a very comforting remark when one recalls that he had had a diplomatic career before succeeding to his father's title, and it was supposedly on the basis of his achievements then that he had subsequently been elevated to an earldom.

It is hard to decide to what extent he himself nursed disappointed hopes along these lines. To Lady Byron he protested that he had neither the temperament nor the inclination for the political career she urged:

[K]nowledge & observation render one more averse to sacrifice one's independence of thought & action by serving anybody or with (i.e., under) most colleagues. The body of the public is so capricious, hypocritical, so timid & at the same time so rash, so confident (without being confiding) & at the same time so stupid that for the most part they are not worthy to be served by any who have not some personal object to be gained by doing it. . . . However, these feelings may alter some time—I think they will.[56]

When they did, it was to no avail. In the spring of 1852 Ada wrote her mother that Dr. Lushington, Lady Byron's friend and chief legal adviser, as well as the Lovelaces' tenant at Ockham Park, had been consulted regarding William's prospects as Lord of the Admiralty (on the strength, perhaps, of his having a son at sea). In Lushington's considered opinion, no one could help him if he would not pursue his object himself, and Ada concluded, "I believe L– has the reputation of being *crotchetty* & *impracticable,* — & one who would not work well with others."[57]

Much more suited to his temperament, seemingly, was a literary career of sorts that Lovelace began abruptly in 1847, only to end it with equal abruptness in 1849. During that period he turned out over a dozen articles and reviews, on a wide variety of subjects, most of which he succeeded in publishing in some of the most prestigious intellectual journals of the time, though not without his share of frustrations and difficulties. Nine of them were eventually bound into a book he titled *Labor Ipse Voluptas,* the King family motto.

The earliest of his publications was "Theories of Population," which appeared in the *Westminster Review* in April 1847. "Population" generally referred to the numbers of the poor; as an agricultural landlord, he found it a subject of great interest. His article consisted of reviews of two contrasting works. The first, *The True Law of Population shown to be connected with the Food of the People,* by Thomas Doubleday, propounded the interesting theory — still occasionally revived — that well-fed populations are less fertile than hungry ones. Moreover, meat eaters were also held to be less fertile than fish eaters or vegetarians. Lovelace took issue with the author's statistics and argued that working-class conditions had improved over the previous fifty or sixty years — the period under review — during much of which meat was actually cheaper than grain. The economic problems that had plagued the country since the Napoleonic Wars were the result of causes other than the high fertility of the underfed poor.

From the excessive prosperity of the war time, full employment, and high prices, we passed at once into a state of pauperism, short credit, and commercial stagnation. From the supineness of other countries, and our own good fortune and courage, we had been till then in the exclusive possession of advantages which we were unwillingly obliged to share with others.[58]

As far as agricultural workers were concerned, Lovelace declared that the more remote country districts were not in any case subject to the law of supply and demand with respect to wages, as were the urban and commercial centers. Then, too, agricultural workers continued to be maintained in winter, when there was little work to do. Given this inflexibility, he felt agricultural strikes should be firmly suppressed, since farming employers did not enjoy high enough profit margins to justify them.

The second book reviewed in this article was *Overpopulation and its Remedy*. Its author, William Thomas Thornton, subscribed instead to Malthus's proposition that population always tends to increase until it presses upon its food supply, whatever the quantity or quality of the latter. He deplored the loss of the commons, since any gain in productivity achieved by enclosing them would be nullified by the ensuing population growth. Against this Lovelace maintained that enclosure did benefit the poor, the use of the commons having been in any case a privilege confined to the holders of certain tenancies only. To Thornton's contention that small farmers were more productive and that landowners should break up the large holdings they were then in the process of consolidating, Lovelace objected that it would hardly be profitable to do so. In this entire discussion of population problems, the possibility of birth control was not once mentioned.

It is clear that Lovelace, despite his liberal political and religious outlook, held paternalistic social and economic views that meshed rather than conflicted with Babbage's technocratic and free-market beliefs. Lovelace approved of "allotments"—plots of land assigned to a laborer to cultivate as he saw fit—even though they were said to increase population and, contrariwise, to interfere with the labor market. A decade previously he had described to his mother-in-law a scheme to provide his tenants and laborers with incentives by remitting part of the rent as a loan, to enable them to make purchases in bulk.

All were very thankful. . . . The farmers would become security for steady men in their employment and we should thus have the advantages of a loan society. Indeed by studying this branch of rural economy, I think we may in a very few years materially raise the character of the Labouring class here and at the same time add materially to their comforts.[59]

In 1844, however, he had observed that tenants who fall behind on their rent seldom have the means or energy to catch up, and should be got rid of; such tenants, in any case, cannot be depended upon to furnish adequate employment to the laboring population.[60] But he never wavered in his conviction that the power to distribute carrots and sticks among the working class should belong to the rich. In late April 1848 he wrote lady Byron a report of a conversation he had had with his stooping son, apparently provoked by recent events. After complaining about his persistent slouching—"he really looks like a navigator—his head is badly set on"—he went on the relate a discussion of history, civil wars, government, and political institutions, in which he informed the boy that "there might be as much despotism in the delegates of the mob as in tyrants by inheritance."

. . . then we talked of Chartism—their claims—showed him that with such an empire as ours any mistake in her councils would be fatal to England—thence the advantages of education—its necessity for those who wished to take part in public affairs—danger of ignorant men attempting to govern—[the] rich might be corrupt—but the very poor must always be indifferent politicians from having no leisure to study or think of ought but their daily bread—that the favour shown to the rich was not on acct of their *riches* but because with riches generally went education—the knowledge of how to use them—From his remarks I think he was interested in the subject.[61]

Political participation was a right reserved for the wealthy, but not a duty enjoined upon them; those who chose to engage in it looked after the class interests of those who did not. But many of the laboring class at that period felt that their interests had been sadly neglected and betrayed by the liberal reformers of the previous decade.

Working-class reaction took several forms, almost all of which were destined to fail or to be transformed into institutions compatible with the political and economic structures they were intended to combat. Chartism was born of disappointed hopes for the electoral reform of 1832; for the lower class this middle-class victory had brought only the hated Poor Law, coarser food, and greater economic insecurity. Then, in the wake of the failure of Robert Owen's cooperative communalism movement—whose adherents had envisaged the transformation of the capitalist system by example and, paradoxically, successful competition with it—many radicals pinned their hopes on redress within the parliamentary system. In August 1839 a "People's Charter" or "National Charter," previously drawn up by the London Working-Men's Association—the group in which Karl Marx would later become active—was adopted by a much wider spectrum of workers' groups. The Charter called for annual parliaments, equal electoral districts, voting by ballot, the abolition of property qualifications for members of Parliament, payment of members of Parliament, and, most radical of all, universal manhood suffrage.

In support of this program large meetings were held, newspapers were launched, and petitions were circulated. After the first petition was, foreseeably, overwhelmingly rejected by a Parliament constituted in the manner the petition was intended to change, the movement became fragmented by disagreements over the extent to which extra-Parliamentary actions should be pursued and, in particular, over the advisability of resorting to the double-edged weapon of violent uprising. Those who actually attempted force were easily crushed, their leaders imprisoned and transported. In years of relative plenty the movement lost ground; in harder times the pinch of hunger revived it. A second petition was defeated in Parliament in the grim year 1842.

By the late 1840s Chartists held the same place in the genteel English imagination that in earlier generations had been occupied by French revolutionaries and Luddite machine breakers, those objects of Lord Byron's compassionate maiden speech in the House of Lords. As the millennium wore on, they would be succeeded by Parisian Communards and eventually, of course, Russian Bolsheviks. In each of these cases the fears inspired by the unlettered hordes agitating for a more equal distribution of political power, and of its associated economic benefits, were far in excess of the actual threat they posed to the hierarchical English social structure that for the respectable classes was the only imaginable alternative to chaos.

In the spring of 1848, fueled by the inspiration on the one side, and the terror on the other, that had been generated by the French and Italian revolutions earlier that year, the Chartists planned a mass demonstration for the purpose of presenting yet another petition, from which, this time, the demand for voting by ballot was dropped. In the event, attendance at the rally was much smaller than had been hoped and feared, the planned march to Parliament was prevented by rain and by the threat of legal sanctions, and the petition, when delivered, was once more dismissed. Moreover, the validity of many of the signatures on it was challenged. The elaborate military preparations taken to safeguard public and private property had been superfluous, and the anticlimactic outcome was credited to the "good spirit" of the middle and shopkeeping classes. Lord Lovelace's account of his discussion of the subject with his son was written a fortnight after this occasion.

Lovelace's next published review was of *The Financial History of England,* by the author of *The True Law of Population.* This time he turned to Babbage—who was on mutually helpful terms with the editor of the *Edinburgh Review*—hoping that his old friend would put in a good word for his article:

I have been writing some observations on Doubleday's "Financial History of England" in the hope they might have found a place in the Westminster [Review] but Mr. Hickson tells me they have an article in prep[aratio]n on that subject. Do you think it possible it could obtain space in the Edinburgh, i.e., could you ask Prof. Empson whether it is open? If it is I can endeavour to get it ready in the course of a few days & if not judged worthy after inspection I shall not be chagrined at its rejection. I have not dealt with the present currency in times of crisis, which requires more knowledge than I have. My observations go chiefly to qualify some of Doubleday's erroneous views & statements & to contradict some mischievous conclusions which he evidently wishes the out-door-masses* to blink[?] at in order to produce & justify a national bankruptcy. I suppose it may be a matter of 20 pages.[62]

Babbage did as requested, and Mr. Empson indicated an encouraging willingness to consider William's paper. He was interested, he said, in "a good Doctrinaire statement of the great general principles based on the dangers from which these principles alone can save us, & showing, by reference to such crises as in all mercantile communities have been brought . . . at one time or another by neglecting those principles, the reality of the dangers." But he also wanted a clear statement of "the damage which may at all times accompany the adherence to these principles. . . . The two sides of this case represent the whole truth; & shd be stated tog[ethe]r I think." The difficulty, he thought, was in finding the right person to write such an analysis.[63] It is evident that what he really sought was someone with a prestigious name who would sign a piece he had essentially dictated. With such definite views, he was not likely to find a contributed piece satisfactory.

Lord Lovelace retrieved his manuscript from his mother-in-law—to whom, in addition to Greig, he regularly submitted his work—and passed it hopefully to the editor. Empson returned it, expressing his transparent regrets that it exceeded the length permitted by his publishers. He then turned around and asked Babbage to commission Samuel Jones-Lloyd, later Lord Overstone, a recognized authority on banking and currency, to produce the article he really wanted. Again Babbage complied.

In the end, Lovelace had to make do with the *North British Review* (probably through Greig's good offices) as the vehicle for his strictures on Doubleday's recommendations to cancel the national debt and reduce taxes—policies Lovelace considered ruinous without qualification. Despite his brave declaration, he was chagrined. To Ada he wrote plaintively, "I rec[eive]d a cheque for £12 from the ed[ito]r of the Nth British, Dr. Hanna—with a complimentary line. I had rather it had been in the Ed[in]b[urgh] *unfeed*. Still it is a douceur that consoles one for the rejection in some measure."[64] The article was also shown

*Probably a reference to the Chartists, an extra-Parliamentary political group.

to John Crosse, who communicated his approval to Ada, a further douceur.

Lord Lovelace never did succeed in lodging a publication with the *Edinburgh Review*, though he tried several times, both on his own and through Babbage and Greig. "I long for the day when I can command instead of beg admittance," he sighed to Lady Byron. His relations with the editor of the *Westminster Review*, who did accept his writings, were also turning sour:

Hickson would not be sorry to have another from me for the West-minster on Lord Nelson, on which I rather long to try my hand, but I do not particularly wish to write for a man who neither pays—thanks—or even dispatches one the copies one asks for until the thing has grown stale. As to compression & *brevity* there are things which cannot be treated thus superficially unless a review is to become a mere magazine of notices.[65]

Despite his reservations, he had a compelling reason to do a review of the book on Nelson's dispatches and letters, and his article duly appeared in the *Westminster Review* in January 1848. It was part of an attempt, in which he and Babbage were collaborating, to help and support the book's editor, Sir Nicholas Harris Nicolas, an antiquarian and bibliographer who had been active in criticizing the Public Record Commission and in reforming the British Museum, activities that had won him some enemies.

In the autumn of 1847 Sir Nicholas found himself a ruined man and fled to France to avoid his creditors. From Boulogne he wrote to Babbage,

I am sure you will be grieved to hear that my affairs are *utterly wrecked*. Everything is sold. . . . I am taking refuge during the storm. . . . So far as I can see my way there is *no chance* of my returning. This is a happy illustration of the reward of devotion to any abstruse pursuit in *liberal* England! If you had not a private fortune my fate would have been *yours too*.[66]

He had barely been able to complete the second volume of his history of the Royal Navy before his hasty departure. Of Babbage's sympathy he could be certain, having rallied to Babbage's aid in the past. In 1843 he had written a statement on the Difference Engine, which Babbage valued so highly he later included it in his autobiography.

Lovelace's review was written in close consultation with Babbage, especially the final section, where he turned aside from the merits of the book to consider the editor:

Besides being the author of several erudite and laborious works . . . , Sir Harris published a pamphlet, pointing out how [the Record Commission] might be rendered really useful and creditable. The commission was abolished very much owing to the honest exposure of its uselessness by one whose pecuniary interests, had he listened to such a suggestion, would have induced him to be silent.[67]

In contrast to Ada's refusal to be Babbage's "organ" in seeking redress from the government through her Notes, Lovelace in this review went on to make a comparison between the sad fate of the editor of Nelson's dispatches and the rich rewards accruing to the editor of Wellington's dispatches (and to his widow), the kind of comparison that Babbage so often made between himself and other men of science. Wellington's editor had "received a pension (was this for literary merit?) of £200 per annum, and after his death his widow received one of £50."

In general, Lovelace added, he disapproved of the practice of pensioning as a recognition of creative achievement because "literary pensions too often hold an awkward medium between recognition of merit and relief of poverty." He felt rather that the government should recognize "Sir Harris" by conferring on him some appointment commensurate with his attainments. Unfortunately, Sir Nicholas died the following year, still in exile, leaving a destitute widow and eight children. Babbage did what he could to help them.

From time to time Lovelace did what he could to help Babbage himself to an appointment or recognition that the latter sought. These included membership in the Railway Commission and knighthoods in the Order of the Bath and the Order of St. Michael and St. George. Here, at least, Ada tried to be of assistance, attempting to introduce him to various members of the aristocracy she thought might be of use or interest to him; Lovelace, encouraging her efforts, told her, "If *you* find an opportunity of being presented to the D[uke] of W[ellingto]n by all means use it—& speak to him at once about Babbage. . . . We owe it to Babbage not to promote his cause by inferior means—now Miss C[out]ts has scarcely mind enough to be charged with such a mission."[68]

"Miss Coutts" was Angela Burdett-Coutts, who, thanks to the eccentricities of two wills, had inherited her grandfather's huge banking fortune. She was a good friend of Babbage's and a philanthropist on a scale that dwarfed Lady Byron's carefully calculated good works. The slur on her mind was undeserved; she eventually attained an enviable social position and considerable influence solely on her own (and her money's) merit.

Despite his willing support of Babbage's interests, Lord Lovelace at times privately expressed his impatience with the loquacious philosopher to Lady Byron:

I hope Babbage may succeed—but one is in a minority when one advocates the philosopher who seems condemned to an orbit of his own—and this too by men who have met him at our house & have been struck with his genius.

Nicholson and his daughter are rather plagued by his [Babbage's] excessive fulsomeness. I do wish he would leave those things, which do not suit his age & exterior.

This must have been a reference to certain of Babbage's humorous anecdotes, some examples of which appear in his autobiography. They were often arch or silly, and sometimes risqué. "Besides which," Lovelace continued, "we know he is not always considerate."

There! Babbage's transgression in attempting to induce Ada to withdraw her Translation and Notes from *Taylor's* was not a thing one could easily forget. Nevertheless, Lord Lovelace went on,

I maintain he ought to be put in harness with the other conrs. [commissioners]—if he kicks (as no doubt he will) it will be awkward for the coachman Strutt [the chairman of the Railway Commission] & for those who have to pull with or against him but if it ends by Babbage having it all his own way perhaps the public service will be the better for it—if Babbage is drafted as an incurable malin [gadfly] it will at least prevent him from declaiming against mankind in general—& his wrath will scatter only his colleagues.[69]

But the prime minister, Lord John Russell, according to Lovelace's network of private intelligence, was determined to fill the Railway Commission vacancy with a professional engineer. This disqualified Babbage, despite his enormous skill and experience, the services he had already rendered the railways in experimentally analyzing the causes of discomforts and accidents, and his successful championship of the wide-gauge track. Lovelace sadly passed word of this new disappointment on to his friend. "I thought you would be as well for knowing what the gov't views now are—but how they will succeed

I cannot imagine. The military engineers are not competent. The civil are not disinterested—strong grounds for reverting to you."[70]

As far as the Order of the Bath was concerned, there too William's delicate inquiries yielded only the response that it was confined to those who had been in the service of the Crown. To no avail were Babbage's eloquent arguments that such recognition of the contributions of men of science—his own contributions in particular—was long overdue. When it came to Babbage's attempt to become Chancellor of the Order of St. Michael and St. George—supposedly limited to Crown subjects who had held high positions in the Mediterranean—Lovelace was somewhat hard put to express tactfully his concern that his friend was losing dignity in his avid pursuit of some sign of government recognition.

Although I should not have advised your application for the chancellorship of the St. M. & G. yet I trust it will not be denied to you. It will cause you I fancy some trouble on certain occasions, and you must take care to avoid heraldic blunders which are *immortal.*

I am rather concerned that the gov't should get a notion of your value being (at your own assessment) lower than it is & lower than they ought to think it.[71]

Lovelace then passed on to a selection of the variety of topics he repeatedly canvassed in his correspondence with Babbage. Most of these were fairly staid, though once he sent an extract he advised Babbage to burn after reading. (In that case, the recipient was more discreet about his correspondence than the sender.) There was the Irish question: "Irish men of taste and genius do not like living in Ireland. They come and spend their money in civilized places but this is a social not a political question." Then on to growing conditions—wet; hobbies—his experiments with the new technique of calotype photography; and, as always in this period, his review in progress—the Gasparin review, aimed at the "leather-gaiter-and-top-boot-mind."

If his opinion of his readers was becoming a bit contemptuous, his feelings for editors were now really scathing. Regarding his work on a book by Quêtelet,[72] he commented, "I should much wish to get it into some other review than the Westminster for whose cobbling editor I have a loathing"; it appeared in the *Prospective Review* early in 1849.

Quêtelet had chosen to examine some of the laws revealed by the statistical regularities found in social institutions. Paradoxically, the effects of chance are found to produce a quite predictable range and distribution of events, and "free will" results in a narrow and limited set of human behaviors. "Nay," asserted Quêtelet, "those very actions resulting from the undoubted, unmistakable will and pleasure of man, appear to recur and to be performed by him with even greater regularity and more constant and exact proportion than do the facts which are the consequence of simple physical laws." The proportions of marriages during the year, for example, are more constant than the death rates. Their very "strength of reasoning power" acts to make men uphold the norms and hence ensure the regularity of behavior.

To this declaration of dreary human predictability Lovelace objected that even scientific activity is compatible with imagination; he cited the *Ninth Bridgewater Treatise*. And creative artists, such as painters, have been known to study mathematics. Age seems to have something to do with artistic production as well, he continued: comic writers are usually older than tragedians; the craft of comedy requires more experience and less passion.

If all this seems somewhat beside the point, Quêtelet too was wandering from the statistical study of social phenomena. He went on to observe quite impressionistically that the pursuit of money cools down patriotic ardor and leads to national decay. Hence, countries, like individuals, have limited life spans. Lovelace allowed as plausible Quêtelet's notion that "the infancy of governments is inclined to the monarchical; their manhood is more or less constitutional or republican; their decline is marked by a surrender up again into the hands of a supreme ruler."

With respect to population trends, Quêtelet followed Malthus. He counseled celibacy and self-discipline, but noted that the aristocracy had a lower birth rate because of "precaution," a bold reference to contraception. Here Lovelace observed that Doubleday's fallacious infertile-meat-eater theory had been influenced by his reliance on birth rates among the peerage as evidence. Lovelace's point, an excellent one, was that the well-fed nobility's use of contraception had made them seem less fertile and had introduced a spurious association between meat eating and low birth rates.

Fearing that the middle and upper classes were failing to reproduce themselves and that illegitimacy (that is, the poor) was on the increase, Quêtelet recommended punishment for the unwed mother. Lovelace demurred at both diagnosis and remedy. The problems of illegitimacy should be met, he thought, not by punishment, but by prevention: better arrangements of "those early domestic conditions on which it has been shown that the preservation of innocence depends, to a very great extent." Here he was referring to the cramped dimensions of most working-class homes: it was common for a large family—sometimes several families—to share one or two rooms. He agreed, however, that it was important for a nation to have a substantial middle class to ensure an informed public opinion.

But vision, as well as knowledge and material ease, were necessary. "As to knowledge," he averred, clinching the argument, "China has long been afflicted with pedants in her administration, and in Germany the scholastic element has just burst out in undeserved, but probably transient, importance." The inferiority of those nations compared to Britain must have been obvious to his readers. In closing, despite his criticisms and disagreements, Lovelace politely hailed Quêtelet as a pioneer in "Social Physiology," the science that treats the moral anatomy of nations.[73]

Several of Lovelace's reviews were on the subject of his most compelling economic and intellectual interest, the improvement of landed property. This included the burning issue of "poor removal," the eviction of that part of the agricultural population found to be unprofitable to the landowner. In an article published in the *Westminster Quarterly* in October 1847, he worried over the backwardness of Britain's agriculture compared to its arts and industry. In the preceding fifty years, he noted, while the population had doubled to 27 million, manufacturing output had multiplied tenfold but food production had remained stagnant. Yet improved farming practices were available that could increase agricultural yields even without enclosures of the common lands. All that was required, he claimed, was more capital, skill, and knowledge on the part of landowners.

He then went on to consider the concept of rent according to various economic theories, calculating the value of the houses and gardens provided for tenants and laborers. He disapproved of the practice of making incoming tenants pay the outgoing ones for any "improvements" they had made in their holdings; he conceded, however, that it did tend to screen out the shiftless. More landlords, he said, should do their own farming. By this he did not mean that landlords should take up guiding the plow, or even supervise their hired hands in a day-to-day fashion, but that they should dispense with tenants and hire managers who would be directly accountable to them. Landowners of substance could easily obtain loans for the needed capital improvements; machinery should be introduced only gradually, however, since such capital investment creates an inflexibility in the need for labor, which would be disadvantageous for the landlord.

One desirable outcome he foresaw from his recommended increase in the number of gentlemen farmers was that fewer landowners would insist on harboring game where it was harmful to the crops. He was no friend of the hunt. Farming, he claimed, would be a healthier and more beneficial activity for gentlemen than hunting, and he wished to see game confined to waste land, "such as much of this kingdom is well fitted to be." In this article he showed himself able to quote figures to refute historical as well as living writers.

The theme of the most suitable use of different regions of the country was further pursued in a review of the "Report of the Select Committee of the House of Commons on Settlement and Poor Removal."[74] He began by observing that the impoverished agricultural laborers who migrated to towns in search of work acquired no rights in the towns on the basis of a few years of employment and residence. During slack periods they were dumped back on the country district whence they had come. But pauperism was the fault, not of the poor, but of others; therefore, the whole of a town should be treated as one parish for the purposes of poor law administration, and the floating population of workers, essential for urban economies, should acquire the right to assistance there.

Thus, Lovelace disagreed with the Committee, who held that the power to remove and send him home served to "stimulate the poor man to exertion" to avoid being deported. On the other hand, the Earl felt that the paternalism of the small county rate areas should be retained. Making rural areas larger, or even national, would lift immediate responsibility for the local poor from the ratepayer, and thus weaken the incentive to employ the weak and the old. The systems the Committee proposed, however, would promote the free circulation of single men, which would have adverse social consequences. Furthermore, the destruction of the homes of laborers, in order to prevent them from acquiring the rights of residence, forced them to walk long distances to work, which reduced the value of their labor.

There were certain districts, such as Lincolnshire, so unhealthy that in his opinion they deserved to be depopulated: "The whole of the rural economy of that district is arranged, accordingly, to get in the crops with the smallest quantity of indigenous labor, relying on the migratory but regular assistance of Irishmen for the hay and harvest, . . . and, as long as such can be had, it appears to be a natural and beneficial arrangement for all parties." What could be more natural than that only the migratory Irish should be exposed to the agues and fevers of the insalubrious Lincolnshire fens?

In still another article, published this time in the *Quarterly Journal of the Statistical Society of London*—and, for once, signed—he dealt with a work by two French historians that examined the effects of the inheritance law in France, which divided a parent's property on his death equally among the offspring. The result was a continually increasing number of small landholders, which the authors considered inimical to productivity. The situation contrasted with the high regard for primogeniture in England and the consolidation of large land-holdings that had taken place there. Lord Lovelace partially disagreed, opining that when many people owned land, there was more respect for it. He admitted, however, that this had not yet happened in France, where the farms were smaller and the people poorer than in England. Perhaps equality was incompatible with maximum wealth. In any case, it was important to have a wealthy class, and that class should be landowning, not commercial:

The industry which supplies the caprices and pleasures of consumers in distant countries, of men whose manners we cannot influence . . . is no doubt precarious. . . . But then, on the other hand, a land in which no luxury is enjoyed, one entirely free from "barbaric splendour and pomp," still more so, one in which property is equalized by law, leaves no margin and store for bad times; it supposes famines unknown, seasons uniformly healthy and propitious; it provides no reserve fund wherewith a rich class, as in England, had occasionally kept alive a poor nation, such as Ireland.[75]

Here he seemed unable to imagine a system of public or state savings, such as had in fact long been in effect in Oriental lands of barbaric splendor and pomp and logically could occur in egalitarian states as well. Thus we find Lord Lovelace, from his own upper-class viewpoint, writing and publishing discussions of the same questions that were then exercising his contemporaries John Stuart Mill and Karl Marx. Indeed, at one point he mentioned to Babbage that he was engaged in a heated dispute with the former over what he termed "Mill's fallacies" in the division of Irish property; he considered Mill very ill informed, except in a bookish way, when it came to agricultural matters. He himself demonstrated that he had a very good understanding of the condition of at least the agricultural poor, as well as of trends in their welfare in the preceding half-century or more.

This intimate knowledge stands in sharp contrast to Babbage's refusal to accept the reality of urban poverty and the inability of impoverished individuals to better their own condition. In their outlooks in this respect, Lovelace and Babbage exemplify two distinct ideals prevalent in England during the first half of the nineteenth century, the period that saw the birth of a self-conscious working class. While Babbage crystallized an impatient entrepreneurial vision, Lovelace embodied a revitalized aristocratic and paternalistic ideology. While Babbage enjoyed recounting anecdotes in which he exposed beggars as lazy or shamming, Lovelace declared, "But the labourer can *not get work* let him try ever so much—we have *not* got free trade yet, and are not getting it as fast as he gets hungry for the want of it."[76] Yet if he was more sensitive and humane on class issues, his views of other nations and races remained narrow and parochial compared to Babbage's more international outlook.

Interestingly enough, too, his paternalism did not always extend to his own servants, for while Babbage in the end was to look after one of Lord Lovelace's servants when he himself had abdicated his responsibility, Lovelace retained the inevitable suspicion and distrust aroused when strangers are admitted into an unreciprocated intimacy. "The insecurity is worst," he complained. "We know so little of what goes on below the surface which they [servants] expose to us—Yet we know enough not to trust them."[77] Ultimately, of course, it was the servants whose position was the more vulnerable and who would have to pay the price of his insecurity.

The articles described so far by no means exhausted Lord Lovelace's interests or knowledge. He pursued in print his concerns with modern history, French medieval literature, and architecture. But he was not content with simply studying and writing about the latter. Before the Institute of Civil Engineers he read a paper describing the method he had devised for shaping by steam the beams used in rebuilding the great hall at his Horsley estate.

The steaming operation was effected by placing as many of the timbers as could be conveniently accommodated in a long, stout, airtight plank chest, into the side of which was inserted the pipe of a temporary boiler. After four or five hours' exposure to the steam driven into the chest, the beams were removed and bent around a gauge formed by strong posts fixed into the ground; after they cooled, they retained the curve they had acquired. They were then lashed together in sets of four and hoisted into place. The beams met precisely at the roof peak and, because of the curve, had no tendency to thrust the walls outward. Furthermore, they could be carved and decorated without loss of strength.

The idea was not original, he said: it had been suggested by the work of a Frenchman, Colonel Emy, in a book on carpentry; but the latter used it only for very gentle curves, with radii of at least 60 feet, while Lovelace's smallest curve had a radius of only 7 feet. There is no evidence that Babbage gave him any assistance in developing this method, which I. K. Brunel, who was in the audience when he presented his paper, considered better than the one he himself had used to roof over the Bristol Station of the Great Western Railway.

There were also one or two papers he never did succeed in publishing. "My dear Babbage," he wrote at one point, in exasperation, "it is plain we ought to have only lodgings in Grub Street or wherever disappointed authors are banished to."[78] At another, he commented wryly, "here I am with two going a begging—a thing Lady Blessington & Mr. James & Mrs. Edgeworth never did in the whole of their honoured & useful lives. . . . A little more persecution will make me as cynical as you."[79]

What part did Ada play in her husband's short literary career, outside of her involvement in the Gasparin review? There is no doubt she approved, writing to Augustus De Morgan (concerning Lovelace's review of "The Agricultural Statistics of France"), "In consequence of yr kind reply to my former note, Lord L– thinks I had better send you the paper. . . . Should *you* be at leisure & disposed to look at it, you will at once see *how much stuff* there is in it, & of what a solid quality. . . . Much of the paper, tho' on a dry subject, is amusing enough."[80] Again, as in her comment to her mother about the Gasparin review, a somewhat patronizing note has crept into her approbation. Yet, despite his disappointments and occasional frustration with editors, Lovelace's attempts at writing for publication had, overall, clearly been more successful than hers, just as his architectural ventures were more sustaining and satisfying than her musical ones. For her, writing could never be a sideline, a steady but not all-engrossing activity; and she seemed to need, to require, a collaborator as wholehearted and pain-staking as she was herself in her most fervent moments—but far more patient and plodding. Such partners were not in overabundant supply.

An aristocrat, a landed gentleman of means, could take up and put down his pen at will. Lovelace had begun his writing, as he said, for his own "instruction and pleasure"; it is not really clear why he stopped it so abruptly after maintaining so profuse an output for almost two years. The correspondence does not support the simple explanation that he just grew tired of the rudeness and arbitrariness of editors and found himself reaping diminishing returns in personal satisfaction. Perhaps his increasingly intense involvement in the practice as opposed to the theory of the improvement of landed property comes nearer to accounting for the close of his writing career. Lord Lovelace not only used every penny of his resources not tied up by his marriage settlement and his mother's jointure in expanding and consolidating his inherited holdings, but had bought himself a large neighboring estate at East Horsley, free of hereditary entail. After 1846 he made it his principal seat, leaving the family estate at Ockham Park, gloomy and dilapidated by his own reckoning, to be leased by Lady Byron's lawyer, Dr. Lushington. It was then that his career as Surrey's most flamboyant polychrome Gothic architect began to flower.[81]

East Horsley, renamed Horsley Towers, was to be transformed from a pseudo-Tudor mansion into a mock medieval castle, complete with subterranean tunnels 140 feet long. Lovelace's projects, however, were not of a kind to afford real comfort and pleasure to his family, who felt rather deprived by them. He was quite willing, for example, to build a harbor at Ashley Combe and even to buy a boat, which Ada enjoyed; but he considered that installing a bathroom would entail too much expense in the pumping of water—although Ada had been prescribed daily hot baths as part of her medical regime. Ada, and Annabella afterwards, complained not only of the miseries and inconveniences of the interminable construction works, which they were endlessly called upon to admire, but further that the resulting alterations were made with an eye only to "architecture and art generally." The floors within the Gothic fantasy at Horsley were of asphalt, without even a carpet, and the rooms were mostly fireless.

Worst of all, the building projects made money an obsession with him; in his letters to her, Ada was more likely to be lectured on household economies than on political economy. With the prospect of her mother's £7,000 income continually brought to mind by Lady Byron's frequent illnesses, Ada found it difficult to accept his stringencies.

By this time it had become natural to confide to Greig what she could not make William understand. One of these confidences involved a sum of £500 that she had borrowed from a neighbor, a Mr. Currie, on the first of May, 1848. Currie was a London banker, brother of the man from whom her husband had purchased East Horsley eight years previously. In a letter acknowledging the loan, Ada thanked him for waiving the usual demand for securities (which, as a married woman borrowing money without her husband's knowledge, she was unable to furnish); she promised to repay the money within three years, and explained,

My present embarrassment originates in the *very small* sum (considering my inheritance and position), which was settled on me upon marriage. This sum has been *totally* insufficient to meet the expenses incidental to my *position as Lord Lovelace's wife*. Very heavy expenses which are at this moment entailed on Lord L. by his Buildings & by some other circumstances [his mother's jointure], have made me feel that I should be wrong if I adopted your suggestion of applying to *him* (at present).[82]

At this point she bethought herself that Mr. Currie might be wondering if he would ever get his money back, under the circumstances, so she added, "But I have the fullest ground for expecting from him henceforward those increased means which will preclude difficulties *in future*."

In fact, she had little ground for such hope, as she revealed early in 1849 in a letter to Greig:

I said *you* advised me to ask for an *increase* of £200 per ann. — which I acted on; but that the party declined to *bind himself* to *any sum, regularly*, — but agreed to pay for Court-dresses & did pay *one* such Bill for last year. —

This is the whole my mother knows & she must not know more. . . . But if she knew of the *debt*, it would do irreparable mischief, in more ways than one.

I have told her that *books & music* are THE two things which have made me *overflow* at times.[83]

Ada never managed to settle her debt to Currie. A note in Greig's hand on the letter to Currie of May 1848 states that £408/15 remained unpaid after her death.

Over the next few months Ada continued to harangue her husband on the subject of money, and at the end of March he sent her another check for £100, though not before she had addressed a further, more vehement plea to Greig, asking him to support her demands:

It is a possible (but I think not very probable) thing, that *Lovelace* may refer to *you* for an opinion upon some (pecuniary) arrangements between him & myself; — and I think I ought to give you notice of my having *suggested* to L— to do so. . . .

I have stated my own opinions & feelings upon the subject of property & money, in a very decided way to Lovelace, in *writing* . . . being entirely determined that the *past & present* state of things must *not* CONTINUE & never ought to *have existed*. — I of course have said nothing unkind or offensive to HIM *personally*. Not only would this be unwise, but I do *really acquit* him of all *intentional* wrong; — & think that ignorance has been the cause of the discomfort & difficulties I have been allowed to experience. —

The point of difficulty is that Lovelace entirely *denies* that anything unusual or unjust has been done; — & indeed I believe he considers the *miserable pittance* alloted to *me*, as MOST *liberal*; (even the *£300 per* *ann.* out of the future £*7000(!) per ann.* Which *he* is to enjoy to my exclusion!)[84]

Ada was receiving the same amount in "pin money" that her mother had in her original marriage settlement, before Byron agreed to an increase. Even after Lady Byron's death, it was Lovelace who would become life tenant of the Wentworth estates, and Ada's allowance would not be increased unless he agreed to it. Nevertheless, it is clear from these letters that books, music, and even court dresses were not all that caused her to "overflow."

Lord Byron, about 1814, by Thomas
Phillips. Depicting him in Albanian
dress, this painting was copied by the
artist from an earlier portrait executed
by him about 1814 (and now in the
British Embassy at Athens), which was
bought by Ada's grandmother, Lady
Noel. All through Ada's childhood it
was hidden from her, but once she
was married she was considered mature enough to stand the sight of her
father, and Lady Byron presented the
painting to her. (National Portrait Gallery, London)

Lady Byron. Inscribed "Engraved on Steel by Freeman, from an Original Drawing," this portrait was "Presented with the Court Journal of 5th Jan. 1833." Lady Byron was over forty at the time. The drawing may have been executed some time earlier and in any case is in an idealized style. (Murray Collection)

Lady Byron, by Benjamin Haydon. A detail from the artist's painting of the World Anti-Slavery Society Convention in London, 1840. Haydon admired Lady Byron and gave her a prominent place in the scene. (National Portrait Gallery, London)

Ada Byron, 1820, engraving after a painting by F. Stone. The portrait of his daughter was painted at Byron's request; the black-and-white engraving was sent to him ahead of the original. Byron then asked for a lock of her hair, so that he could see its color, and sent some of his own to be made up into a locket for her. That locket now belongs to descendants of John Crosse. (Murray Collection)

Ada Byron in childhood, by an unknown artist. Possibly painted at the period of her first serious illness, at age seven. (British Museum)

Ada Byron, dressed up for her coming
out in 1833, by A. E. Châlon. (British
Museum)

Ada as Lady King, 1835, by Margaret Carpenter. (Her husband did not become Earl of Lovelace until 1838.) The ornaments in her hair are presumably the King family diamond parure, which she later twice pawned. Ada did not find Mrs. Carpenter's picture flattering; she remarked to her mother, "I conclude she is bent on displaying the whole expanse of my capacious jaw bone, upon which I think the word Mathematics should be written." (Murray Collection)

Ada, Countess of Lovelace, 1838, engraving after a sketch by A. E. Châlon. (British Library)

Charles Babbage, 1845, by Samuel
Laurence. (National Portrait Gallery,
London)

Model of Babbage's Difference Engine.
It was this that Ada viewed soon after
meeting Babbage and later called the
"gem of all mechanism." She never
saw a model of the Analytical Engine.
(Science Museum, Kensington)

Drawing for Babbage's Analytical Engine, 1841. On the left is an elevation of one of the devices that presented the linked punched cards so that the information on them could be fed into the mechanism. The device shown related to the variable cards, on whose operation Babbage gave Ada instruction while she was preparing her Notes in 1843. (Science Museum, Kensington)

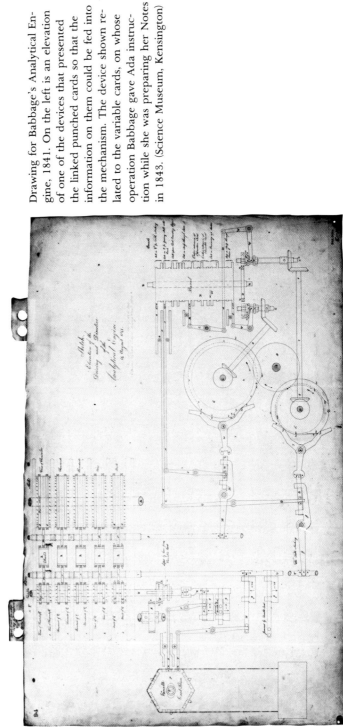

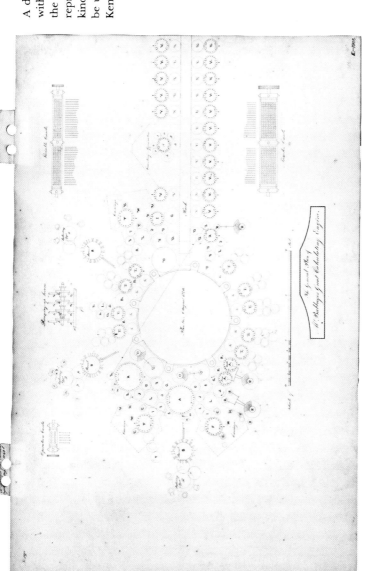

A drawing of the Analytical Engine with the parts arranged according to the plans of the early 1840s. Note representations of the three different kinds of punched cards that were to be used as input. (Science Museum, Kensington)

Mary Somerville, 1832, by F. Chantrey. (National Portrait Gallery, London)

Augustus De Morgan. This photograph was the frontispiece of *Memoir of Augustus De Morgan*, by his widow, Sophia De Morgan, published in 1882. It was probably taken in the 1850s.

Michael Faraday, painted 1841–1842,
by Thomas Phillips. (National Portrait
Gallery, London)

Andrew Crosse as a young man.
(Somerset Archaeological and Natural
History Society)

Fyne Court, Somerset, where Ada visited the Crosse family in 1844; within the walls all was "chaos and chance." (Courtesy of Audrey Mead)

Ada, Countess of Lovelace, 1852, by Henry Phillips. In August 1852, when she knew she was dying, Ada several times posed for two portraits by Phillips, sitting at the piano with her hands prominently displayed. The sessions were often interrupted by excruciating pain. (Courtesy of Doris Langley Moore)

William, Earl of Lovelace, daguerreotype taken in the 1850s, possibly before Ada's death. (Private collection)

Horsley Towers, Surrey: the tunnels. Lord Lovelace was an amateur architect and engineer. The purpose of the tunnels with which he embellished his estates at great expense and effort remains obscure. (Courtesy of Stephen Turner)

Ashley Combe, Somerset, the Lovelaces' second estate. (Private collection)

Byron, Lord Ockham, 1862, by Gustav
Rafe. (Private collection)

Lady Anne Blunt. From her clothing
and hairstyle, this photograph appears
to have been taken in the late 1860s,
when she was about thirty years old.
(British Library)

Ralphe Milbanke, about 1860. Ralph
adopted his maternal grandmother's
maiden name, as requested in her will.
(British Library)

6

I Begin to Understand Death

On 2 September 1845 Ada wrote Babbage from Brighton, where in the latter half of the decade she made a practice of retreating from her family, and where Lady Byron had furnished her with a house. "I enclose you a note I have had from Mr. John Crosse the author of the article in the Westminster. He is in great distress at the infamous manner in which it has been printed."[1]

The article in question was a long review of two works of natural history. The first, and the chief subject of the review, was Alexander von Humboldt's *Sketch of the Physical Description of the World* (the first volume of Humboldt's five-volume *Kosmos*), published earlier that year. The other book Crosse discussed, *Vestiges of the Natural History of Creation*, was published anonymously (in two volumes, 1844 and 1846) because it contained what the discreet author considered to be very daring speculations concerning the evolution of species. The attention it received is thought to have helped prepare the ground for the acceptance of Darwin's theories when they appeared fifteen years later. The author was eventually revealed as Robert Chambers, publisher with his brother of *Chambers' Encyclopedia*; he had feared that owning up to a naturalistic view of creation might injure the family firm, even though he had ventured no ideas concerning the mechanism of evolution, such as were the key to Darwin's revolutionary hypotheses. In the meantime, the speculation over the identity of the author did nothing to injure the sales of his book. Late in 1844 Lord Lovelace had mentioned it to Lady Byron:

I am quite delighted with the "Vestiges" which Greig has sent as a most acceptable present to Ada—Babbage recommends her to read it "if she has not written it"—a high testimonial to her in the eyes of those who know the book without knowing her—and to me who do know her, an ad[ditiona]l recom[men]d[atio]n of the book itself. The chapter on the mental faculties of animals is most curious.[2]

Lovelace took Babbage's gallantry as high praise indeed, for he seriously reported it to Hobhouse, who remarked in his diary that he had never before heard of either her mathematical reputation or her reputed authorship.[3] How close Ada's intellectual promise had already come to illusion among her family and friends is clearly indicated by such inordinate pride in the false imputation of this achievement. (The list of contenders for the reputed authorship of the *Vestiges* was a long one and included Prince Albert, whom Ada's friends at that same period hoped to make her disciple.)

The note that Ada enclosed with hers to Babbage is the only letter of substance from John Crosse to Ada to survive. It read,

I have just glanced through a portion of the article in the Westminster — the printer seems to be some sarcastic dog who rejoices in the most mal à propos distortions & displacements. Thus he makes me (or rather Humboldt who is far from being so original) say that the time of the earth's revolution around *the sun* (instead of her own revolution) depends on her volume! Proper names of eminent individuals are defaced without pity — Commencements of paragraphs are made to end them. *Condensations* of the original are given as *quotations* — Greek words are transmuted into forms that would puzzle Bopp, Pott, Grimm or any other monosyllabic philosopher (the very names of these fellows are most truculent roots) — Mr Babbage's note, which belonged to the Oceanic part of the argument (where I requested its insertion) is placed in solitary dignity amongst all sorts of things that don't belong to it — And to crown all, the notice of "the Vestiges" (added to please Hickson) comes in as a weary straggler instead of being dovetailed into its proper place.[4]

This was very different from the epistolary lectures she was accustomed to receive from William. If Crosse's character and the nature of his relationship with Ada are difficult to make out, it is not because, as in the case of Lord Lovelace, he was reserved and uncommunicative. Rather it is because almost all his letters to Ada were destroyed. (It is difficult even to construct an outline of his personal history, since his name was later changed, following, as usual, the terms of an inheritance.) This sole surviving letter reveals a personality as extroverted as William's was retiring. Interestingly enough, however, like Lord Lovelace, he complains of his ill usage as an author in the *Westminster Review*. But the bumptious jocularity of his grievance contrasts sharply with Lovelace's wry exasperation.

John Crosse's view of "the Vestiges," as well as of Humboldt's book, was that it was a work of popular science. As such, he contrasted it with the work of Mary Somerville and with Whewell's treatises on inductive science, which, he said, "serve as sign-posts to the student, but afford no broad and beaten track on which the multitude can travel." There was great need among "the half-educated masses" for understandable explanations of the natural world and for "some account rendered of the learned leisure which is made possible by the labor and ignorance of millions": "Give us bread, is the cry of the multitude, and they will not be satisfied with a stone, even if that stone mark the most interesting transition state of two most interesting strata, and be held out by the very hand of the very president of the Geological Society, and be accompanied by his most affable smile."[5]

Toward the end of the article Crosse dropped his sardonic facetiousness and addressed seriously the evolutionary philosophy that underlay "the Vestiges." The author, he pointed out, believed that all development took place by means of the modification of each successive generation in such a way as to adapt the organism more successfully to its external environment. Crosse was prepared to be even more comprehensive in suggesting a universal law of change:

That there be some common law we have allowed, but it must take in inorganic as well as organic existence, and link them both to the world within. Is it not possible that the "high analysis" (its nature once understood) may lead in this direction? . . . The [author of "the Vestiges"] concluded by representing *gravitation* as the possible one and comprehensive law of the inorganic, *development* as that of the organic world. We do not think it is possible to effect the radical separation of these two.[6]

The mention of "high analysis," by which Crosse presumably meant mathematical description and analysis, is striking. It is exactly the phrase that Ada used in her letter to his father to indicate her scheme to unravel the mysteries of the nervous system and the molecular constitution of matter—the letter that preceded the visit during which she made John's acquaintance, shortly before this review was written. It is possible that she inspired this speculation of his (if she did not actually write it), and its inclusion suggests that she had after all found something of the intellectual communion and partnership she longed for—a suggestion strengthened by the presence of notes on Humboldt's *Kosmos* among her papers.

After abandoning the study of law, Crosse seems to have taken up the life of a "gentleman"—that is, without income except for the allowance his father chose to give him. In 1855, upon his father's death, he succeeded to the family estate, with its untidy manor house, Fyne Court. Early in 1859, in accordance with the will of his mother's brother and in consideration of an inheritance of £10,000 he assumed his uncle's name of Hamilton and discarded that of Crosse completely. Before coming into these legacies, however, he must often have been short of funds to maintain himself and a clandestine family.

It is not certain when Crosse's relations with Ada went from "stretching my mind" to stretching her pin money, but the date of her loan from Currie, together with Greig's later anxiety to recover Ada's letters of 1848 to Crosse, suggests that within a single year Crosse managed to involve himself in serious and long-term relationships with two women. And the records of the tangled financial affairs that Ada left at her death, which included the purchase of furniture for Crosse's house, hint that his connection with one woman was used to help support the other, with or without the knowledge of the first.

There is no evidence that Ada's attempts to raise money by gambling began earlier than 1850; the little notes and slips of paper that were found among her papers, issuing from tipsters and often addressed to her maid, Mary Wilson, date from 1851. Nor is there any evidence that Babbage was in any way involved. Even Greig, who eventually came to know all there was to know of her racing affairs, never accused Babbage of complicity; the "harm" Greig claimed his influence did her must have proceeded rather from his general tolerance of free-thinking and of the confidences she reposed in him.

There is also no evidence that Ada used any sort of mathematical system in placing her bets, although she may have implied as much to her betting partners in order to inspire them with confidence in her judgment. A paper of Babbage's entitled "An Examination of Some Questions Connected with Games of Chance" cannot have been of much assistance to her, for it only formalized in mathematical terms a rule of thumb commonly employed by gamblers—and it presupposed an unlimited number of trials. All of the trials, furthermore, were supposed to involve similar (and known) chances of success. Needless to say, none of these conditions obtained in Ada's ventures. Furthermore, according to Babbage's paper,

The first and most simple plan is that of doubling the stake whenever a loss occurs. This is well known, and has . . . acquired a peculiar name; it is technically called a *martingal*; it requires for its success, that the person who employs it have the power of leaving off whenever he pleases, and that he have the command of unlimited capital. If the chance of events happening is one third instead of one half, the stake must be tripled.[7]

Babbage, characteristically, went on to generalize the situation from there. At one point he arrived at the result that, out of 800 trials, it was necessary to win 390 of them just to break even. There was not much incentive for reckless gambling there.

Ada's career in horse racing began openly and innocently enough. Indeed, it was only the size of her losses that was ever much of a secret, and this only from her mother, and temporarily. In May 1850, on the occasion of Lady Byron's fifty-eighth birthday, a message from Ada mentioned casually, "I am afraid you will take no interest in what interests *me* just now, — viz. the running of the *Derby* (May 29th). Indeed I am in danger of becoming quite a sporting character. The 2 horses I care about are Lord Zetland's *Voltigeur*, & Lord Albemarle's *Boling-broke*."[8] The day after the Derby she wrote again, very puzzlingly:

I have such a horrible *head* today (owing of course to despair at the great pecuniary losses I have sustained by betting) that I can't write the long letter I intended. I hope to do so tomorrow, for I have *very* interesting things to tell you. —

You will however see in the Papers: — "total *ruin* & *suicide* of the Earl of — — — owing to the gambling of the *Countess* &c &c &c" — What else could you expect?

We are delighted at Zetland's triumphant victory yesterday. *Voltigeur* galopped to the winning post gloriously & with perfect ease. No one expected it. . . .[9]

Either she had not followed her preferences in the matter of Voltigeur, or her losses in the other Epsom races had much exceeded her Derby gains—unless of course she was simply playing with her mother's unsporting admonitions.

The issue of Ada's gambling did not become a matter of vituperation on the part of her mother until much later, and then only because it revealed the extent of Ada's alienation and provided Lady Byron with a weapon against her son-in-law. At the beginning she covertly aided and abetted Ada in her new enthusiasm.

In the late summer of 1850 William and Ada, ready "to stir from our den at last," took a tour of the north country, during which they visited a number of the grander houses of the counties they passed through. They were guests not only of Lord and Lady Zetland, owners of the wonderful Voltigeur, but also of Colonel Wildman, who had bought Newstead Abbey from Ada's father. In anticipation of the trip, and of the racing meetings they planned to attend, Ada secretly accepted a gift of money from her mother: "I have *not* told L– anything about the fund for travelling because I am not sure it will be desirable to touch it *this* year. . . . Of course I *must* tell him, if it *is* touched, — because he would wonder how *I* came to be *so rich*. Indeed he would know that I am not in such circumstances at present."[10] Ada called it a "fund for travelling," but both parties must have been well aware of the uses to which it might be put.

What stirred and disturbed Lady Byron about this trip was not the prospect of gambling losses but that, in the event, both Lovelaces were so taken by Newstead and expressed such poignant regret at its loss. None of the three considered for a moment that even if it had not been sold to pay off the consequences of Byron's youthful indulgences and to secure his wife's jointure, it might, by customary practice, have passed to the seventh lord along with the title. To his mother-in-law William confided, "Nothing however that I have seen interested me half so much as Newstead. I felt quite melancholy at times in conducting Ada to the halls of her forefathers forfeited by their misfortunes — a sort of fate — it would however have been impossible for the last Byron to have kept it, except from extraneous sources."[11] He had to admit that the great crumbling pile had been restored with taste, reverence for the Byron name and family, and more expense than the family could possibly have afforded. In Ada's reaction both admiration and regret were framed in more extreme, thrilling, and romantic terms. Later she would admit that Colonel Wildman had revived the Abbey, and that she herself felt resurrected; but her first impression was of desolation and decay.

I feel as if, however, it ought to belong to *me*; & altogether horribly low & melancholy. . . . I seem to be in the Mausoleum of my race. What is the good of living, when *thus* all passes away & leaves only cold stone behind it? There is no *life* here, but cold dreary *death* only, & everywhere, the death of *everything* that was! . . . I shall not be sorry to escape from the *grave*. I see my own future continuing visibly around me. *They were! I am*, but shall not be. . . . They tell such tales here of "the *Wicked old Lord*" as he was called commonly, — my father's *predecessor* & great-uncle, the one who killed Mr. Chaworth.
    . . . I feel as if I had become a stone monument myself. I am petrifying here.[12]

This was Gothic and strange enough, but the most curious part of this letter was her sense of deprivation over money. "We ought to have been happy, rich & great," she said, "but one thing after another has sent us to the 4 winds of Heaven." The civil war and the Round-heads, she declared, had taken most of the Byron fortune; only a small portion was ever restored. The truth is that, enmeshed in a spreading web of debt, Ada was prey to increasing resentment over her exclusion from control of both her mother's and her husband's property; some-how she imagined she would have been in command of her father's, had it not been alienated.

If only she had left well enough alone. In her next letter she chose to torment her mother a bit by admitting that her first melancholy had "gradually changed to quite an *affection* for the place . . . as if it were an *old home*." She had been reluctant to leave. Then, after a generous tribute to the sensitivity, taste, and expense with which Colo-nel Wildman had restored it, she began to hint that her mother might wish to buy it back for her, repeating a "prophecy" that it would be returned to the family in her generation. She expressed no concern at all over how her kind and attentive host might feel about being asked to part with the home he had rehabilitated with such lavish care. "I do love the venerable old place & all my *wicked forefathers!*" she exclaimed.[13]

Lady Byron's anguished reaction is attested by the presence of two copies of her response in the family papers. "If the Mythic idea generally entertained of your father affords you satisfaction, do not forget, dearest Ada, how much of it is owing to my own line of conduct," she cried. By this she referred to her having refrained from publishing his mon-strous conduct to the world and suffered in consequence of her noble silence, since the fashionable world in which Ada now moved might not understand what a devoted saint she had been. What if such misconceptions should damage her standing with her grandchildren?

It often occurs to me that my attempts to influence your children favourably in their early years—will be frustrated & turned to mis-chief,—so that it would be better for them not to have known me,—if they are allowed to adopt the unfounded popular notion of my having abandoned my husband from want of devotedness & entire sympathy—or if they suppose me to have been under the influence, at any time, of cold, calculating, & unforgiving feelings,—such having been his published description of me, whilst he wrote to me privately (as will hereafter appear) "I did—do—and ever shall—love you!"—

But from the whole tenor of my conduct . . . you must be convinced of my repugnance to anything like Self-justification,—only the strongest conviction of its being a duty towards others could force me to enter into it—

I do not judge for you & him as Parents—but surely you could not *as such*, come to the conclusion that your Children had better,—for the sake of Lord Byron's Fame—believe me *what I am not.* Write to me what you think & feel openly.[14]

Even if the Lovelaces had been inclined to heed this last injunction, it was soon canceled by a subsequent letter informing them that the agitation over Newstead had damaged her health. Both did their best to soothe her with assurances that they had heard neither approval of Byron's conduct nor aspersions at her own. Colonel Wildman had known Byron only during their school days. William exclaimed over the horrific notion of parting her from the children. He had always found invaluable her taming influence over them—and over their mother as well, as he had pointed out years before: "all the latter are more or less docile from it—and the bird too—is now becoming so gentle and tractable on points which formerly ruffled her wings."[15] "The bird" wrote even more comfortingly that she had heard ill of Augusta Leigh, and that she regarded her father's conduct toward herself as *"unjust & vindictive"*—presumably because he had willed his money away from her in favor of Augusta's odious brood. She also carefully noted the similarities between her new fashionable acquaintance and Lady Byron's friends.

Little dreaming of the implacable resentment their offense had bred, the Lovelaces turned their attention away from the past in favor of the far more interesting present: their tour continued, and in due course they headed for Doncaster, where there was to be a three-day race meeting. Ada sped on ahead of William, eager to attend the entire session. She dispatched an ecstatic report to her mother, promising her a picture of Voltigeur.

The SECOND victory of Voltigeur on Friday, when he won the Cup, beating the *invincible* Flying Dutchman—whom it was supposed *no* horse could equal, is the greatest triumph ever achieved in racing!— It was a magnificent struggle between these two *greatest* of Horses. Like single combat between two heroes—I am glad to have seen this greatest of Races, which will be *historical.* I believe that no horse has ever yet won the Derby, the St. Leger, *and* the Cup—especially against such a champion as the Flying Dutchman.[16]

The Flying Dutchman had won the big event at Ascot in June, as well as the Derby and the St. Leger in 1849.

Ada was now in the grip of a new enthusiasm, as great as her previous ambitions to be a singer and a scientist. William was not nearly as much of a racing fan; he had lingered behind to admire more architecture and did not appear at Doncaster before the final day of the meet. In reporting back, he adopted a tone of amused detachment, admitting, "Half only of that cup of pleasure fell to my drinking as I took care only to arrive on Friday to see the last." As to Voltigeur, the poor horse needed protection from

his indiscreet admirers . . . pressing upon him and helping themselves too freely to parts of his Tail which has been much diminished by their ardour to have a hair. The bird has distinguished herself . . . — she was out about Volt[igeu]r however. The worst of racing however is the company—both of legs among men, & fashionable among women that it brings.[17]

His pocket had been picked of three or four pounds as he walked through the crowd. A few days later William complained, still good-naturedly, that "an eager ardent Avis" was trying to convince him to go into the business of breeding racehorses, arguing that it was a more profitable line than training them or raising other types of horses. She must have been lucky in at least some of the races, for she shortly wrote that she would not that year be spending the "travelling fund," which, it turned out, had amounted to £30.

By 1851 Ada was deeply involved in horse racing, and her correspondence indicates that she was the ringleader rather than the dupe of her small group of associates. A series of letters to her from Richard Ford, one of this band, written in January 1851, concerns arrangements for a dinner party whose purpose was to discuss their betting strategy; it gives us a glimpse of the kind of sporting life in which Ada had immersed herself without her Lord. The party was to comprise "your Ladyship, the sporting Nightingale & the charming Fleming"; both the latter, like Ford and Crosse, were accepted friends of the family. Ford was a traveler, a writer, and an art collector; though twenty years older than Ada and a man of some prominence, he wrote to her in terms of easy familiarity such as Babbage never used. Where their shared racing interests were concerned, he was brusque and efficient. "I have written to Nightingale to come up Thursday for the *Triple* event," he wrote on the sixth; "if you will please to let me have instructions touching the bank, I will attend to it tomorrow afternoon."[18]

But Nightingale did not answer, and on the thirteenth Ford wrote again.

I cannot think what has become of Nightingale; I have just written him to come up, to talk over the wonderful combinations in your letter. I could not make these bets without much consultation: the very largeness of the odds shows the little probability there is of winning & I imagine the safe rule is to bet, rather than take the odds, but I confess all this is Greek to me.[19]

Ford was well advised in his caution; apparently Ada was suggesting the extremely risky option of making book. There is some indication here, but nowhere else, that she might have been trying to dazzle her equally inexperienced colleagues by implying that she had the mathematical resources to improve their chances. The communications from tipsters and touts elsewhere in her papers show how far this was from being the case. The other men involved—except possibly Crosse—were just as timid and compliant as Ford with regard to Ada's bold suggestions. When Nightingale eventually surfaced, he sent a note to Ford, which was in due course passed along to Ada:

Absence from home has prevented my answering your two Notes. . . . No doubt the thing can be done without her Ladyship's being known in the affaire.
    I am a poor Authority in Racing matters! I would never risk my judgement for more than 5£. . . . However I will do all I can to further her Ladyship's views—but it is like having to move Napoleons Army to Russia for a Man who has only exercised a Brigade of Guards in Hyde Park![20]

This note was accompanied by a cover letter from Ford, yet more explicit about her role in their confederacy:

I think your wishes will be complied with. . . . I have had a visit from Crosse touching the wine, & will give him the benefit of my vinous & gustatory experience. . . .
    What you say about the Flyer [Flying Dutchman] is very perplexing. I admire your nerve & discipline. Making a book seems to me to be living on the brink of a precipice.[21]

Among the proposals being considered at this point was Ada's belated determination to back Voltigeur against the Flying Dutchman at the York Spring Meeting, which was to take place in early May.
    All that spring Ada was in a state of euphoria similar to that previously inspired by Science. To her mother's attempts to win her over to her own latest religious enthusiasm, she responded that she was too busy living and that *"that filial* relation" was "always hanging like a Millstone round my neck."[22] On 17 May the anticipated York Meeting took place. Afterward Lady Byron jokingly condoled with her on her "defeat," though she was probably unaware of just how much had been staked. Ada made an uncomfortable attempt at humor in reply,

embellishing the date of her birthday letter to her mother with the word "Doomsday": "The Voltigeur defeat distresses me less than *yr age.*—I was a good deal prepared for it—expected that the Flying Dutchman *would* have it this time & in fact strongly *expressed* this opinion to some of the party."[23] Even the fact that she was part of a group who shared her interest was no secret.

But Ada was passing through one of her agnostic phases, while Lady Byron was at the moment entranced by a consumptive young Brighton preacher named Frederick Robertson. "I am afraid," declared Ada, "Mr. Robertson thinks me a *desolation* & a *negation* and I think HIS 'love *to God*' ditto!" Moreover, she told her mother, "In *all* respects at this moment you appear to me to be a lamentable example of *my* blackest views of life."[24]

"Doomsday" that year struck Ada not on her mother's birthday but some days later, when she and her partners met to settle their losses; Ada's debts amounted to £3,200. In addition, she managed to obtain a further £1,800 from Lovelace, to lend to a less fashionable member of the "party" named Malcolm, who was in reduced circumstances; his income amounted to a middle-class £300, out of which he was obliged to support a family. Lord Lovelace was in no position to refuse Ada's demand: it later came to light that he had actually given her a letter "assenting to and authorizing her betting proceedings."[25]

At first Ada insisted that Lady Byron be kept in the dark about the magnitude of her losses, and perhaps William too had reason to be reluctant to confide in her. Then, at the end of May, Ada, who had been experiencing profuse menstrual periods since February, had a severe hemorrhage. A Dr. Lee, who was consulted, suspected uterine cancer as the cause. Sir James Clark was called in for another opinion. Twelve years earlier he had precipitated a scandal in Queen Victoria's court by mistakenly diagnosing an abdominal cancer in a lady-in-waiting as an illegitimate pregnancy. Now he was not so positive in his judgment, yet he too was pessimistic. Ada's own favorite physician, Dr. Locock, was characteristically optimistic and, just as characteristically, wrong. As usual, Ada's trust in him was unshaken, and she placed herself in his hands, just as he demanded.

The news that Ada transmitted to her mother concerning these medical developments referred delicately to her illness as "the local state." Lady Byron saw no reason to be alarmed. Her own opinion, she said, was not as unfavorable as that of Dr. Lee. Ada was prone to simulating a variety of diseases without actually having them. Witness her heart attacks. "You know," she recalled, "I have always held the same opinion as to the necessity of an *equable* life & the avoidance of stimulants with a brain like yours." Ada had blithely disregarded her advice all that spring; it was no surprise that she should now be taken ill. "I have long anticipated some illness in you from overexciting habits."[26]

Ada protested that latterly she had been living an extremely quiet and regular life. She attributed her various symptoms that year to "the miserable East Winds," which had so often been a source of distress to her; "& then there is also the *high pressure* of the present age & epoch & state of society." Dr. Locock, she thought, was managing her present condition admirably.

Everything turns out exactly as he predicts . . . although the treatment *appears* to do harm sometimes.—He was anxious to effect the first *sloughing*, & this has been effected most satisfactorily, tho' followed by a Haemorrhage which would have alarmed me *much*; but for his absolute confidence & assurance that it was all as it *should* be & as he *wished*.[27]

After examining her internally—a fairly unusual gynecological procedure at the time, and a measure of the severity of the hemorrhage—Dr. Locock explained that "tho' now there is still an extensive & deep-seated *sore*, yet it is a *healthy sore*."[28]

The fact that Ada had been subject to "heart attacks," in addition to everything else, appears at a surprisingly late date in her correspondence, surprising because it is accompanied by the claim that she had been subject to them for twenty years past. It appears in a letter dated "Thursday Evening, June 14," indicating that it was written in 1849, an indication supported by her London address at the time, also heading her letter. Ada had apparently been stricken at the home of Mr. Murray, the publisher, to whom she now addressed a note of gratitude, apology, and reassurance. "I must thank you for y[ou]r kindness & *excellent judgment* yesterday," she said. "It is 'spasms of the *heart*' that I am subject to;—& I have been so at times, more or less, for about 20 years; but *this last year* much more. It is of course (*to the sufferer*) like a mortal struggle between life and death, tho' I believe not really dangerous under proper treatment." (Even in such a letter of embarrassed apology to her host over having spoiled his dinner party, Ada could not resist philosophizing that she was unable to decide whether life was more "desirable or alarming.")[29] The twenty-year period she alluded to would place the onset of these attacks about 1829, or at the same time as her mysterious paralytic illness; this is corroborated by a statement of Lady Byron's early in 1830 that "she has lost the tendency to spasms."[30]

Ada had another attack while she and Lovelace were on their northern tour, staying with the Zetlands. There she was tended by the Zetlands' physician, Dr. Malcolm (presumably no relation to her racing colleague), who thought he had identified the cause in a strange intermittent matutinal fever. He dosed her with quinine and nitrate of silver and prescribed a healthy, restful, outdoor life for the next two years. Above all, he thought, writing was harmful; presumably he was referring not to normal letter writing but to her attempts at serious intellectual work.

For a time Ada was quite taken with Dr. Malcolm; she tried to follow his directions for several months. At all events, the spasmodic attacks seemed to cease, but Ada continued to recall them with a horror beside which the pain and weakness connected with her "local state" seemed quite bearable. And so she could assure her mother quite solemnly that she had been living a quiet and regular life for some time past.

It was at this point that Lord Lovelace decided it was impossible or inadvisable to conceal from Lady Byron the extent of Ada's racing debts any longer. No doubt, with his own resources tied up in building works, he hoped for assistance from his mother-in-law, who, as she had declared to Greig, always had sums idling in the bank, and whose manner of living was as frugal as her income was lavish. But Ada still feared that once her mother realized that something of importance had been withheld from her, she would not rest until she had devoured the last crumb of secrecy, including much that William himself did not know. Her mother, she knew, was of a far more suspicious turn of mind than her husband. If he must go and make confession to Lady Byron, Ada begged, would he try to protect her from the reckoning she knew would be demanded?

The result was an irrevocable shattering of the alliance between mother and husband that had encircled her life since her marriage. Impetuous and unannounced, Lovelace presented himself before his mother-in-law on the twentieth of June, a month after the Derby Day disaster, and poured out in person the seriousness of both Ada's losses and her illness. Consequently, only Lady Byron's account of it, duly delivered to her solicitors, survives.

I distinctly recollect the comments I made on her to him in the Interview at Leam[ingto]n. That she had been deserted by him in circumstances of temptation & exposed to the influence of low & unprincipled Associates, from whom he ought to have protected her—that Genius was always "a child" & needed such Guardianship—that *I* felt, & the *World* felt he had not afforded it from the moment he let her go alone to Doncaster—that she was not only easily excited but had "an overweening confidence" in her own powers, that these dispositions required to be met firmly & without the apprehensions expressed by him as to her violent emotions for such emotions passed away & did no harm—

I lamented that she had used dissimulation with me, & said it was not less hypocrisy—tho' she had reason to know from experience how willing I was to help her—& I said if she had but confided in me, she "might have had as many thousands as I gave *hundreds* to her last Spring"—

Did this refer to the previous year's "travelling fund" of £30?

I added with reference to both—that it was yet time to take another position—but it was necessary they should see clearly the dangers of *continuing this course* & I described them in very strong language—the degradation, the loss of Children's respect, of private & public character—saying "I know what I tell you is not very consolatory, but I have never yet interfered in your affairs, & now they are placed before me I cannot mask the truth & give up my principles."[31]

Her lawyers, of course, were paid to believe her.

When Lovelace arrived, Lady Byron was herself, as usual, ill, which was one reason for her residing at Leamington Spa on the occasion. She was too ill, she wrote Ada, even to attend her stricken friend Miss Doyle (one of the Furies), who, indeed, died the following day. No doubt she was angry to discover that she had been excluded from some of the Lovelaces' affairs and that her daughter had been evading her. But what she found truly unforgivable was that when she demanded to "come to an explanation" with Ada, Lovelace incautiously admitted that Ada did not feel up to such a catechism. Her faultless reputation as a mother was in question; she must now prepare to spend the rest of her life canonizing her maternity, as she had lived the previous thirty-five years tending the shrine of her wifehood.

She suffered agonies over the estrangement. When she visited Dr. Lushington at Ockham, the realization that she was not at liberty to call on her daughter was humiliating in the extreme. To Ada she uttered a cry of pain that was almost a supplication.

I suffered so much from the *proximity* at Ockham under circumstances which prevented my going to E. Horsley that the question of future intercourse must be settled at once. I cannot endure such a position & should take means to prevent its recurrence.

As to any apprehensions on your part of what might be too exciting in your present state of health, I trust that,—as you once told me I had never in your life uttered a word which wounded you—you would be able to say so still, after we had met.

If Ada had ever made such a statement, it must have been under duress. In any case, Lady Byron was choosing to ignore the many occasions on which Ada had confessed her resentment and fear. But it was as near as she could come to pleading.

As I pass thro' London tomorrow I shall send to your house for a reply. Whatever it may be, it will make no change in that which *has* undergone no change & never *will*—O Ada had you but known what a word from *you* could have done.[32]

She made it clear that William was to bear the full brunt of blame for the estrangement between them. "I cannot—will not—be on the same terms as before, with your husband." In response, Ada admitted that she had been unwilling to talk "on the subject of recent occurrences"; now (however reluctantly) she agreed to a meeting.[33] She did see her mother once or twice during the ensuing year; but the visits were stiff and formal—and controlled by Ada.

Over the next months Ada's physical condition deteriorated to the point where even Dr. Locock admitted it was "worse than Heart attacks." As long as it remained relatively painless, however frightening and enervating the loss of blood, Ada was more concerned over the possible effects on her mind. In August she reported the opinion of Dr. Cape, yet another physician:

for months previous, there had been a continuous *current drawn off from* the Brain. I often felt great CONFUSION, & difficulty in *concentrating my ideas.* . . . I WAS (partially) *dead*, there is no doubt;—but I have come to life again. . . . But what a fearful thing is an *insidious* painless disease, that undermines before one knows it.—Locock was *very* blameable in *Feb.* I believe *none* of the things that *have* happened *would* have happened had my state been discovered.[34]

The reference to a "current drawn off from the Brain" and the mention, in the same letter, of two manuscripts of Rutter's reveal that she had returned to her interest in mesmerism and, along with Dr. Cape, was considering its possible relevance for her case. Yet Dr. Locock still had not owned, and would not for another year, exactly what her true state was.

It is a matter of speculation whether Ada's life might have been saved had she been instead under the care of Dr. James Simpson, the Edinburgh obstetrician who had treated Greig's wife Agnes for an "inversion of the womb" (prolapsed uterus). Dr. Simpson became Queen Victoria's physician in Scotland, and it was he who persuaded her to have her last two babies while under chloroform sedation—a turning point in childbirth anesthesia. But beside his work in anesthesia, he made pioneering contributions to gynecological diagnosis and the control of bleeding, and it was his opinion that cancer of the cervix should be treated by a partial hysterectomy, which was then within his capability to perform. Once the cancer had spread beyond the uterus, however, the case was hopeless. Cervical cancer may take up to twenty years to develop, so it is quite possible that by the time her symptoms proclaimed themselves unmistakably, Ada's fate was already sealed.

In any case, she was under the care, not of the queen's Scottish physician, but of her English accoucheur, Dr. Locock, who, though kind and sympathetic, accepted consultations with other authorities only very reluctantly. Even after he had so clearly failed to control her disorder, he wrote to Ada testily, "I wish people would leave us alone—so you will understand, that if you wish to drive me away, propose a consultation.—When I think one would do good, I will tell you so."[35] "Damn the doctors!" she had once said, and might well have said again.

She was suddenly seized by the realization that she might not have much time or strength left to "use my *brains*." She turned again to writing poetry, reporting to her mother, "It is so entirely & utterly DIFFERENT from what you, or anyone would expect."[36] At this point too she suddenly found sympathy growing up between herself and her previously neglected and almost ignored younger son Ralph. There were many similarities between them, she found; they were both Byrons and geniuses: "Set a *comet* to catch a *Comet*. Set a *Genius* to warn a *Genius* and set a *Byron* to rule a *Byron!*" But even better, "He will have the *manly* temperament, which I adore & envy, (& the *more*, from being under the influence of a WOMAN'S *complaint*."[37]

In the meantime, while trying to disarm the still-guarded, evasive Ada, Lady Byron pursued her quarrel with her son-in-law. He answered in terms both submissive and bewildered, acknowledging all her previous generosity to him. However, he was as yet unbroken enough to add, "I do feel that your condemnation of me without any definite cause being assigned is not in accordance with justice."[38] This only brought on a further blast: he had kept the existence of his letter consenting to Ada's betting a secret from her right up until their meeting on 20 June. This letter was now labeled the "chief cause" of his "embarrassments."

By late autumn Ada's lack of progress was so patent that even the previously complacent Lady Byron wrote perceptively, "I observe however that all medicines & changes are very successful at first & for a time—misleading in fact some of those about you to think that the permanent improvement is much more positive than it is—and your own temperament leads you to exaggerate both the first favourable symptoms & in some measure at times, those that are less so." She also observed that Ada had always lacked "persistence" in her undertakings, and was glad that she contemplated a long period for the writing in which she was now engaged.[39]

When Dr. Locock at last admitted that Ada's "healthy sore" was in fact a tumor, he once more took too optimistic a view of the prognosis. Toward the end of the year the hemorrhaging began to be accompanied by severe pain. Dr. Locock then explained that this was simply part of the "curative" process. "There is a good report to make of my *complaint*," wrote Ada. "The contraction & softening progress quite steadily. But the *pain* of the curative state is far greater than when disease was actually gaining ground. I hear that the *contracting* of the vessels is often very painful." Already she found the pain unbearable without the intervals of relief and rest provided by narcotics. Trying to put a brave face on her new agonies, she consoled herself with the reflection that at least hers was a condition that did not require an operation, and she still compared her present state favorably with "those horrible HEART attacks." Yet, she reflected, "I am more & more convinced of the HORRORS of human existence. My own (by NO MEANS one of the worst samples) is a *struggle* & an *endurance*. What then may *others* be?—I cannot make head or tail of it."[40]

Lady Byron had often (though inconsistently) disapproved of the use of narcotics; to relieve pain too easily, she felt, was to frustrate its mission. Now she attempted to persuade Ada to resort to mesmerism instead of opium. Ada once more canvassed the opinions of her doctor, Faraday, and Babbage. Dr. Locock responded conditionally and diplomatically, and managed to remain in Lady Byron's good books; Babbage and Faraday were skeptical enough to antagonize her for life. Lady Byron's letters frequently referred to the experiments of Dr. Rutter, whose discoveries were foreshadowed by "a passage in Newton's Principia." Murchison, another eminent scientist, had been inquiring into Rutter's Magnetoscope and was quite satisfied. Faraday really ought to go along and be convinced too. Ada, however, found that mesmerism was not as effective as opium for her "REAL pains." She was also trying cannabis, which she had heard about from her friend Gardner Wilkinson; she found it soothing, and helpful in inducing sleep.

Babbage was so uneasy about Ada's condition that he consulted his friend Sir James South, who in turn demanded another opinion. But Dr. Locock, after another examination, assured Ada once more that all was well: the pain was only the result of the contraction and return to normality of her long-distended "vessels." The tumor was quite gone.

Great pain and her near brush with death had made her much more subdued. It was dreadful to know the magnitude of suffering that was possible. And to what end? She tried "*not* to think about it, since I can alter nothing." Now, having convinced herself that she was on the way, if not to recovery, then to stabilization of her condition, she was able to reflect, "What most disturbed me, at the time when I certainly apprehended a fatal termination, was the idea of the length of illness & suffering there would be. Dying by inches is a horrid way, is a dreadful fate, & did appall me when it seemed impending."[41]

Toward the spring she began to feel that if her disease resulted only in the loss of "the uterine function," she could be reconciled to all she had endured and could even look forward at last to a period of sustained intellectual endeavor, albeit as an "incurable invalid." After all, there were many other steady and productive habitués of the shawl and sofa around for her to emulate. "This *would* be a compensation probably for this formidable illness. My mind, nerves, everything about me, have been as uncertain as the 4 winds of heaven, ever since that cursed function first began. It has been to me an unmixed evil."[42] If only the pain would cease.

When I find that not only one's whole being can become one living agony, but that *in* that state, & AFTER it, one's *mind* is gone, more or less,—the impression of *mortality* becomes appalling. . . . If my wits & feelings were not *crushed* as they certainly are by great illness & pain, I could hope that there is some ulterior object in individual existence.

But alas! everything spiritual & human *goes* with some miserable alteration in the material tissues. The more one suffers, the more appalling is it to feel that it may all be in order to "*die like a dog.*"[43]

Except for narcotics, most of the treatments simply added more suffering to that caused by disease. Still she tried to convince herself of the benefits she derived from them.

I never mentioned *leeches* to you, thinking it might give you anxiety. I have had them very often. *Three* times BEFORE the great haemorrhage. Other measures were tried first & did no good. . . . & yet local depletion seems good, & indeed requisite. I take Bark & Port Wine 4 times a day always, & unless this & *very high diet* is kept up, the SORES become threatening, & indeed I break out *all over*. And yet, local bleeding is necessary. One thing: there has been no PERIOD now for some months. . . . Many symptoms look like the departure of the periods altogether; — particularly the *haemorrhages* I had for a year, previous to the present *total stoppage*.[44]

But the pain continued and intensified. "To be dead sleepy, & absolutely *torn* awake by agony, & obliged to use every imaginable resource again & again for hours!"[45] The resources now included chloroform, to which she had submitted only out of desperation. In April she reported to her mother that she found it "effectual but unfortunately only transient in its power. . . . I was terribly *frightened* the first time it was tried, & indeed had *for weeks* refused all resort to it, till it became scarcely a matter of *choice*."[46]

Heartrending as her communiqués were, they could not deter Lady Byron, who, convinced that Ada was still concealing some of her affairs from her, was once more clamoring for a meeting. Now housebound, Ada continued to reassure and elude her:

If I were in any URGENT way in difficulty or compromised, I would *at once* & without an instant's delay tell you so, & beg to be rescued.

A year ago, I would have *died*—literally & really—sooner than have owned to you difficulty, or have asked for even the *merest trifle*. So that there is *indeed* a difference in my feelings, at all events.[47]

But this admission brought only renewed demands, and, grasping for other means of fending her off, Ada hit upon her mother's hostility to the hapless William. "I am greatly vexed that I can't see you," she wrote. "My lord is a *Fixture* at present, but I will let you know the moment I am at liberty."[48] She was now permanently in London, no longer able to travel there for her medical treatments; but Lovelace continued to commute to the country to attend to his duties as principal landlord and Lord Lieutenant of Surrey.

Ada's life had always been beset by fear. Even the periods during which she had reveled in horseback riding were punctuated by long stretches of time in which she feared to ride at all. Occasionally she had given voice to her fears, both general and specific ones; but to these confessions Lady Byron always gave short shrift, at one point retorting, "You will certainly not 'die' of a prolonged fright, for your life for some time past has been nothing else!"[49] Ada's greatest fears, however, were of her mother, and these she now began to reveal, hesitating and imploring.

> I have a GREAT DESIRE to see you, & yet a *great dread*. I fear yr not being fully aware of the *extremely* delicate condition I am in, & how *one breath* that was painful to me might have a severe effect on my much-shaken health.
>
> I think it is best to say this to you plainly at first. You certainly do *not* know *all* I have suffered, from occurrences of the last year, & how these many painful feelings have left me a comparative *wreck* as to what I can now stand.[50]

A few days later she tried once more, making no less than three drafts of her appeal.

> In yr last, 3 or 4 days ago, you say, "*There is much virtue in plain speaking*." But on the other hand: *What is once* SAID *never can be* UNSAID, & *this* is a formidable drawback from resolutions to be perfectly frank. It is *kill or cure*; & it requires much courage to run the risk (if one *sees* it). . . .
>
> There are unhappily *associations* now connected specially with *you*, extremely painful to me, —& which have weighed hard on me throughout my illness; —so that I cannot but feel some agitation in first seeing you.[51]

She begged her mother to consult Dr. Locock before she came, to hear from him how dangerous any agitation might be. But at this point Lady Byron could bear her exclusion no longer. She was determined to force a complete confession of all Ada was still withholding, to bring her back to a suitable belief and trust in religion, and to place her physical condition under the influence of phrenomesmerism. Before Ada's carefully composed letter was sent, Lady Byron left Brighton, where she had been staying, and, mastering her own trepidation, presented herself at the door of the Lovelaces' house in London.

> I am alive & have *done it*—seen Ada this morning with cheerful self-possession on my part, but on hers, feelings that seemed as if they might destroy her—one eye was almost invisible from the squint—

there was not much light—it looked like the first act of a death-bed, but perhaps a second interview may make another impression.

She had written to prevent my coming *now* it seems—but the letter went to Brighton after I had left it—Something urged me to risk the Journey yesterday tho' Miss Montg[omery] told me I was unfit for it.[52]

The final stage of the struggle for Ada's mind and soul was now joined, and Lady Byron decided that the day-to-day saga of her preparations, alarums, and excursions should take the form of a series of letters to one of her idolizing lady friends, a Miss Fitzhugh. These letters, a kind of journal, were meant to be, and eventually were, reclaimed from their first admiring audience, to be kept as a record of her campaign. She probably never dreamed with what a jaundiced eye future readers, for whom they were preserved, might regard these accounts, in which minute descriptions of Ada's deteriorating physical state alternated with technical discussions of mesmeric matters, self-conscious attempts to gird herself for battle, and self-congratulatory reports of her speeches and triumphs.

I am conscious of the power to destroy, even by a word—not to save—She lost her voice from the nervous shock on our first interview—I must not be cold—I must not be effusive—I must not shun confidence, yet I must not force it. . . . [I]f her state made it possible for her to move, I would take her away & then she would be open. . . . Babbage wont hear of Rutter *because* Dr. Leger uses his Inst[rumen]t for Phren[omesmerism]! . . . I shall go to Faraday today as this Sun gives me strength.[53]

But in less than a week her patience was exhausted and she "forced it"—driven, she complained, to such uncharacteristically crude tactics by the defects of her daughter's character.

Persuasion & tenderness failed—Delicacy was not understood. I was obliged to take the tone "You *must*—you will be happier afterwards—*out* with it—"

It is pain to me, because against my nature, to do things in this unpoetical rude way—but the lesson to me is that no illusion favouring Self-deception must be spared.

Certain pursuits have been called "speculation"—I always call them by their common discreditable name. I see that this does good.—There is so strange a mixture of Trust & Mistrust—or rather a *misplacement*—for confidence, withheld almost *to the death* from those who deserve it & would use it kindly, is given to any professing Stranger, nay, to the known Deceiver!

My mistake has been to allow the existence of illusions, not to tear away roughly every sort of fallacy or self-flattery. . . . From tenderness toward the Living or the *Dead*, I have perhaps in a measure neutralized my own influence & suffered a worse one to prevail—But my position has been made most perplexing by the unfairness & counteracting influence of a Third.[54]

Most of the veiled allusions in this passage are fairly easy to penetrate. The "speculations" must refer to Ada's racing ventures. The "Living" and the *"Dead"* designate the dying Ada and her dead father, on whom Lady Byron had lavished such thankless tenderness. The "Third" can only refer to the now despised Lovelace, since neither Babbage nor John Crosse had yet been openly accused and excluded from the Lovelace household. These gentlemen, however, may very well have been caught up in the references to the "Stranger" and the "known Deceiver." The forced confidence, extracted from a cornered and desperate Ada, was that she had been obliged to pawn the heirloom diamond parure belonging to the King family. Lady Byron, through her solicitors, took prompt steps to retrieve the diamonds, and the matter was for many months kept secret from the "Third," who happened to be the rightful owner of the jewels.

The knowledge she had acquired in this fashion served well to tighten her hold over Ada's affairs and confidence. Ada's relief and gratitude that her mother not only did not "destroy" her but entered into a conspiracy to keep her husband in the dark ranged her firmly on her mother's side and against him. Lady Byron now felt free to begin reordering the entire household, with the object of eliminating any other prop or source of support outside herself, and any other claim on Ada's sentiments. The first to go was Annabella's governess, who had been so unfortunate herself as to develop a breast cancer, and with whom Ada showed signs of developing an alarming sympathy and understanding. The letter just quoted continued immediately,

I have had to save poor Annabella from sleeping in a poisoned atmosphere & her health *is* affected. The cancer under which the Gov[erne]ss has been suffering broke out some time ago & is but too perceptible—The change will take place next week—I have to find the Substitute to manage the important changes in a Household where sad things have been going on unchecked.[55]

Meanwhile, she pursued her researches into mesmeric cures.

I spent last Ev[enin]g in the fullest investigation of cases like Ada's in which Mesmerism had been of value & cannot have a doubt that in ulcers & Wounds externally the cure is more rapid than any Surgical measures but I was looking for internal diseases—There are two not very dissimilar to Ada's—where considerable enlargements were re-moved—I learnt the passes.[56]

A mesmerist was found to treat Ada, but she succeeded in giving the patient only transient relief at best:

After she went away a fit of pain threatening to be very severe was coming on. I stopped it by application of Rutter's principle—not of the common passes. She was writhing about—unable to be still for a moment—Under this influence she became perfectly still within 5 min—complained of my making her head hot (I did not touch, only pointed my thumb to her N. Pole & held her left thumb with my other hand, the left)—as I continued for 1/4 of an hour the composure encreased & the eyes were less wide open—When I stopped she said the pain had relaxed—In a few minutes it was returning—I [was] obliged to come away. . . . If I had been at leisure I might have written a curious Memoir of the Magnetoscope & its struggles with prejudice.[57]

Despite the strong and increasing hold she had gained over Ada's condition and affairs, she was still not satisfied with her position. She complained to her friend that her visits were still limited, that Ada did not like her to be there except at a fixed time. Her own freedom was curtailed by having to wait for Ada's assent to their audiences; and when there, she was forced to make minutes do the work of hours. If she could not bring her daughter to a wholehearted belief in mesmerism, she was still determined to bring her to religion. In any case, the two were connected; a Miss Goldsmed claimed that prayer was an imbuer of mesmeric power. Eventually she would be forced to swallow her distaste for Lovelace and install herself in his house.

By the middle of July Ada was never out of pain, and even the mesmerist came to the conclusion that her case was incurable. Lady Byron took comfort, however, in observing that Ada was drawing the proper lessons from the biography of Margaret Fuller, "in whose life she traced the evils of letting Imagination gain the full ascendancy."[58] Several weeks later, when Dr. Locock finally brought himself to reveal how little time she had left, it was to her mother that Ada turned and asked, "Can nothing save me?" only to be told, "Nothing."[59]

The tone of Lady Byron's reports was now almost complacent. Lovelace, overwhelmed as he was by misfortune, was viewed by both mother and daughter as an obstacle in the preparations to be faced. "I shall *never* get thro' all I have to do," said Ada to her mother, "unless I have more leisure. Lovelace *will* not let me alone, tho' I often send him away. But he returns like a *dog* to his master. I shall be quite glad when he goes away."[60]

Lovelace too had begun to keep a journal of Ada's illness, which by the end of July he expected to last only a few more weeks. Although lonely and bewildered, he was far from realizing that he was being sent away like a dog from its master. On finding her arranging her papers and being told she wished him gone about his business to East Horsley, he assumed she was moved only by the wish "neither [to] see nor cause my sadness. I cannot bear to be away from her let alone the terrors of solitude for me when she is in this state."[61]

Like the journal kept by Ada's governess, Miss Lamont, thirty years previously, her husband's was written for perusal by Lady Byron, in the hope that she would be struck by his devotion to her daughter and soften the stony face with which she regarded him. Consequently, there is almost no mention of a new matter that had arisen in the previous few weeks. For one reason or another, Woronzow Greig had chosen that moment to place in Lovelace's hand the end of a thread that would unravel one of Ada's most difficult secrets.

John Crosse had, over the years, become as regular a guest in the Lovelace household as Babbage. In fact, he and Babbage were often there together; on one occasion, Ada reported, they had "both been plagued by her impudent maid," Mary Wilson, who had once worked for Babbage. Lovelace, faithfully relaying the story to his mother-in-law, added that Miss Wilson "seems to be a baggage—or hussey—entering into conflict with mechanics & metaphysics."[62] (Crosse, as early as 1847, had apparently superseded Babbage, in Ada's eyes, in the more prestigious calling of metaphysics.)

Still a frequent visitor at Ada's sickbed, he called on her on 25 July, though he himself was recovering from a serious illness. On the same day, while walking about the grounds at East Horsley, Lovelace and Greig took refuge from a sudden thundershower in a shed. As they waited for the storm to subside, Greig casually mentioned Crosse's wife and children, of whose existence he had learned the previous Christmas. They had been brought to his attention by an acquaintance at his law offices, who happened to be a neighbor of Crosse's where he lived in Reigate. Greig, as he explained to Lord Lovelace, had been surprised to hear of Crosse's family at the time; but in raising the subject now, he assumed that it was no secret from his listener, the Lovelaces being Crosse's "most intimate friends."[63]

The news astonished Lovelace, who requested that Greig, in his role as family solicitor, undertake a preliminary investigation. Soon

Greig was able to report that Crosse's name had appeared on the voters' register at Reigate for at least two years and that his wife was "a young and charming person." Should he look any further into the matter?[64]

After mulling over the information for some days, William questioned Ada about it. She must have denied any knowledge of the marriage and agreed to take it up with Crosse himself when he visited in the evening. The answers she received, or at least those she relayed back, were far from satisfactory, for he decided to interview the gentleman on his own. Once more at East Horsley, he wrote to Ada, who was gripped by anxiety lest this new development result in Crosse's visits being discontinued.

I told Mr. C. that in saying I did not wish to withdraw my countenance from him, I did not thereby imply that I desired to receive him as formerly — that discussion and explanations of the sort that had passed between us rendered it less agreeable for the respective parties to meet often — and that the knowledge with which I was fixed with regard to his proceedings at Reigate rendered it undesirable that he should frequent my house — but that of course your wishes were sacred in my eyes.

He told me that his position was most painful & that your wishes were very strong — & rather asked for advice & suggestion from me. I replied that I could give him none — that my position & feelings were at least as painful & he must judge for himself — but that your wishes must be a paramount consideration with *me*.[65]

Ada had practical as well as emotional reasons for clinging to John Crosse's visits. She was still attempting to make various secret payments to and through Crosse, including, it later appeared, those on an insurance policy on her own life. These transactions required some confidential intermediary with the outside world, since she could no longer even post her own letters. The number of possible agents was being remorselessly whittled down by her mother, who had already banished the governess and was now threatening her devoted and trusted maid, Mary Wilson.

In a panic, Ada took several desperate steps. During their brief interview on 3 August she gave Crosse the diamond parure that had been redeemed and returned by her mother in secrecy, to be pawned by him once more. A week later she appealed to her old friend Babbage, attempting to make him her executor. She confided to him a number of papers and keepsakes, to be distributed after her death. Then she gave him a letter that she hoped would enable him to settle some of her financial affairs without scrutiny or interference by her mother or her husband.

Dear Babbage,
In the event of my sudden decease before the completion of a will I write this letter to entreat that you will *as my executor* attend to the following directions;
　1stly you will apply to my mother for the sum of £600; to be employed by you as I have elsewhere privately directed you. 2ndly you will go to my bankers Messr's Drummond's and obtain from them my account & balance (if any) and also all of my *old* DRAFTS. 3rdly you will dispose of all papers & property deposited by me with you, as you think proper *after full examination*.
　Any *balance* in money at my bankers you will add to the 600£ above named to be similarly employed.[66]

Babbage had grave doubts that either Lady Byron or Mssrs. Drummond would accede to any such requests; but Ada was in such great agony, and reiterated so piteously that both mother and husband would abide by her arrangements, that he felt obliged to accept the responsibility she thrust on him. She was relying heavily on the assurances, so often made simply to comfort the dying, that her wishes would be sacred.

　The account that John Crosse gave Lord Lovelace of his "proceedings at Reigate" was later summarized by Woronzow Greig, no doubt in a far more orderly fashion than that in which the information was elicited. (The practice of repeating information back to the clients who had imparted it was, fortunately for the biographer, common among the many lawyers involved in the denouement of Ada's sad story.) Crosse had explained that for the first year he had shared his house at Reigate with a married cousin on his father's side. Upon the cousin's departure, an uncle on his mother's side, said Crosse, had installed his (the uncle's) mistress and children—who, oddly enough, happened also to bear the name Crosse—in the unused part of the house. Finally he admitted that the mistress and children were in fact his own, living under his name, and that they had now been removed to Godstone, at some distance from Reigate.

　The possession of a mistress and children would not in itself have been sufficient to bar Crosse from the Lovelaces' friendship, as long as he made no attempt to introduce them into respectable society. Despite Greig's informant's claim to the contrary, Crosse assured Lovelace that he had never attempted to do this. Indeed, verbally winking and nudging, he expressed surprise that Lovelace had not known of his worldly arrangement. The possession of a legal wife and family who were not taken into society was quite a different matter. The only inference to be drawn from such a state of affairs was that there was something shameful about either the wife or the intentions of the husband.

The situation was, and still is, very puzzling. John Crosse's younger brother Robert was a clergyman who early in 1852 had been appointed rector at Ockham—a living at the disposal of Lord Lovelace. (A survival of an ancient feudal arrangement by which the priest was a vassal of the lord upon whose land the church stood, the privilege of appointing clergymen to particular livings had become a piece of property that could be sold or alienated separately from the land itself.) Lovelace then had some proprietary interest in the moral rectitude of the whole Crosse family. Furthermore, the Lovelaces were still, in the early 1850s, in sufficient contact with Andrew Crosse for descriptions and discussions of his second wife, Cornelia, to appear in their letters. It is highly improbable that they would not somehow have heard of John's marriage had it been known to his family. The fact that Robert Crosse remained rector of Ockham through all the revelations concerning his brother, and for twenty years thereafter, is further confirmation that John Crosse's acknowledged family were as ignorant of his unacknowledged one as any of his acquaintance.

Burke's Landed Gentry lists John Crosse as having been married in 1847 to a Miss Susan Eliza Bowman, daughter of Charles Bowman of Ipswich. This "young and charming person" was born in 1818 and died in 1916, in full possession of the Crosse (then Hamilton) estate at Broomfield—her husband having died in 1880. The birthdate of their first child, John Hamilton Jennings, is given as 26 March 1848. There is, however, no record of Crosse's marriage in the marriage registers of England and Wales. Even more puzzling is the fact that there is also no record of the birth of John Hamilton Jennings Crosse. The birth of a daughter, Mary, was recorded at Reigate in 1851, and the birth of another daughter, Susan Hamilton Crosse, was recorded in 1854 at Godstone, where, as Crosse told Lord Lovelace, the family had relocated. Possibly the marriage and the birth of the first child occurred abroad. In any case, if the marriage existed in 1852, it seems to have been kept a secret even from Crosse's own father and brother.

Lord Lovelace decided to accept Crosse's clandestine family as respectably illegitimate and wrote to Greig that he had no wish to pursue the subject further. Yet it refused to go away. On 19 August he recorded in his journal that he and Ada had had a "sad discussion about Mr Cr." Apparently Crosse had complained to Ada about the cross-examination to which he had been subjected, and she took her husband to task over the matter. There followed an acrimonious exchange between husband and (as yet undisclosed) lover, the result of which was further to jeopardize John Crosse's visiting privileges.

Now frantic, Ada begged Dr. Locock, who might have been particularly amenable to his patient's wishes because of his signal failure to preserve her life and health, to intercede for her. There is a copy of the doctor's response, in Lady Byron's handwriting, among Woronzow Greig's papers. Somewhat disconcerted by her request, the doctor nevertheless assured her, "I have written Lord Lovelace in such a way as not to make it appear that you have written to me on the subject, but that we had discussed it when I was with you." He enclosed a draft of the letter he had composed, with the hope that she would find it answered the purpose:

I know that for several years Lady Lovelace has had a great friendship for two or three gentlemen to whose society she had much devoted herself. I allude especially to Mr Crosse and Mr Babbage—I know also from recent conversations with your Lordship that from not personally liking one of these gentlemen, you might be led, either by word or manner, to prevent him from being so much with Lady Lovelace as she herself would anxiously desire.

I beg particularly to urge your Lordship not to needlessly disturb Lady L. by depriving her of so much of Mr Crosse's society as she wishes or has the strength to bear—for I know the dread of such being the case has often much harassed her & I think in her present deplorable state it would be both cruel and mischievous to debar her from what has been such a source of comfort & happiness.[67]

The original of this letter was dated 20 August. By that time it was too late, for her mother had already assumed command of the Lovelace house and of admission to Ada's sickroom, and was already privy to her secret. Lady Byron's copy of Dr. Locock's letter was sent off to her lawyers on 30 September, doubtless as evidence of the need to control Ada's visitors. From 16 August the door had been barred to Babbage as well. He was the bearer of liberal religious views of which Lady Byron had long since ceased to approve. He had dared to dispute her belief in mesmerism. But above all, Ada had confided in him and attempted to make him her executor, and this was not to be borne. On 22 August Lady Byron herself became a member of the household, and from then on, no one was admitted who had not been approved by her—except the wretched William, with whom she felt obliged to assume the appearance of a truce. Three days before, however, Fleming—one of Ada's betting partners—had called and managed to see Ada privately, carrying off a check for a payment on the insurance policy, of which he himself was now the beneficiary. It was the last communciation she was to have with the outside world, other than a letter she wrote to John Crosse on the very day her mother installed herself in the house.

The insurance policy was but one strand in the complicated web of transactions by which Ada and her partners borrowed and reborrowed to cover their own and each other's losses. A life insurance policy was, and is, a common method of guaranteeing the payment of a debt; in setting up his London house at Great Cumberland Place in 1850, for example, Lord Lovelace had insured his life to cover his obligation to his mother-in-law for money loaned to buy new furniture. In this case, the policy was intended to secure a sum of £600 owed to Richard Ford, and Fleming seems to have been an agent acting for yet another party who had advanced the money. Although a lawyer himself, Woronzow Greig, who later assumed the responsibility for winding up the details of Ada's turf adventures, was to find himself lost in the intricacies of this particular transaction.

There may have been blackmail involved, and Ada would certainly have been a most likely and vulnerable victim; but the evidence is strong that John Crosse, at least, was not blackmailing her. Among the scraps of paper covered with penciled scribblings left by Ada in her final bouts of agony is a list of the titles of her poems, to two of which are added the parenthetical initial "T."; and equally revealing of the tenderness with which Ada regarded "T." in the last months of her life is a bequest that may have been among the papers given to Babbage but returned by him when he found himself unable to carry out his charge. Wrapped in a piece of paper marked "To B." on the outside is another headed "To T.":

My gold pencil-case (Morden's Patent) having my name & Coronet on the top as a seal; with the request that he will use it habitually, in remembrance of the many delightful & improving hours we have jointly passed in various literary pursuits. Also I request him to select from my Books, any 12 WORKS (not vols) which he may prefer. . . .

Also for the sake of the same pleasant memories, my very shabby dark-coloured leather writing *Box* which is always on one of my tables, & which I have had for *very many* years & have habitually used, I wish T. [written over the lightly crossed-out word "him"] to have this with all its contents ins[ide] = some *seals*, minerals, etc etc *exactly as they happen to be found.* There is nothing in it of any value, excepting for old association's sake; & in fact it always contains much rubbish, — odd bits of paper, pins &c &c. But I wish it delivered to T. as *it is* with it's contents, & no attempt made to clear it of anything or to put *it tidy*. It's value would cease, if it & its contents were in any way altered after I had last used it.[68]

There can be no question that this is an affectionate and sentimental bequest, an attempt somehow to remain almost physically with the loved one after death. It was hardly the sort of legacy to leave a blackmailer who was threatening to expose her as she lay in extremis. But is it certain that John Crosse was "T."?

That Babbage was not "T." may be safely concluded from the form in which the bequest was made. The "B." on the cover sheet, so clearly separated from the inner sheet addressed to "T.," and the deliberate repetition of the "T." twice in the body of the bequest, once over the crossed-out word "him," make the distinction not only clear but emphatic. Then too there is a complete absence of any other reference or address to Babbage as "T." among scores of letters and references, many of them in playful or affectionate terms.

The only other possible candidate for "T." is Field Talfourd, the painter and younger brother of the much better known judge and writer, Sir Thomas Noon Talfourd. In the autumn of 1850 Lord Lovelace mentioned that there was an artist visiting them for the purpose of making landscape studies of the area. This artist was probably Talfourd, who is better known for his portraits, particularly that of Elizabeth Barrett Browning, but who also occasionally exhibited landscapes. He was born in 1815, and his work was occasionally seen at the Royal Academy from 1845 onward. Some half-dozen letters from him to Ada are found in the Lovelace and Somerville papers, most of them responses to her invitations. Those that are dated were written in the autumn of 1851 and, considering the other emotional involvements Ada had at the time with men and horses, indicate a surprising possessiveness in their intended recipient. Her own letters must have threatened the bounds of propriety, for in one of his he reassures her, "Your caution with regard to any *letter* you may bestow upon me is unnecessary."[69]

At then end of September 1851 he was in ecstasy over a sonnet she had sent him. "I have seldom if ever read anything more graceful," he rhapsodized, "so evidently a strong earnest feeling struggling into poetry—just what a sonnet should be." He was certain she had composed it. "I scarcely know whether most to envy the writer or its object—Have you not been misleading me? I cannot doubt the writer to be a *woman*, who could its *object* be but Mr. B. and who the *writer* but yourself?"[70] That he could jump to the conclusion that Ada had written a sonnet to Babbage is somewhat startling; what she had actually asked him was whether he thought the poem had been addressed to *her*. But he thought it not altogether applicable.

There is only one sonnet among Ada's papers that could possibly be the one in question. If it is, Talfourd's speculation was needlessly coy and deliciously wide of the mark. Babbage was certainly not its object, nor was Talfourd himself, and it was clearly not written by Ada.

Sonnet to — — —

There's one to whom thy presence is delight
    Who humbly loves, yet claims no thought from thee.
    Thy few but gentle words from malice free,
Thy many graceful deeds so mildly bright,
Thy pencil's touch that lends each scene new light,
    Thy placid form—all are one harmony!—
    Thy beauty hath a holy symmetry
That sways the heart with sweet and silent might;
And while in melting hues all virtues blending
    Unconscious of the many who extol,
And meekly to the will of others bending,
    Thy life breathes but the music of thy soul—
Oh! All who own thy magic grace transcending,
    Depart the purer for thy soft control!—[71]

We may compare this poem with the one polished piece of poetry that is certainly of Ada's authorship and finished to her satisfaction. It is also a sonnet, as it happens, and was composed within a year after the Talfourd letters were written—that is, during the final year of her life.

The Rainbow

Bow down in hope, in thanks, all ye who mourn,
Whene'er that peerless arc of radiant hues
Surpassing earthly tints,—the storm subdues!
Of nature's strife and tears 'tis heaven born,
To soothe the sad, the sinning and forlorn,
    A lovely loving token to infuse
    The hope, the faith, that pow'r divine endues
With latent good the woes by which we're torn.
'Tis like a sweet repentance of the skies,
To beckon all by sense of sin opprest,
Revealing harmony from tears and sighs,
A pledge,—that deep implanted in the breast
A hidden light may burn that never dies,
But bursts thro' storms in purest hues exprest![72]

Ada's didactic style and philosophical preoccupations, so insistent here, are totally absent in the "Sonnet to — — —" above, where instead the hand of the master is so suggestively present.

"The Rainbow," which in some copies was subtitled "Good out of evil," encapsulated her last "intellectual-moral" mission, her last metaphysical query. Significantly, the line "To beckon all by sense of sin opprest" is a vaguer, less open substitution for the more revealing "To all who cry, who sin, who fall, addrest," which was later crossed out. She was so proud of this poem that she not only signed it but left instructions for its inscription on a plaque to her memory.

Soon after Talfourd penned his admiration of the mysterious sonnet, he removed to Torquay, where he was attempting to establish himself when Ada sent him an urgent summons to visit her. Tactfully, he refused, claiming the press of work. A few days later he wrote again, protesting this time that he was too ill to travel, and furthermore,

There is something so *mysterious* in your request that I am lost in speculation [as to how] my presence for so short a time could (which pardon me I cannot conceive) conduce to your well-being & at the same time . . . such a journey (400 miles) would prove almost fatal to my success this season. . . . As you give *no clue* to the object of my presence—I know not if the occasion is pressing.[73]

Whatever Ada replied to this, whether she gave some account of the sudden crises that had descended upon her—her own frightening illness, that of her younger son Ralph, who had come down with a serious case of scarlet fever, the near loss of her elder son's ship in a hurricane—or whether she reassured him that the urgency was past, he replied only, "I *thought* you had written under some degree of excitement."[74]

On balance, it is safe to conclude that Field Talfourd was not "T." The evidence of "delightful & improving hours . . . jointly in various literary pursuits" is sparse compared with that relative to John Crosse, and there is no further mention of Talfourd in the last turbulent year of her life. His speculation about the "object" of the sonnet shows he was not on familiar terms with Babbage. Crosse, on the other hand, was well acquainted with Babbage, to whom Ada had confided documents that must have revealed their relationship, had he not known of it before. It would not be at all strange that she should have entrusted to him her mementos to Crosse. Finally, the note on the Reichenbach review mentioning "T. C." does provide some positive evidence that John Crosse was known to Ada as "T."

The presence of Ada's testamentary letter to "T." in the Lovelace Papers indicates that Crosse never did receive the pencil case and writing box. Before her death, however—perhaps even before her illness, but more likely through Babbage's offices once more—Ada did manage to pass on to him three other keepsakes: a gold ring that had belonged to her father, a locket containing a lock of Byron's hair, and a miniature portrait of his "Maid of Athens."* Although Crosse was later to prove himself quite capable of blackmail, Ada's gifts, both attempted and completed, must be taken as testimony that her anxiety to keep him with her proceeded from her attachment to him and not from extortion on his part.

By comparison, her mother's presence was fraught with ambivalence and the shadow of coercion. As the summer of 1852 advanced, Ada continued to write to her in terms both affectionate and placating:

I am so greatly comforted & composed by the degree to which I feel *you* a *prop*, that I do think you will either enable me to *get well*, or else to depart *very comfortably* & it seems to me very strange how I ever *could* have regarded *you* with *suspicion*!

But I dread beyond anything the idea of yr being far AWAY. If I could not see you often, I should feel so lonely now. So you must get a place *near*.[75]

Yet somehow she still sought to postpone Lady Byron's installation in her house, and even wrote with some urgency to forestall it on the day before it was finally effected: "I begin to understand DEATH, which is going on quietly & gradually every minute, & will never be a thing of one particular *moment*. I shall send for you at any minute but we have still a *little* time before us."[76] But Lady Byron no longer waited to be sent for.

Lord Lovelace's journal during this period contrasts sharply with the evidence of Ada's frantic efforts to extricate herself and protect her lover, as well as with Lady Byron's seething resentment and inexorable invasiveness. He tried instead to present a tender, almost placid picture of a Good Death: the stoic sufferer, reflecting deeply on the meaning of life, pain, and illness, gathering her loved ones about her, ordering her effects, and arranging her affairs with calm fatality.

By her desire I had written to Col[onel] Wildman to tell him of her condition—I read this to her & she approved of it . . . —an hour or

---

*John Crosse (then Hamilton) in his will (1880, Somerset House, London) left the locket with the hair to his daugher Susan, Byron's gold ring to his son John, and the "Maid of Athens" to his daughter Mary.

two later she called me in to talk to me of her wishes about her remains being gathered to those of her father the poet. She had had she said this strong wish ever since visiting Newstead & Hucknell in Sept[embe]r 1850. I was not aware at the time that she had conversed so earnestly with Col Wildman on the subject . . . but it seems that *He* had in his life time indicated & even prepared a tomb at the Abbey itself—his ex[ecu]tors however had determined on Hucknell as for some generations the Byrons were buried there—but Col Wildman appeared rather to have wishes that Ld B[y]rons own wishes had been acted on. She then gave me a letter to read which she had written to Col Wildman expressive of some wish of the kind, but so delicately worded that it should not place him in any difficulty supposing it should be determined ultimately to rest at Hucknell.[77]

The delicacy displayed to Colonel Wildman in this matter was noticeably absent from her dealings with her husband; for in wishing to lie next to her father, she implicitly refused to lie next to Lovelace. If he felt the snub, he resolutely concealed it. Her mother, more intricately slighted, was not so forbearing. While Ada's preference of Byron's celebrity over Lovelace's obscurity might be countenanced, her choice might equally imply a preference for paternal profligacy over maternal sanctity. Some days later, when Lady Byron got wind of the plan, she at once attempted to take control of it by proposing it as her own idea, inspired by Ada's dread of burial at Ockham. Lovelace noted that she was "rather pleased that Ada had already anticipated her," but to her friend and confidant Miss Fitzhugh she admitted she had been "secretly wounded by the *reserve*."[78]

Relieved, Ada set about giving elaborate instructions for her interment and for the plaques to be placed to her memory in a variety of locations. There were to be endowments distributed in recognition of the nursing services she was receiving as well. Lovelace forbore to speculate on where the money for all this was to come from, noting only the strange pleasure she seemed to derive from her plans. "She walked about the room on my arm for a time, speaking almost with satisfaction of the posthumous arrangements & simple inscription to the effect that she was placed by *his* side by her own desire."[79] It was to be her last feeble defiance of her mother's authority.

She was also making very liberal use of the privilege granted to the dying of ordering the behavior of the living, in both present and future. If she could no longer receive John Crosse or Babbage, she could still command the presence of several others, including Charles Dickens. She had been struck by the death scene of little Paul in *Dombey and Son*. Obediently William wrote to Dickens "to hasten if he would see her alive." Dickens arrived that evening, "spent an hour with her & was wonderfully struck with her courage & calmness."[80]

Her most high-handed commands involved the children, for whom she had for so many years expressed indifference and a belief in the value of benign neglect. Here, in contrast to her arrangements for the disposal of her remains, her wishes were very much to her mother's liking; they were set down in pencil in a paper dated 15 August, and traced over in ink by some other hand:

It is right you should immediately be made aware, my dear Mother, of the wishes I have finally expressed to *Lovelace* regarding my children in the event of my decease.

1stly—that *Byron* should under no circumstances be permitted or induced to leave his Profession, which I regard as essential to his welfare both moral & physical

2ndly That you should be the female Superintendant & Director of Annabella—& that she should *reside with you* during not less than 6 weeks annually. It is almost unnecessary for me to add that I desire & hope *she* will consider her *Father* as her first & great object in life & that she will in all things endeavour to supply *my* place to him.[81]

Ralph, of course, was to continue under Lady Byron's "guardianship & management"; any change there would be "ruinous" to the boy. There was also a fourth clause in this most curious document, which insisted in the most vehement terms that her children should be kept rigorously from contact with their father's side of the family (as she herself had been kept from her aunt Augusta). Lovelace had always been on bad terms with his mother and brother (who made something of a name for himself opposing the privileges of primogeniture in Parliament). From time to time, however, there would be signs of rapprochement between the brothers, usually fostered by Woronzow Greig. Lady Byron had always reacted with rampant hostility, and no reconciliation ever came about.

If the harm that might have befallen Byron's grandchildren as a result of associating with their paternal uncle or grandmother is a matter of speculation, it is even more intriguing to speculate on what Ada (or Lady Byron) might have intended by proposing to substitute young Annabella for her mother in relation to her father. In any case, Annabella did soon develop her mother's concealed anger and resentment toward him and entered into a similar surreptitious alliance with Lady Byron against him.

Lady Byron mentioned Ada's disposition with respect to the children in one of her self-congratulatory dispatches to her friend: "She is coming silently to a true feeling of what my *Mother-life* has been. She had pointed out to Lovelace that I had reason to feel this marriage 'a disappointment' & has made me guardian of the younger children."[82] But in Lovelace's report of Ada's instructions only the arrangements for Annabella are mentioned; Lady Byron's continued custody of Ralph was taken for granted.

Byron, or Ockham, the Lovelace heir, had returned from sea earlier in the month. Ada was overjoyed to see him but found herself in too much pain to be able to entertain him for long with the necessary composure; for most of his leave he was kept away from the house, either at one of the Lovelace estates or with friends of the family. Apparently, however, he was with his mother long enough to make known to her how unhappy he was in the navy and how he longed to leave it. Her answer (and her mother's) was a flat refusal.

The object of Ada's yearnings after this was the return of her younger son, who had been sent to Switzerland to imbibe the Fellenberg educational system at first hand—an experience he would later look back on with revulsion. Once arrived, however, Ralph too was largely banished from his mother's sickroom. Only the girl seemed able to withstand the emotional ravages of their mother's wasted appearance, agonized writhings, delirious outcries, and drugged torpor. Lovelace recorded the following scene:

Annabella & I went to her: awake but slightly febrile—we sponged her face & hands, which she enjoyed. The water gurgled a little reminding her of a cascade—we went back to some of the scenes of our Tour on the lakes . . . on which she dwelt with great pleasure & interest, telling me I must at some time take Annabella.[83]

(Lovelace kept this promise, taking Annabella to visit Newstead, just as he carried out almost all the dying wishes Ada made known to him.)

Ada now regretted having once been disappointed at the birth of a daughter. She could envisage the entire moral burden of "the Family," especially the disconsolate father, falling on the shoulders of a girl who was not yet fifteen. Annabella tried bravely to shoulder this responsibility, writing to her younger brother to reconcile him to their bewildering mother:

She is *so* beautiful now, and so gentle and kind—I think her like what I should imagine an angel to be. . . . I think dear R. that you *do* know now how *really* Mama *does* love us—only, as G. M. [grandmother] told me, "*she has so little idea of her value to others.*" Do you know, she really

said to me one day, that she thought she must be a *"nasty creature."*
I of course contradicted that assertion. . . . She has been playing duets
with me these two last evenings.[84]

Lovelace himself seemed scarcely aware of his children in his preoccupation with his stricken wife and her mother, the intended reader
of his journal:

We talked again of her strange destiny. Her mother had said it was
hard at first sight. — Yes — she said "we both take the same view. She
calls it *hard*. I call it *difficult*." In the one the view of feeling predom-
inating, in the daughter the intellectual. She did not see why so much
suffering was necessary, & yet she bowed in submission to its infliction.[85]

Up to this point Ada had borne her ordeal with astonishing patience
and fortitude. But now she became incoherent and distraught at times.
She was terrified of being buried alive and repeatedly pressed her
doctors to reassure her that they would not permit it to happen. And
there was something else troubling her. As William hesitated at her
bedroom door one day, she called him in and dismissed him several
times. He ordered flowers to be sent for, "a hydrangea, fuchsia, ge-
ranium &c growing in pots so that she might see them in the doorway."
Once more she called him in. "It was very kind indeed the thought
of the flowers; very kind & I thank you for them — but now go, I am
not strong enough for anything else, not for the whole truth just
now."[86]

Before she could tell him "the whole truth," she fell into convulsions
that lasted for two days. On 30 August her pulse stopped for about
ten minutes. Yet, against all expectations, she did not die but gradually
recovered consciousness and reported a strange dream.

She spoke but hardly recognized me — "Surely it was a great mistake —
which excessively vexed" her — but whether this alluded to the
trance . . . we could not determine. About 5 there seemed to be more
return of consciousness — but mixed with some vague terror of some
one out of doors who was to do her a mischief. She told her Mother
& me that the suffering she was to undergo would last 1,000,000
years — that it had been revealed to her in a dream with experience
[?] — thus evidently confusing the future with the recollection of the
agonies she had gone through. She knew me & kissed me. She asked
her Mother to pray God for her & me also later — with one hand in
each of ours, she said "Therefore, my Mother & you my husband
will ask God's mercy for me." We said both all that would console &

comfort her—At times the strong fear of someone from without, & of some precautions that ought to be taken, returned.[87]

Lady Byron's picture of the same scene differs materially and characteristically:

The prevailing impressions were that she was dying—that she was guilty towards God—& should have "a million years of the horrible pains she had suffered in this disease." . . . My answer freed her from all her fear—Have I lived in vain? She then turned meekly to Lord L. & said—"Will you forgive me?" This is the best moment she has had, & for that she must have been kept alive.[88]

Later she added to this passage a note that revealed that what had made this moment the "best" was not so much Ada's meek plea for forgiveness for sins as yet unconfessed, as "her strange illusion, thinking a dream reality: that her *father* had sent her this disease, & doomed her to an early death! She spoke of it as cruel, & unjust of God to allow it."

Lord Lovelace's journal broke off at this point. He founded it too painful to continue, for the following day Ada confessed to him her adultery with John Crosse. It was by no means "the whole truth," but it was enough to shatter the picture of brave and innocent affliction he had clung to. It was not news to Lady Byron. Her account indicates that Ada's relationship with Crosse had been revealed to her on 20 August: on that day a confession was made that she refused to elucidate to Miss Fitzhugh. Possibly this was the final precipitator of her decision to "become an Inmate" of her despised son-in-law's household. Until she did so, too, there is no indication that Ada felt the slightest guilt or shame over her affair with Crosse.

Lady Byron had been willing to enter into a conspiracy of silence over the matter of the pawning of the diamonds, for the time being at least. A confession of adultery was another matter. Certain that the revelation would be as painful to William as to Ada and convinced as usual that truth, if sufficiently wounding, must be beneficial to the hearer, she relentlessly goaded Ada to confession, seeing nothing in this at odds with her earlier pronouncement that "my one object will be to make myself the medium of Christian influences, elevating and cheering."[89]

To that end she hoped to forestall hovering death, "& yet there are reasons why one should not wish it long deferred . . . the suffering, & the probability that she might survive *me*—which would be dreadful to her."[90] Nevertheless, when the confession came as Ada was emerging from her coma, still with large gaps in her recollection, it may have taken Lady Byron somewhat by surprise. Although the subsequent exchange of letters between her and Lord Lovelace makes it quite clear that he first learned from Ada of her infidelity on the first of September, the record that Lady Byron eventually filed with her lawyers postdates the event by three weeks—unless another and better-rehearsed performance was staged at the later date:

I know that I am essentially necessary to her—I have determined to bear and forbear—I came to this house when I ceased to have intercourse with Lord Lovelace on account of his grievous injuries towards me, with the express stipulation that I should not be considered to have prejudiced my own case, as regarded him, by accepting his hospitality *on the ground, distinctly stated, of being of use to her*. It has been acknowledged by him that I *was* of use. . . .
  Many things in his conduct as related to her, have given me pain—but this day I am perfectly horror-struck by his temper, shewn in an interview with her, before which he was made fully aware that he must hear what she might wish to say, but reply as little as possible & as calmly—
  She spoke to me afterwards with anguish & tears of his "bitterness" towards her—I saw his angry countenance after he left her—He was unwilling to speak of the Interview to me, as he usually does—I enquired—he said "it was very painful"—then added sarcastically (or bitterly at least) "it was impossible not to contradict her"—but that he had said before he came away "God be merciful to you!"—
  God be merciful to *him*!—Pharisee as he is, from whom his poor penitent wife could draw no kinder expression by her complete submission & humility for such was her spirit—I *knew*. If I remain in this house, it is for the reason assigned. AINB[91]

To put an aggrieved Victorian husband, learning of his wife's infidelity, in such a position that any attempt to justify himself can be construed as ill-tempered is a legal and moral achievement of impressive dimensions. At what point in Ada's recital did he find it impossible not to contradict her? He later complained that he "was forced to listen to the most degrading excuses."[92] Did she blame the inception of her wayward career on his miserliness?
  To the catalogue of his other crimes—concealing the truth about the racing disaster, estranging Ada from her mother, abandoning her to low associates—Lady Byron had recently begun to add that of harassing his wife on the subject of his financial difficulties. To Miss Fitzhugh she contrasted Lovelace's meanness in evicting his poorer cottagers "in order to *diminish the rates*" with Ada's generous deathbed

distributions to the needy (out of money supplied by her mother, to be sure). But where Ada herself was concerned, she did Lovelace an even greater injustice than usual. She wrote to Miss Fitzhugh, "She said he had been in tears when he told her that he found he was going down more & more as to his income & should soon not have the command of £100.—All this of course tho' *she* said nothing was a dagger to *her*—I soothed her by talking of his low spirits."[93]

In actual fact, since his wife's huge racing losses had come to light when her health was already failing alarmingly, Lovelace had done his best to relieve her mind of anxiety and guilt, attributing their embarrassments instead to general economic conditions and his mother's jointure. "[D]o not be disquieted dearest about pecuniary matters," he had written her, "for I hope & trust that some day we may be relieved from our present difficulties. . . . I am certain that all persons deriving income from land will be so much affected by these prices, that they *must* give up a good deal but it cannot stop there, & the distress will come to be felt by the other classes in the country who till now have had all the benefit of free trade & low prices."[94]

He was quite unaware of the virulence and extent of the antagonism daily committed to paper and dispatched to Miss Fitzhugh. Indeed, he was under the impression that their joint ordeal in witnessing Ada's suffering, and their joint efforts at comforting her, had melted Lady Byron's anger and rewelded their kinship. Gladly, he placed the management of his household in her hands, even supplying her the written authority she was in the habit of insisting on.

As for Lady Byron's more general object, the provision of elevating and cheering Christian influences, Ada at first resisted with scorn the notion of "death-bed repentance." Had she not for some years past stoutly maintained her Crosse- and Babbage-backed agnosticism against her mother's ecletic mixture of unfocused Christianity, mesmerism, and spiritualism? "The more I REFLECT on the subject," she had once written, "the *less* I can believe. I am more disposed to believe even in the error of my *own senses*, & in being *self-deceived* than in what is so contrary to all previous experience & philosophy."[95]

But now the older woman won all the arguments. "Can you prove what I feel towards you?" she demanded. "Yet *can* you disbelieve it?" Writhing in pain, Ada was forced to admit that feeling was not demonstrable.

She then went on to say that she *would* "try experiments"—that experience never had any weight with her—that she entered upon every experiment "innocently"—tho' she got wrong in the course of it—She said—as a reason why longer life was *not* to be desired, that it would only "show her more of her errors"—I told her I had another fear for her, were she at this moment to be freed from her Disease, that she would renew those experiments—We then agreed together that the conditions of this life might have done their work for her, & that she would be mercifully released from struggles & trials for which she was not fitted—a sense of her extraordinary folly in being *independent* of counsel & guidance when she was so incompetent to act with judgment in the affairs of life.[96]

For a time it seemed as if Ada's convulsions and subsequent partial amnesia might rob Lady Byron of her ultimate victory; but as soon as she began to rally, her mother lost no time in consolidating her position. Written in a shaky hand, a scrap of paper among Ada's effects asks her friends to return her letters to her mother after her death. One such returned letter, dated 12 September, is a flatteringly worded bequest to Mary Millicent Montgomery, followed by just the postscript to warm her mother's heart: "I feel that *Miss Montgomery* would have been my *best* FRIEND & I wish her to know it."[97] Miss Montgomery had been the first among the Furies.

Furthermore, if Ada, in her pain-racked, drug-blunted progress toward dissolution, could not resist taking her mother's friends as her own, neither could she refuse her enemies. Soon Lady Byron could happily report that Ada was realizing the "conspiracy" against her.

The Idolatry in which she was encouraged by all about her (& under the sanction of one especially) for her father as a man not less than a poet & that his aberrations were no longer felt to be such, with the halo thrown around them—She particularly said that—excused the *worst* of his known (to an intimate few) immoralities. Gradually *I* was thus placed in an unfavorable light, without any direct censure being passed upon me—as if I could not enter into the Byron nature. . . . She told me that the person whom she considered most her friend "felt animosity" toward me—I had after one interview declined further acquaintance. . . . A Swindler & a Brute—"O my prophetic Soul."[98]

So much for John Crosse. Lady Byron could now afford to dismiss him; her own apotheosis was at hand:

Have I lived to hear it?—She has just said "There is but one reason why I could wish for life—*to live with* YOU entirely["]—She believes that with my "protection" she might have gone through life happily for herself & others—What a change!—I keep saying to myself "Wonderful God"![99]

In the ecstasy of her triumph, new strength poured into her, almost transforming her chronically frail constitution. "Nothing agitates or hurries me,"she reported; "I command sleep when I choose— . . . & am equal to anything but standing long."[100]

The convulsions of the last days of August had left Ada with an impaired memory for events of the preceding weeks and months that was only gradually fully restored. Toward the end of September her discovery of a memorandum relating to the second pawning of the diamond parure suddenly brought a recollection of the whole transaction flooding in upon her. This time she lost not a moment in reporting it to her mother, who immediately ordered her lawyers to redeem the jewels and confide them to Dr. Lushington for safekeeping. Not a word was to be vouchsafed to William until the heirloom was restored to him after Ada's death.

Once the diamonds were secured and Ada had made every submission to her mother's vanity and will that could be desired, not even Lady Byron could understand why she continued to live. Yet her now restless mother, her crushed and spent husband, her exhausted doctors, nurses, and servants, were still unremitting in their efforts to prolong the torture that Ada herself found senseless and interminable. "I believe I have suffered," she scribbled—though now to no one in particular, simply covering scraps of paper in the intervals of agony and incoherence—"as much as a human being *can*, but I still hope tho' I do not perceive it that it is for some good here or hereafter— that it is for some good."[101]

Surprising as it seems, so far from bringing her final relief with an overdose of narcotic, her attendants employed every possible precaution to extend her life. Care was taken, for example, to send Ockham back to his ship unnoticed, because "the parting would kill her."[102] So successful were these efforts that it was not until 27 November that Lady Byron could at last describe to Miss Fitzhugh the final torments, which she did with astonishing composure and a hint of disappointment:

For some hours the last agonies. . . . Faintings & fierce pains alternating—I think this may last all night but it *is* Death—like nothing that has come before. . . . I am there constantly but only *physical* relief is sought. . . . Since I wrote this I went & told her the Truth—to believe & trust God as her support & only hope or words of that kind for I can't recollect—

On the following day she amended her report, adding, "At ten O'clock last night, the last *hour* tranquil. . . . "[103]

But now that it was over at last, or should have been, Lady Byron herself was far from tranquil. Once more she adverted to her sole meeting with John Crosse, insisting that she had stood between him and Ada "in the *very last hour*."[104] Almost at once she removed from the hated Lovelace's house to a hotel, gathering up and taking with her, or forwarding to her lawyers as possible useful additions to her armory, the lists and scraps of paper that Ada had left in her last few weeks and months. Some were dictated, some penciled in a trembling hand. Among the latter was one bearing only the date, September 1852, the inscription "To my mother 'Malgré tout,' " and a reference, "PS:17—v 8."[105] Did she recognize it at once, or did she open her Bible, then or later, to read Psalm 17, verse 8: "Keep me as the apple of the eye, hide me under the shadow of thy wings"?

Hide Me under the Shadow

To Lord Lovelace, Ada's long-expected death brought not the finality he anticipated, nor the reconciliation and mutual consolation with Lady Byron he hoped for, but only renewed pain. Before the month was out, she mounted a fresh campaign against him.

She wrote to Woronzow Greig to try to range him among her allies. Lovelace, she complained, had wished to visit her in her hotel; but she insisted that before she would see him he must submit a written retraction of the statement he had given Greig in an attempt to defend himself against her charges. Above all, he must admit that he alone had been the cause of Ada's revulsion against her mother in the year that followed the disclosure of her racing losses.

This may seem like an odd way of recommending her cause, but Greig had already shown himself almost hysterically eager to act as peacemaker between his difficult clients. Moreover, in the past she had not stinted in showing the whole Somerville family those sensitive fiscal attentions that had so won the gratitude and admiration of Harriet Martineau, Anna Jameson, and other of her illustrious but impecunious middle-class friends. When a loan to Greig himself proved unnecessary, she delicately offered to dower his two sisters, should the need arise. As it happens, both would die unmarried and in straitened circumstances, but her kind thoughts had won her extravagant praise in Greig's communications to his mother.

The revelations that had come to Lord Lovelace during his wife's final months, her long, harrowing illness, and, worst of all, the new disclosures concerning the family jewels had by this time darkened her memory and dimmed his devotion to her. Yet, strangely enough, his need for and devotion to his mother-in-law seemed only to intensify. His long antagonism toward his own mother, as he himself admitted, made him cling more fiercely to this adopted mother. He was willing to make every retraction and submission, except the all-important one of owning himself the sole cause of Ada's estrangement from her mother.

For the past year Greig had added his own entreaties to those of Lovelace. "Remember," he wrote Lady Byron, "*he never had a Mother*, and let your heart pity while your judgment condemns him."[1] Now he who, as well as Lovelace, knew from Ada's own lips the truth of her feelings about her mother was ready to make all the necessary declarations that still stuck in his friend's throat. He assured her that Lovelace regretted any slur he had cast on her "lifelong, unwearied and devoted affection" to her daughter. "He is wax in your hands," he went on encouragingly; "you may remodel him to your will."[2]

Perhaps under other circumstances she could have, but in the dissolving chaos that his world now seemed, with "every cherished conviction of my married life . . . unsettled,"[3] he obstinately sought reason and justice from the woman who had always presented herself as truth and virtue incarnate. Now she rejected his apologies as inadequate and found his countercharges inflammatory. She crushed and disarmed him by revealing her rescue of the diamonds, but most wounding of all was her implication that *Ada* had claimed he had caused the rift between mother and daughter, while he was only acting on her express entreaties. (Indeed, he had nursed the forlorn hope, right to the end, that Ada would vindicate his conduct to her mother and persuade her to drop her hostility.) "The pain with which I read and understood that part of your letter," he wrote now, "has hardly been surpassed by that inflicted upon me by her revelations of the 1st September or your later one touching the Jewels."[4]

She conceded him no just greivances whatever. As far as "September 1st" went, she said, it only proved that he had not adequately protected his wife. "You also bring forward," she continued, "in reproach to her who is gone, the matter of the 'Jewels'—As all the trouble and cost of recovering them were *mine*, *your* interests are not injured by the fact of your property having been endangered for a time through the villainy of one who abused the *unlimited* confidence you placed in him."[5]

Uselessly, he pointed out that he had confided his property not to John Crosse but to her own daughter. And if Ada was not to be trusted, why then, after redeeming the jewels, had Lady Byron restored them to her? He begged for an interview to clear up their mutual misunderstandings:

For the past week I have been prey to the utmost wretchedness of mind—Every cherished conviction of my married life has been unsettled. Perhaps in conversation you may be able to give me some relief. I will lay my whole heart open to you . . . if you will but be patient and listen, & correct & convince me wherever I may have misconceived either you or as far as you know her. Do not treat lightly a request which is so important for the recovery of my peace of mind—for you know not the state to which I have been reduced by your last two letters.[6]

His request was refused as often as it was made. A complete retraction, worded exactly to her satisfaction, was a precondition to any meeting. She even sent Greig a suitable statement for Lovelace to sign.

Greig was playing a most curious part, having been privy to all the details respecting the jewels that his friend was so mortified to find had been kept from himself, the most closely concerned. Now urging surrender, Greig wrote, "You know that she is very peculiar in her views, opinions, and actions. These you cannot expect to change. You must therefore take her as she is."[7]

Much as Lovelace needed his mother-in-law, he wanted justice more; denied a meeting, he poured out more and more of what might otherwise have been hidden away, becoming in the process first piteously reproachful, then bitterly sardonic. "Few mothers," he wrote her, "have been so ardently beloved by their own offspring as you have been by me"; yet now she did not scruple to aggravate his distress by her views of "truth." As to his marriage, he reflected sadly,

Few have endured in that relation graver wrongs or forborne more that I have—and if after her imperfect confession her dying hours were embittered by no reproaches from me can you not in turn spare the infliction of them upon me? If I remitted much to her, and in part for your sake, should you not also forgive a little for hers? Should you not have compassion on your son-in-law as he had pity on your daughter?[8]

She had had genuine grievances against Byron, and her adamant refusal to see him or consider a reconciliation at the time of the separation was probably sensible in view of the inordinate privileges and favorable treatment accorded to husbands in law. Against Lovelace she had only the festering knowledge that Ada had at some point turned to him for protection against her—that and the competition she was determined to promote for the allegiance of his children. She accused him of poisoning their minds against her. Wearily he reassured her, perhaps to an extent he would later regret; but at that point he was too beset by his woes to think about them. "Do in respect of the children what you think best, for I cannot now pretend to suggest anything, however much I may value every intervention of yours."[9]

The correspondence grew ever more vituperative. "They are beyond control," observed Greig to Lushington.[10] In the vehemence of their exchange a number of matters emerged that might otherwise have been referred to more obliquely, if at all. When Lady Byron flung back his allusion to Ada's "imperfect confession" by retorting that he had no right even to recall what he had pledged to forgive, he only enlarged upon the allusion:

I am precluded by no pledge given in or out of your presence from remarking freely on the wrongs committed against me as you have done on those you think me a party to. I have said the confession was imperfect. One grave crime only was made known to *me*—it was palliated as "a mistake"—and I was forced to listen to the most degrading excuses of some of which only you are cognizant. Of the robberies no confession was made to, or *for me*—nor do I suppose I should ever have known the extent of [her] guilt in that direction but for your apprehensions that the former accomplices might again be successful in their attempts.[11]

"One grave crime" must refer to the "palliated" adultery—which he forgave, he now said, partly for her mother's sake; "the robberies" alludes to the double pawning of the diamonds; and Ada and Crosse are now "accomplices." The "apprehensions" that eventually led to the disclosure of Ada's guilt were those that made Lady Byron deposit the parure with Dr. Lushington—the latter had only recently enlightened Lovelace completely as to Ada's full culpability and her mother's blamelessness. More than twenty years later, when it fell to him in his turn to enlighten his daughter, he accused Ada of "infidelity to him, of swindling & of forgery, matters which came to his knowledge during her last illness."[12] Unfortunately, the details of these charges are lost, the relevant pages having been circumspectly cut from Annabella's diary. But whatever his terms referred to, his use of words like "crime," "robberies," "guilt," and "accomplices" makes it clear that his once "beloved Ada, dearest bird," had come to be regarded as little different from a criminal.

His affection and devotion to his mother-in-law were more difficult to renounce. For several weeks he had continued to profess his love and admiration for his relentless tormentor, despite her "harshness"; but eventually she worked herself up to a pitch of grandiose hyperbole and self-praise that he could no longer tolerate. She reiterated her claim that he had no right even to mention Ada's confession:

You say that you gave no pledge. *I* understand forgiveness to be a *pledge* that what is forgiven shall be no more brought forward *against* the person forgiven, neither to injure that person living nor to blacken the memory of the departed. . . . Her *spontaneous* effort, in opposition to every worldly interest & at every risk, to obey the voice of conscience, has I trust been accepted by a Judge more merciful than Man. The time may come when under sufferings less severe than hers, you will be taught to feel how cruel it would be to weigh your words with the coolness of the stern Magistrate.

Did I not know that there is in your Imagination a rigidity which makes it hardly possible for you to enter into another's position & feelings, I should look at that passage of your letter with horror—& at best anyone would consider it an outrage upon the Mother whose child you depreciate & condemn *beyond measure* not taking into ac[coun]t what *led to* her aberrations.[13]

What she had said at Leamington was only the truth, which ought not to have alienated—though sadly, she knew, it sometimes does:

It is then that the tender Parent or devoted friend, will go through the martyrdom of uttering it [truth], as I did in that hour of agony. The Past & Future, as regarded Ada, came distinctly before me,— Whilst you, who had looked less deeply into the human heart & its issues, knew not what you were doing. With the fearful prospects which you then unconsciously opened to me, Disease itself was to be looked on as a blessing to my Daughter.[14]

He was surfeited at last, but even then forbore to point out that she had once more repeated and confirmed the slighting of her daughter's illness that she had so indignantly denied. Nor was Lovelace the only recipient of such sentiments; Miss Fitzhugh had received her share. Some time later Lady Byron went even further and confessed to another friend that her conduct at Ada's bedside had not been as "impeccable," nor her presence as necessary, as she claimed in the depositions she filed with her lawyers. "The staff on which she leaned had not been equal to support her," she admitted; "it was the more necessary that she should be taken away."[15] These occasional confidences make it clear that Lady Byron's interferences with fact were not self-delusions but perfectly deliberate. She could not disguise from herself that she had longed for the death of the daughter she had rejected at birth, who had so often rebelled against her and who had left so sordid a legacy. This might now have occurred to Lovelace too, for he brought himself almost to sneer:

In spite of your clear view into the future as well as the past which you describe yourself to have taken at Leamington, I not being a seer & not being able to share in it remained under the more obvious & literal impression which the then unexplained course of your observations engendered. . . .

In my former letters I have gone to the *utmost* limits of concession that truth would allow. Those admissions have only been acknowledged by you as the ground work for reiterating reproach. . . . She is now where your posthumous praise or support can be of little consequence. And I can with humble but earnest confidence appeal from your unmerited condemnation of me to the judgment of Him who knows what have been throughout my devotion to her & my affection for you.[16]

With this letter, each of the combatants having appealed over the head of the other to a Higher Judge, the correspondence ended. Strangely enough, the illusion he had kept alive the longest was that Ada, had she lived, would have felt compelled to absolve him of blame and substantiate his assertions that the distrust between mother and daughter had existed long before Leamington and had had far deeper roots than his acquaintance with them—that he had acted only as Ada's instrument of defence against her mother, and at her insistence. Somehow this belief survived not only his affection for his wife but even his faith in the justice and benevolence of Lady Byron.

As usual, she had the last word, and in her own inimitable fashion. Attached to his last letter is a statement to the effect that she read only the first two lines, since he "continues the argument instead of acquiescing in my wishes." Her health, she added, had been affected by the quarrel. Appended to this is a statement, notarized by a friend, recalling her various actions over the preceding six months and how they had been witnessed and approved by the lawyers Greig and Lushington.

Possessed, as she was aware, of a legal mind that equaled or surpassed that of any of her advisers, Lady Byron had long held her counselors in contempt. "Is it not strange," she had written Lovelace in happier days, "that in almost all the cases where I have employed Solicitors *they* have been completely deceived & the client has had to point out the rights of the case.—I have come to the resolution of being my own Lawyer in future as far as possible."[17] For a woman actually to represent herself was of course not possible. Instead, at the slightest sign of difficulty with friends, servants, tradesmen, or family members, she worked through a complex hierarchy of solicitors, whose actions and communications she monitored carefully and revised in accordance with her own superior judgment.

In her dealings over the previous year with Ada and Lord Lovelace, she had used Dr. Lushington as a go-between in rendering, to the one, financial assistance and, to the other, moral disadvantage. She had also done her best to ensnare Greig, Lovelace's sole friend, agent, and adviser. To deal with those outside the family, she employed the firm of Messrs. Wharton and Ford, who worked under her own close supervision and that of her prime minister, Lushington. Greig, in his turn, worked for Lord Lovelace through a Mr. Karslake but kept in close and harmonious contact with Dr. Lushington.

Between Lady Byron and Lord Lovelace—though not, curiously, among the solicitors—there developed a clear-cut apportionment of postmortem matters to be disposed of. To Lord Lovelace fell the duty of arranging the funeral and memorials, which he discharged with meticulous attention to Ada's instructions. He saw to it that Ada's coffin was placed by her father's side, in a ceremony of unstinted pomp and splendor. Even more woeful was the task of dealing with the racing fraternity, who now came forward clamoring for whatever spoils his extreme vulnerability to publicity might yield. This labor he delegated entirely to Greig, at the latter's insistence.

Lady Byron for her part appropriated the more congenial and re-spectable task of settling the more openly recognizable claims and distributing a number of small gifts in Ada's name; she also took upon herself the satisfaction of dealing with Babbage and with the servants dismissed from the household at Cumberland Place. (This was now to be broken up, as a result of Lord Lovelace's disinclination and financial inability to continue as his mother-in-law's tenant; he was determined to avoid living in London until his daughter's "progress in life"—that is, toward marriageability—should require a settled res-idence there once more.) Discharging her tasks in her own fashion, Lady Byron peremptorily overruled any of Ada's arrangements with which she disagreed and set up others in their place more to her liking, insisting all the while that she was merely carrying out her daughter's wishes. To one favored creditor, who could not possibly appreciate the irony, she volunteered,

Thank you, my dear friend, for enabling me to settle your claim. I must explain that Lady L. having no power to bequeath anything could have neither "Trustee" no Ex[ecuto]r. *I* placed £5000 of my own at her disposal, & took upon myself besides debts to the amount of between £7000 & £8000 so that Lord L. has been completely exonerated from such burthens (none of them Gambling debts) & I could not allow him to be troubled—It is happy for me that my retired, & as many think *parsimonious* way of life, has enabled me to give comfort to the Departed, & relief to the Survivor. From the one I had thanks which I shall never receive from the other.

You have had Law-expenses also I fear, & the least I can do is to send you the enclosed.[18]*

It is difficult to guess what debts Ada might have left behind, other than gambling debts, to amount to over £7,000. In any case, a statement by Lady Byron of the "Sums given by me for the benefit of Lord Lovelace" contains no mention of such an amount. Instead, it lists £1,500 "given to Lady L. Dec 10 '50," presumably a birthday present, and £6,500 "from that time to Nov 52 & applied to the payment of claims which would otherwise have fallen on Lord L."[19] The two sums do indeed add up to £8,000, but they were undoubtedly used at least in part to repay gambling debts. The only other item of substance that might have been covered by Lady Byron's munificence is some furniture sent to John Crosse at Reigate, an account of which was among the incriminating papers deposited by Woronzow Greig with Farquhar and Company, Bankers, in June 1853. (Other items included "Papers relating to the family jewels"; "Papers relating to the turf"; and letters from Ada's racing associates, including Richard Ford and John Crosse, the latter written during 1848, 1850, 1851, and 1852. Could the gap in 1849 be related to Ada's learning of Crosse's "marriage" and a temporary estrangement on that account?)[20]

Babbage, whom Ada had attempted to make her executor and whom Lady Byron had turned from the door the previous August, caused some difficulties when he submitted copies of Ada's testamentary letters. Far from complying with the requests so confidently expressed in them, Lady Byron summarily demanded the surrender of all of Ada's papers in his possession, including her letters to him. Her demand was of course transmitted through one of the subsidiary attorneys in her employ. Stung at this brusque treatment, Babbage refused to deal with the deputy, demanding to speak only to Dr. Lushington, who, he somewhat naïvely believed, as "a man of honor" was incapable of such high-handed behavior. Nevertheless, he delivered

---

*The intended recipient of this letter was clearly in on the secret of Ada's gambling, yet personally innocent of involvement in it and obviously not intimately acquainted with all her financial transactions. Hence, it was not one of the lawyer-friends, Lushington or Greig. Possibly it was Sir William Molesworth, the radical politician, owner and editor of the *Westminster Review*. Ada had certainly borrowed money from him—£300 in August 1849—and his sister was married to her confederate Richard Ford. Alternatively, it might have been John Crosse's brother Robert, who was involved in a lawsuit at this time. However, Lady Byron's acquaintance with the latter at the time of this letter could only have been very recent.

up the packets of bills and receipts that Ada had entrusted to him, asking that they be returned to him after inspection and settlement.

Mr. Wharton, the solicitor whom Babbage had found so objectionable, drafted a response: "Pkg of all papers were that day forwarded to Lady NB, who of course will not part with any receipts or bills of her daughter . . . whether they were made out in the names of Mrs or Miss Wilson or Mrs or Miss Taylor [another servant]." Lady Byron inspected the draft and applied her more subtle brand of legal craftsmanship. "Dear Mr Wharton," she wrote, "I think great caution is necessary with Mr Babbage, & I have therefore suggested a modification of the first paragraph in your proposed letter." The wording she proposed to substitute was as follows:

Whether Ly Noel Byron, who was authorized by Ly Lovelace with Lord L's sanction, to settle all bills & claims on Ly Lovelace at the time of her decease, would be justified in surrendering the receipts (in whatever name made out) *necessary for that purpose* & obligingly forwarded by Mr Babbage, will be for her consideration, when she is informed of the grounds of his request.[21]

At last, disgusted by the treatment he was receiving at the hands of his old friend, Babbage composed and supplied to Dr. Lushington a deposition of his own. It must have created a sensation, since copies of extracts from it appear in several places in the archives. It reads, in part,

With respect to the collection of papers & letters given by Lady Lovelace to Mr. B during her life as well as an extensive correspondence carried on with him for years, many parts of which are highly creditable to her Intellect, Mr B feels entirely at liberty to deal with them in any manner he may choose. The conduct of Lady Lovelace's Relations to Mr B. has released him from any feeling of delicacy towards them; & Lady Lovelace's testamentary letter gives him full authority.[22]

"Lady Lovelace's Relations" may have included her husband as well as her mother. If so, it is surprising that Babbage should have resented the rupture of this friendship, which must have been inevitable as soon as Lord Lovelace learned that Babbage had accepted the confidence of matters kept secret from himself. The "old Drafts" Ada had instructed him to obtain from her bankers included payments on the insurance policy and installments on her various secret debts; the bequest to "T." was made under a cover addressed to Babbage; he had preceded her "Relations" in his knowledge of her affair with

Crosse. Just when he learned of it and what his reaction was is impossible to guess; as early as 1843 Ada had playfully discussed an earlier flirtation with him. Undoubtedly, he had not reacted with the censoriousness of Woronzow Greig or James Phillips Kay, and at the end he had agreed to be her channel of communication with her lover.

Of the remainder of the papers given by Ada to Babbage, only a few dozen of her letters have come to light. Most of these were deposited in the British Library by Babbage's son, presumably including those "highly creditable to her Intellect." If so, then one can only conclude that Babbage, like De Morgan, Faraday, and others of Ada's intellectual acquaintance, was as impressed by her speculative, metaphysical "first queries" and her "overweening confidence in her own powers" as he had been entirely convinced of her mother's moral rectitude. Except for the profuse exchange that took place during the drafting of the Menabrea Notes, the letter that came closest to an intellectual discussion was written on Tuesday, 23 July 1850:

I am anxious to learn the progress of yr *nebulous* theory. I suppose it depends [on] rather mathematical laws of *condensation* & *expansion* of *gases*;—the gas being supposed subject to certain INITIAL CONDITIONS, for instance *gravitation, rotary motion*, &c?[23]

Once more, the vagueness and tentativeness with which Ada's inquiry is phrased mark her uncertainty and ignorance of the subject matter she is attempting to discuss. (Here, the "nebulous theory" refers to the formation of planets from gaseous matter rotating about the sun.) In this her letters are in sharp contrast to the specificity and clarity of the inquiries on scientific topics from Babbage's other correspondents, such as Mary Somerville and the Duke of Somerset. Even more significant, perhaps, is the fact that, except for the correspondence during the Menabrea period, none of Babbage's letters to Ada that have been preserved contain references to scientific matters in the technical terms that mark his communications with his peers.

Ada's letters to Babbage may also be examined for evidence of the extent of Babbage's knowledge of and involvement in her gambling. Some writers have suggested that a number of their letters contain disguised references to bookmaking schemes and racing tips. Once such a surmise is entertained, almost any ambiguity, pun, or unclear reference in the correspondence may be interpreted as lending it support. Yet it is quite common for friends who meet fairly frequently to correspond in such a way that many of their allusions are not

immediately intelligible to the outsider. In the case of Babbage and
Ada, while a number of the references must inevitably remain obscure
and baffling, almost all, when examined closely, turn out to have
perfectly commonplace alternative explanations.

For example, a note from Ada dated Friday, 1 November (hence,
presumably written in 1850), has been cited as referring in a medical
metaphor to her gambling losses, sustained in spite of helpful hints
from a tipster recommended by Babbage. The note reads, in part,

I had better delay no longer in letting you know that the invalid is
certainly *better* from Erasmus Wilson's medicine. But the health is so
utterly broken at present, that I wish to follow the plan you suggest,
& to have the examination & enquiries by yr medical friend—as soon
as return to Town shall admit of it. I think this of great importance.—
Some very thorough remedial measures must be pursued,—or all
power of getting any livelihood in *any* way whatever, will be at an
end.[24]

Twelve days later another note followed, containing another cryptic
reference to "the invalid," who is now reported to be "in *statu quo*,"
but over whose condition Ada is still pessimistic. Clearly, neither Ada
herself nor Lady Byron, that other inveterate invalid of her intimate
acquaintance, could possibly have been worried about losing "all power
of getting any livelihood." Once the hypothesis of a coded message
is suggested, however, it seems easy to interpret these communications
in the fashion proposed by Doris Langley Moore:

If we take "the invalid" for Ada herself, the message rather trans-
parently conveys that, while she had recently had some luck, her debts
were still serious, and that the "medical friend" was a person credited
with having inside knowledge from which she might benefit.[25]

When we recall that Ada's notes were written a few months after the
Doncaster meetings, at which she apparently had enough success to
fire up her enthusiasm for the turf to a fever pitch, the explanation
seems to fit quite well.

It turns out, however, that Erasmus Wilson really was a prominent
physician whom Ada consulted on at least one other occasion; among
her papers is a prescription signed by him for the treatment of her
younger son. (It recommends cod liver oil, cold baths, sponging, and
exercise.) Since it is dated 3 July 1852, Ralph was probably not the
invalid of November 1850. Who was it then?

On 2 January 1851 Babbage wrote Ada, "I have only seen the
invalid once, but Sir J. S. called yesterday and had seen her."[26] Thus,

"your medical friend" is revealed as Sir James South, also a physician, whom Babbage was to consult in his concern for Ada during her last illness. On 12 January Babbage wrote again; in this note the identity of "the invalid" is disclosed:

Sir J. South has called today with the accompanying book for your maid with full directions for her use. Perhaps it would be useful if you would open it and read the directions in order to use your influence in causing them to be followed.[27]

Ada's maid at this time was Mary Wilson, who had previously been employed by Babbage and who occupied a position as near to friend as a servant could attain. (Ada's daughter referred to her as "Miss Wilson" and complained about the pet dog she was permitted to keep in her employer's household.) The first of Ada's notes mentioning "the invalid" was written from Ashley Combe, where Ada and William had decided to spend the autumn months of 1850 without the children, accompanied only by maid and manservant. The phrase "as soon as return to Town shall admit of it" reveals that "the invalid" was with Ada when the note was written. The fact that Mary Wilson was Ada's confidante and go-between in her betting does not necessarily mean that Babbage was also involved. Neither Greig at his best-informed nor Lady Byron at her most malignant ever accused him of this, and certainly there was never any hint that Sir James South was implicated.

Another letter that has been cited as evidence of coded communication between Ada and Babbage is a four-page note (two sheets, each with writing on both sides) dated Tuesday, 2 September, and hence written in 1845. If it were a coded message, it would be the earliest evidence of Ada's gambling, by five years. The writing seems to break off abruptly in the middle of the third page and continue again on the overleaf. The last words on the third page begin a sentence: "About the"; the first words on the fourth page are the continuation of a sentence: "some day put up in the *pendant* archway (with *red* instead of *purple* cloth upon it)."[28] The letter seems to make no sense at all unless interpreted as a coded message. Close inspection reveals, however, that the bottom of the third (and fourth) page has been torn away just after the words "About the," and that someone has carefully glued a blank piece of similar, but slightly newer, paper to the torn edge. In other words, the end of the sentence on page three and the beginning of that on page four (and half a page in between) are lost. Of course, it is impossible to tell why part of the letter was torn away in the first place; what is left provides no sign of a premature betting conspiracy but rather a good example of the importance of inspecting the original documents in historical research. (In fact, a number of torn documents in other volumes of the British Library Additional Manuscripts have been similarly "repaired.")

Another type of cryptic communication that is sometimes cited as evidence of collusion is the number of references in Ada's letters to Babbage to matters better discussed in person than by letter. These often seem to receive significant emphasis from Ada's characteristically lavish use of Victorian-style underlining: "I think the matter we yesterday talked of is of *real importance.*"[29] However, on those occasions when circumstances later force her to explain herself on paper or when other evidence discloses the mystery, the revelation is invariably anticlimactic. In one instance the "matter" related to some firearms, said to have belonged to Byron, that were now offered to her for sale; would Babbage please examine them and give her his opinion?[30]

On another occasion Ada sent Babbage "a piece of humbug," a sheet of paper covered with what, it has been darkly suggested, was a cipher resembling Greek letters.[31] It was, in fact, one of two begging letters in bad Greek sent at an interval of a year (when Ada was aged sixteen and seventeen respectively) by a trio of Greek sisters who were all that remained of a family Byron had once befriended. The other such letter is in the Lovelace collection; it had been sent for translation to Mr. Frend, who advised Ada to ignore the appeal.

Finally, the "coded messages" by which Ada supposedly communicated with Babbage about her racing affairs contrast markedly with the openness with which such matters were discussed in letters to her from her undoubted confederate, Richard Ford. All in all, Babbage may safely be pronounced innocent of playing the horses with Ada.

The nature of his behavior in connection with the maid Mary Wilson was perhaps less innocent. Needless to say, Lady Byron's malice toward Miss Wilson knew no bounds. She claimed to have dismissed the maid as "worse than incompetent,"[32] but she treated all the servants just as humiliatingly. She insisted that each, before receiving the year's wages that were customary on such occasions, sign a paper acknowledging that her generosity was a fulfillment of her daughter's wishes and in no way deserved. All possible evidence of good and loyal service was recalled or withheld. Possibly as one such sign of appreciation, Ada had left her favorite dog, together with a grant of £5 a year for its maintenance, to the coachman. Lady Byron permitted him to keep the dog but denied him the maintenance. When he persisted in his claims to it, and to a silver tankard left to him as a further token of Ada's favor, Greig advised Lady Byron to ignore him: as an outside servant the coachman was not in direct possession of the sort of knowledge that could create unpleasant publicity.

Mary Wilson posed more of a threat, especially since she had turned to her old patron Babbage for advice and support. Babbage included in his deposition to Dr. Lushington, entitled "Observations by Mr. Babbage," his knowledge that after 16 August, the date of his exclusion from the house, Ada "was entirely deprived of control *even* over her own servants." His source of information on that point was easily inferred. That he had a good understanding of the only incentives that might move the solicitors appears in another passage in the same paper: "The family of Lady L.," he declared onimously, "are still indebted to the forbearance of Miss Wilson under provocations which might produce & would justify the severest retaliation."

Greig reacted to this barb just as he was intended to, muttering to Lady Byron, "I cannot understand B's game," but speculating that "B" intended to sell his silence for £100 a year to be settled on Miss Wilson.[33] Lady Byron, to Greig's intense distress, was not nearly as afraid of exposure as a lady of her position should have been. As for Lord Lovelace, he was past caring about anything but the lacerating emotional blows that Ada and her mother had combined to inflict on him. In the event, the day was saved by Babbage's own tenderness toward Ada's posthumous reputation. In his will he settled £36 per annum on Mary Wilson out of his own estate, and she never did sign the humiliating concession that would have been the price of a year's wages. A year later the lawyers were still fussing over her unclaimed check.

The incident of the begging letters from the Greek sisters illustrates how Byron's family in general, and Ada in particular, remained for many years peculiarly vulnerable to importunities by real or pretended former associates of Byron's murky exploits in foreign parts. This is doubtless the true explanation of an obscure letter from William to Ada that has been pointed to as evidence of his knowledge of a blackmail threat to her over her racing debts during her lifetime. From internal evidence — such as references to political events in Italy — the letter seems to date from 1848 and concerns an abortive rendezvous of Ada's:

Dearest Bird
I opened your letter this m[ornin]g with some anxiety to know the result of Sundays app[ointmen]t with the extortioner. He may have got scent indirectly that no good would come to him — may have had an accomplice in distant observation for some time before — and seeing

someone e.g. Fleming [?] or some other person on the watch—may thus have been deterred—if so, he will trouble us again in his way, me perhaps as you suggest. It would certainly be far better, as however disgreeable it would be far less inconvenient for me to come into court & prosecute than you. But the state of society which permits & encourages such crimes is full of evil. The law it is true affixes a punishment, but it or the practice of juries render the *proof* almost impossible. For instance—the handwriting—who can say he ever saw this rogue write save his accomplice?

. . . Commerce at Genoa is a sad bathos after all the dreams of Italian independence and of Ch[arles] Alberts suzerainty.[34]

The word "extortioner" and the uncertainly deciphered name "Fleming" were enough to make Ada's biographer Doris Langley Moore assume that the letter pertained to some attempt to blackmail Ada on account of her racing debts. But it is clear from the context in which Fleming is mentioned (if it was Fleming) that he was acting in this instance to protect Ada and, if possible, to apprehend the "extortioner." Furthermore, it must be remembered that the Fleming who was Ada's racing associate was acceptable enough to Lord Lovelace in 1852 to have been admitted to visit Ada on 19 August, after Babbage was excluded. Finally, the complacent righteousness of the tone with which Lovelace decries the "state of society" indicates that on this occasion he considered Ada innocent of any legal, moral, or social impropriety. In fact, the letter is very similar to another, dated only 24 August, that reveals that Ada had been pestered by a person calling himself "G. Byron": "[I]t would have been better to have done nothing in reply—he might perhaps have written again & more violently which would have given me a hold over him—as it is I may have the opp[ortunit]y of inflicting a thrashing on him," wrote Lovelace on that occasion.[35]

"G. Byron" was the subject of an extended exchange between Lord Lovelace and Woronzow Greig, who investigated a plea addressed to Ada on his ("G. Byron's") behalf in September 1847. He claimed to be the son of Byron by a marriage to a "Countess De Luna" in Spain, during the poet's first, lighthearted trip abroad in 1809. After attempting to make his fortune in America, the "son" surfaced in 1843, writing to Byron's publisher, John Murray, and to various Byron relatives, including the Lovelaces. From the latter, he complained, he received "only—to say the least—a cold, if not a sneering reply.—To a sensitive mind this is worse than death." Nothing daunted, he made several subsequent attempts, and an encounter with him was even mentioned by Ada's son Ralph many years later. By that time he had gone on to become an accomplished forger of both Byron and Shelley manuscripts.[36]

One letter from him that came into Lovelace's possession, dated 22 September 1847, gives us a glimpse not only of his literary style but also of Ada—for once not mystic, philosopher, genius, child, nor patient, but conventional court lady:

There is no other alternative—I must court either *death* or *publicity*.—Oh mockery!—The very paper on which this sheet rests, seems to deride me!—A paragraph containing a description of the Countess of Lovelace's dress, at one of the Queen's levées stares me in the face.—I do recollect that very day as one of intense suffering—when the *"sole daughter of HIS house & heart* rolled in her gilded carriage to St. James' Palace—*his* grandchildren were *à jeûne* [starving, i.e., "G. Byron's" children were hungry].[37]

Publicity was the great dread and great threat, the only weapon that members of the lower orders might use to extract benefits, advantages, or even, occasionally, justice from their betters, whose privileges seemed otherwise so secure. Yet "G. Byron" was comparatively easily disposed of; only the softhearted but impecunious Augusta gave his sensitive soul any comfort at all. After Ada's death, however, the more difficult task of dealing with her racing colleagues and settling their claims—which were of course labeled extortion—devolved upon Greig, in part because of Lady Byron's worrisome insouciance over the dangers of public exposure. As Greig wrote to his opposite number, Lushington,

I think that with your assistance we might possibly have protected our friends from what you and I consider one of the greatest of evils—publicity—were it not for the entire disunion—nay, antagonism—which subsists between them. . . . I can manage for Lovelace well enough by not allowing him to interfere except when called upon. But how to keep Lady N. B. quiet? I cannot understand what she would be at! She is dissatisfied with the only proposal she is aware of as regards the [insurance] policy—viz, our idea of sending the strongest protest to the Company that they would pay the money at their peril. *How* can she twist this proceeding into offering hush money? Can she desire to force us into court? What else can be her object? However, my mind is made up . . . as the responsibility will be mine in whatever manner I act, I must exercise my own unhampered judgment, and do that which I consider most *expedient* as well as beneficial.[38]

And what he did in the end was to permit the insurance policy to be paid as hush money. Greig was shouldering this responsibility not simply to shield his old and much afflicted friend but perhaps also because he did not wish either of his clients to learn the full extent to which he himself had been involved in Ada's debts, flirtations, and confidences. Possibly it was in the hope of its affording some relief from his burden that he decided to commit Ada's secret history to paper, a rather odd thing for him to do, considering the frantic efforts he made to conceal her affairs and to cover up her tracks.

Each claimant, as he came forward, was deftly handled with due consideration to his social standing and to the potential danger of his knowledge. Among the other matters disposed of was that of another servant named Wilson: Steven, the butler. He was probably not related to Mary, since, upon finding that Lord Lovelace had refused him the all-important "character" or recommendation, without which it was almost impossible for a domestic servant to be reemployed, he repaired not to Babbage but to Greig, and threatened legal action. As Greig reported it,

I told Wilson that Sir I. Kemp had acted a most unjustifiable part in delivering to him your letter which was a confidential communication and could not form the basis of any legal proceeding. I then got him to admit, that he had informed you that Johns [another servant] reason for leaving your service against your wishes under the peculiar circumstances of the family [Ada's illness], was his intention to go to Australia—and that he Wilson had never undeceived you of the false impression which he had given, although he knew that John had never left the country, but on the contrary had slept in your house and stable for some time immediately after he left your service and had frequently been in your house up to the time of Lady L'[s] death. I asked Wilson if it was possible under these circumstances to come to any other conclusion than that he had intentionally deceived his master. He replied that he certainly did not intend to deceive you; and thus the conversation ended, I undertaking to speak to you.[39]

At that point Lovelace was disposed to punish severely any failure of candor in his servant. But after cutting Wilson's legal legs from under him and maneuvering him into a suppliant position, Greig began to worry. The butler might know too much; it was not wise to make him desperate. Fortunately, Agnes Greig, from whom her husband had struggled to withhold the unsavory details of the Lovelaces' affairs, somehow got wind of his dilemma and was able to suggest a graceful solution for both wretched master and anxious servant:

My wife has been talking to me about Wilson the Butler whose appearance here she of course knew, and whose motive in coming here she heard from the servants. She tells me that according to practice which prevails in London, a character is never absolutely refused to a servant except under very extreme circumstances, but that a Character is always given upon those points *only* to which the master can conscientiously speak in the servants favor. And she strongly recommends you to adopt this, to prevent Wilson from talking & complaining. She has so completely satisfied me by her argument—altho she is wholly unacquainted with the strongest reasons in support of it—that I recommend you to authorize me to tell Wilson that he may refer people enquiring about his character to Karslake who will give such a general character as will not compromise you, but which will stop Wilsons mouth.[40]

Ada's racing partners were not so easily to be silenced and disposed of, although the same arts of legal intimidation were skillfully deployed against the most vulnerable of them. This was Malcolm, whom Greig described as "a poor man having an income of only £300 a year out of which he has to make an allowance for his wife." Nevertheless, he dared to claim that Ada had forced him into the practices of high betting and borrowing. Greig refused to believe that Ada's influence upon such a man of the world could have been so compelling, although her correspondence with Richard Ford does lend Malcolm's assertion some credence. Greig scoffed, claiming that a letter of Malcolm's to Ada boasted of previous winnings amounting to £2,000, and continued his narrative to Lord Lovelace:

He [Malcolm] had been engaged in making bets for Lady L. on the race for the Chester Cup, and had on that occasion received from her as a present for his services one quarter of her winnings! He [had] got alarmed as Lady L.'[s] engagements increased, and attempted to dissuade her, but she turned the tables on him and he became her *agent* in making bets for her at the Derby of 1851.[41]

It is interesting that Greig should have relayed to William this account of the fatal Derby of 1851, which was the occasion on which Ada had extracted the letter of authorization from him. Malcolm, it seemed, was claiming that he had lent Ada money, including the £1,800 she got from William to lend him. (Perhaps it was this feat that Lovelace later referred to as "swindling.") Since Ada had been able to pry £1,800, supposedly to lend to a "poor man," out of the husband whom she had bitterly accused (to Greig) of unjust miserliness toward her, and had been able to make Richard Ford and Nightingale do her bidding, it is quite believable that she had been able to overawe the socially inferior Malcolm into complying with "her Ladyship's wishes" as well.

But Greig brushed this claim aside. Since Malcolm had paid back £200 of the loan while Ada still lived, it was in law a personal obligation of his, and his assertion that it was in fact used to pay Ada's debts could not be sustained in court. Malcolm in desperation offered to get statements from Fleming, Crosse, and Ford to corroborate his story, but his erstwhile friends refused to help him, and by 22 January Greig's tone was gleeful as his trap began to close: "I shall enter into a subsidiary correspondence with him for the purpose of getting him to commit himself in writing with reference to certain facts—and then I shall be down upon him!"⁴²

Ada had paid Ford the £600 owed to him before the insurance policy was reassigned to Fleming to secure a new creditor, and now he refused to be drawn into the proceedings. That left Fleming and Crosse to be dealt with. Greig could then menace Fleming by pointing out that when the policy was reassigned to him, Ada's life had become uninsurable, and that, in any case, since the policy had been taken out without her husband's knowledge, it was not valid. Fleming presumably saw the force of these arguments and retired from the contest.

The insurance policy, on which the company remained so curiously willing to fulfill its undertaking, was then held in trust for Lord Lovelace by Greig and a Mr. Robins—none other than the attorney who had helped to assign it to Fleming and had been negotiating on his behalf for payment. Since Robins's association with these shady transactions had been in a purely professional capacity, Greig had no hesitation over collaborating with him now. Nevertheless, the payment of the policy, as long as it was unresolved, was something of an embarrassment for Lovelace and his lawyer. To collect it themselves would associate them somehow with the other claimants, who had been beaten off. But there was a suitable receptacle at hand.

John Crosse claimed the payment on his own behalf. He had in his possession a number of Ada's letters and, even worse, the letter William had given her consenting to her betting practices. Greig wanted that one destroyed beyond anything else, since its existence was an obstacle to the final crushing of the luckless Malcolm. A series of letters to Lushington shows the excitement with which he pursued its recovery.

I have instructed Karslake to open negotiations for the purpose of getting *all* the letters destroyed by C. including the fatal one written by L. in the event of our allowing C. to receive the money assured by the policy. . . . Lady N. B. knows nothing of this . . . C's attorney strongly approves . . . and will afford every assistance in carrying it into effect. Until the result is known I have directed [the] Malcolm [action?] to stand over also—as if *the* letter is destroyed, there can be no possible ground for staying the proceedings against him. Lady N. B. had been with my wife this afternoon and *I fear* MAY have been too communicative.[43]

Crosse agreed to the ceremonial burning of Ada's letters—a ritual reminiscent of the burning of Byron's memoirs—in the presence of Karslake the lawyer, and also to supplying a list of them. When the list was sent, however, it contained only 18 letters, although Greig knew from having read Crosse's letters to Ada, which were then in his possession, that he had received at least 108; furthermore, Crosse had written that he preserved them. (To Greig's surprise, Crosse did not ask for the return of his own letters; perhaps he thought they had already been destroyed.)

Being satisfied from these facts, not to mention others, that I am dealing with a man destitute of honor and principle, I have determined to allow him to receive the money under the policy, provided the 18 letters are destroyed in Karslakes presence—one of these letters being the most injudicious one referred to from *him* [Lord Lovelace] to *her*— and another being her letter to C. of the 22nd of last August.[44]

Greig included Ada's last letter to Crosse—written just ten days before she confessed and "palliated" the affair as a "mistake"—along with Lovelace's letter consenting to her betting as the two most requiring to be destroyed. As for the latter, his eagerness had some practical justification: it stood in the way of proceeding against Malcolm. The significance he attached to the former was emotional—he wanted to shield his friend, who had already been devastated by learning of "the additional act of treachery, perpetrated by the delivery of his confidential letter to the ruffian who is now causing so much vexation."[45] Her final letter destroyed the pretense that her infidelity had been a "mistake" and that she had given her lover the jewels to pawn the second time "*only under compulsion*," as Lady Byron maintained.[46]

Greig went on to reiterate the advantages of his tactics: if Crosse agreed to sign a statement that the eighteen letters listed were the only ones that to the best of his knowledge and belief existed, and that no copies or extracts had been made, it would prevent him from using any he retained in the future. But as it turned out, Crosse balked at signing such a demeaning statement. And, for his part, Greig refused to countenance the document Crosse demanded: "a letter on behalf of Lord L. and Lady NB and their friends exonerating him from all blame in respect of the racing transactions in which he had been engaged with Lady L."[47] Suddenly, the explanation for Ada's having insisted on obtaining from her husband the "confidential letter" authorizing her betting, and then delivering it to "the ruffian" John Crosse, is revealed! It must have been Crosse who had demanded a letter absolving him from responsibility for placing her bets; now that he was faced with returning it, he required another document to safeguard him from legal responsibility.

At this point in the negotiations, the truth is that Greig and Crosse (who, it must be recalled, had had some legal training himself) found themselves in something of a stalemate. Either proposed document, if ever produced, would demonstrate that both Crosse and Lovelace had something discreditable to hide and that both had been engaged in what amounted to blackmail. Crosse was even more sensitive to adverse publicity than Lovelace: his status as a gentleman was still partly dependent upon his personal reputation. In the end, he permitted Ada's letters to be destroyed and never attempted to use any he retained.

While Lady Byron did not object to certain kinds of notoriety, she was meticulous in building and maintaining a potential brief for her moral fitness, just in case it might sometime be challenged in court. While the delicate negotiations with Crosse were in process, she suddenly wrote to Mr. Wharton, the senior partner of the legal facade she retained, to announce that her son-in-law was demanding the return of several articles of his property that she had taken with her when she ceased to be an "Inmate" of his house. Specifically, she referred to Ada's books, which she claimed the right to distribute, and to the journal Lovelace had kept during Ada's illness.

After he had ceased to keep it, he brought it to me saying that he thought it would interest me, and added, "indeed I wrote it *for you.*" He said he found it too painful to continue it. . . . I had never read it through . . . now I have done so and find that Lord Lovelace gives *most important* evidence as to my character & conduct—This, I fear, is the reason for his wishing to withdraw the book. The question is— Have I a right to keep it? An authenticated copy would I suppose answer my purpose of producing the evidence in my favour, which would bear particularly on my having the charge of my grandchildren. I apprehend that the next hostile step will be to divide them from me—It will not be well for those who take it.[48]

It did not really matter what Mr. Wharton thought of her plan. She made and had witnessed extensive extracts from the previously misprized diary, agreeing to return the original on condition that he furnish her with a statement to the effect that he had always intended to have it returned. The legal principle behind this condition was that a recipient might be held to have forfeited a gift if her subsequent conduct proved unworthy; hence any demand for the return of a gift could be taken as implying wrongful conduct. Litigious in the extreme, she surrounded herself with defences.

It was not Lady Byron but Lord Lovelace who was eventually divided from the children, although he faithfully attempted to carry out all of Ada's wishes concerning them, including taking his daughter to visit Newstead. Ada's dispensation with respect to the children had unhappy results in all three cases.

Byron, the eldest, hated the navy. Instead of reporting to his post when he slipped unobtrusively away from his mother's bedside, he deserted and attempted to take ship to America. His manner and accent aroused suspicion, and he was soon apprehended. His distraught father, preoccupied as he was, pressured him into returning to the service for another year, after which time he could presumably decide whether to remain or leave permanently. But Lovelace was determined to keep him on an even tighter rein, and he was reassigned to a ship commanded by Captain Stephen Lushington, the nephew and namesake of Lady Byron's lawyer. Captain Lushington kept him under strict surveillance and sent regular and critical reports on his behavior to his father. The predictable result was that Ockham became even more difficult and rebellious. His side of the story was reported to his sister:

Captain Lushington tried to frighten me, but did not succeed; I took it all very quietly and whenever I spoke my mind to him he would become exasperated and raise his voice as if he was boiling the main top but he could not get anything out of me that way. I shall have my discharge out soon though; for I have asked my father to try to get it for me; if he cannot obtain it, I shall take it and leave it to chance whether I am caught again or not.[49]

At the end of this letter he noted that it had been written aboard the *Inflexible*. Shortly afterward he did lose patience and run away again. This time he was not found for several months. According to a memoir by his brother's widow, written many years later, he took ship as a common seaman, eventually working his way home from the Black Sea. Landing in Hull, ill and in rags, he was forced to make himself known to some friends of his grandmother's, "who took him in and sheltered him, until relatives intervened and insisted upon some provision being made for him by his father, who never forgave him."[50]

His grandmother never forgave him either, but she felt it wiser to pretend otherwise, in her efforts to win over all her grandchildren. Without the legal control she had had in Ada's case, she set out to catch her flies with honey, and the younger two, at least, remembered her with more fondness throughout their long lives than Ada had during most of her short one. Ockham, however, was fiercely independent and refused to choose between father and grandmother. Wishing a plague on both their houses, he ran away from the tutor who had been engaged for him, who happened to be a son of Dr. Arnold, the headmaster of Rugby. He became an ordinary workman in Scott Russell's shipyard and nursed some dreams of his own. Learning that Ralph too had had a row with the stern earl, who had turned him out of the house, Ockham wrote to his brother that a change of scene was always an improvement; he signed his letter "Jack Oker."[51] He and a friend, he said, were saving up their money to buy a steamboat in which they intended to travel around the world. Meanwhile, he corresponded with Babbage, even daring to challenge, from a position of experience, the latter's serene view that the interests of employer and workman must always be in harmony (and always those of the employer). Babbage, in reply, considered sending him a copy of *The Economy of Machinery*.[52]

His father's spies, among them Greig, continued to file reports on his conduct; sadly, he never sailed his own boat around the world, but succumbed at the age of twenty-six to tuberculosis, the scourge of all classes in the nineteenth century.

Ockham's sister, Lady Annabella King, was horrified by his opinions and actions, but went to him and nursed him in his last illness. Aristocratic though her outlook was, her life was scarcely less marked by discomfort and privation than her brother's. Upon moving into the house at 6 Cumberland Place in 1850, for example, she had written Ockham excitedly about the "little convenances" her new schoolroom provided, such as a clock over the mantlepiece. And on returning to East Horsley the week following her mother's death, she wrote Lady Byron,

Thanks to Mrs. Clarke's arrangements, the schoolroom is beginning to look and feel quite comfortable. A *Carpet* had been put down over the matting, and it makes the floor a great deal *warmer*. The whole of this end of the house is being attended to much more than it has ever been before—fires are lit, and the rooms aired—I do not think they have ever been fairly treated until now, for I don't remember that any except the schoolroom have ever had a fire all day long.[53]

A year dragged on for her, in the company of a silent, embittered father, who nursed his griefs and wrongs but was unable to confide in a daughter who felt so warmly toward his inexplicable enemy. His only interest and relief now was in his building projects.

I wonder Papa is not tired of always, always making alterations & always residing in the same place. He is constantly going to see how everything is getting on—& *we* are always asked if we have been to see either the garden wall, or some wheelbarrows being wheeled about with earth in them. . . . However there seem to be so few things in which Papa takes an interest that I am glad he does care about these things. Do you know really Papa will hardly even talk about anything but Lodges and building and art in general? Please do not judge of my handwriting by this letter, for my fingers have been so cold that I could hardly write.[54]

Her grandmother tried to make life more cheerful by giving Annabella a pony and phaeton of her own. Yet still, after her yearly visit with Lady Byron at Brighton, returning to the dreary construction site her father called home seemed almost like imprisonment.

Although I have so long contemplated the dire necessity of returning to Horsley, I can hardly recover from my astonishment at being here again. It is like having had a long sleep with a beautiful dream & then being unpleasantly surprised at waking & finding that it was a dream—too enchanting to be real.[55]

Like her younger brother, Annabella complained bitterly about the limitations of her education. In later life she laid the blame on an "ignorant governess" (who, she neglected to mention, had been hand-picked by "G.M.," as she called her grandmother). But at the time she poured into G.M.'s receptive ear a stream of charges against Papa, who refused to take her to London, or at least to Brighton, where proper music and drawing masters were to be found. It was still unthinkable for a girl who was of the social as well as "the intellectual aristocracy" to be sent to school.

Papa kept promising that they would travel abroad, but they never seemed to go. At last her grandmother revealed that the reason for the repeated postponement was the uncertainty surrounding Ockham's second disappearance, about which her father had not even told her.

She had one more reason to be grateful to G.M., who alone guided her through the gloomy, anxious mystery of her brother's fate. When Ockham was rediscovered and "provided for," Papa at last took her abroad, where she became fluent in French, German, Italian, and Spanish. Later she studied drawing with John Ruskin and acquired a famous violin, becoming a competent musician and an accomplished artist.

A dozen years after Ada's death Lord Lovelace married again. His new wife, the daughter of a Calcutta auctioneer, had lost her own first husband. The younger generation referred to her snobbishly and unkindly as "the Indian Widow," but the new countess returned good will for ill and did everything she could to bring about a reconciliation between the old earl and his alienated children. In Annabella's case she was at least partially successful: after much effort on her step-mother's part, she was summoned to East Horsley, later recounting the interview with her father to Agnes Greig, now widowed herself:

What I found so difficult was the part in which he spoke of my Grandmother, for in many things I do not understand her character & yet to me she was really like a guardian angel, & I owe almost all the good I have learnt in some way or other to her, & yet it would seem she must have been mistaken towards my Father. Then as to my Mother, I must believe all that my Father tells me. Do you know, I think all mysteries are quite wrong & that *however painful* it would be better I & my brother should know exactly the *truth* as to my parents & my grandmother. My Father says he has been silent so long, & that he has been thought of & treated unjustly—How can one know what to believe if the circumstances on which one must form an opinion are purposely shrouded in mystery?[56]

Annabella was almost thirty years old when this was written, and still unmarried. Lady Byron had died seven years before, in an odor of sanctity, and had left her independently wealthy. Since her father's marriage she had lived mostly abroad. Ralph, set free to travel by his keeper's death, could not provide her with the settled home she and her friends felt she should have. A well-planned marriage seemed the answer; and, with the mixture of imagination and pragmatism that in previous generations had been exercised by Lady Melbourne and Woronzow Greig, a friend helped her into another and comparable disaster with a handsome but moneyless young man of suitable birth.

Wilfrid Scawen Blunt tried self-consciously to model himself on Annabella's illustrious grandfather. A poet, if at that point still un-published, he was as erratic a diplomat as Byron had been a legislator.

The object of his life was "romance," the search for which could not long bear the confinement of monogamy. As a husband, if he was less given to drinking, threats, and vandalism than his model, he was equally slighting and abusive, and even more self-indulgently promiscuous. By a strange quirk, Annabella, or Anne, as she was now known, actually was as dutiful, selfless, and devoted a wife as her namesake had pictured herself. She accompanied her husband in his restless wanderings in the East, and even found life with him happiest in a tent. She mastered Arabic and kept a diary from which Blunt edited two charming books on Eastern travel. Her fortune was used to rebuild his own and to buy him leisure to pursue his interests. The consequence was that their separation was long delayed—their matrimonial misery belonged to Anne alone.

Ralph was twenty-one when Lady Byron died, leaving him her maiden name, the bulk of the fortune she was free to dispose of (outside the estates that had been settled for life on Lord Lovelace), and the consequences of her attempts to isolate him from the influences she was convinced had depraved his grandfather. It was very soon thereafter that he began to interest himself in penetrating the mystery that had surrounded the Byron Separation. The vast archive that his grandmother had hoarded, which seemed to hold the answer, had been left in trust, guarded unofficially by Dr. Lushington (who might have revealed more than the papers, had he chosen) and officially by Lushington's sister-in-law Miss Carr (one of the Furies, now turned Cerberus), a Mr. Bathurst, and Miss Mary Carpenter, a younger woman who had been involved in Lady Byron's charitable works. It was many years before this faithful band could be persuaded to deliver up their trust, or died in office. In the meantime Ralph became obsessed with amassing and transcribing all the papers relating to his ancestors that he could get his hands on.

In 1869 Harriet Beecher Stowe published a magazine article entitled "Lady Byron Vindicated"—provoked, as she claimed, into the unseemly action of publicizing what had been told her in confidence by the appearance of Teresa Guiccioli's memoirs. Mrs. Stowe was concerned that La Guiccioli's account might present too favorable a view of Byron. To redress the balance, she made the incest story public for the first time, claiming it had been told to her by Lady Byron during their rather brief acquaintance. An uproar predictably followed, in the course of which Lord Lovelace incautiously expressed his doubts in an interview with a journalist. (Lovelace not only disbelieved the tale of Byron's incest; he also claimed to have discovered that Byron's marriage was bigamous, according to his son-in-law.)[57] Ralph (now

Lord Wentworth, a title he had inherited on the death of his elder brother) determined to revindicate his grandmother on the basis of the evidence in the immense hoard of papers she had compiled for that very purpose. Eventually, his careful study of the papers enabled him to publish a book that, appearing shortly before his own death, he hoped would establish his grandfather's guilt once and for all. His numerous, meticulously accurate transcriptions reentered the archive; they show plainly that his interest had strayed beyond his grandparents to the enigma of his mother's story, even though he left but little written commentary on it and published not a word.

The mystery of Ada's life, which she herself probed so painfully, remains to puzzle us: why one so rich in promise, opportunity, and aspiration should have left behind so little in achievement. It cannot be said, short though it was, that her career was cut off in its flower; her nearest brush with achievement was eight years behind her, and she had drifted in a downward spiral for some years before her fatal illness became incapacitating. The obsessive gambling in which she engaged late in her life was not for any purpose, as has been suggested, such as that of financing the construction of Babbage's Analytical Engine or even that of testing a mathematical theory, but was rather an expression of the resentment she felt against a mother and husband who seemed so wealthy, free, and powerful at her expense.

A single case, of course, can reveal nothing about the general effects of the various factors that may or may not influence the course of individual development. On the other hand, if any doubts remain that mathematical talent, as well as interest, can be manifested in women, the case of Mary Somerville can lay them to rest. And unlike Mary Somerville, Ada was strongly convinced that she had inherited genius of some sort from her father; the spiritual tasks that most absorbed her throughout her life were first, to find the most exalted channels through which that genius could be developed and expressed, and second, to account for the failure of her gifts to flower. Again and again throughout this study, these issues have been examined explicitly and implicitly; from within, in her own words and in those of others about her, and from without, in terms of a series of contrasts with contemporaries who addressed or embodied the same problems of worthy endeavor in a privileged social context. It is time to sum up these issues and their apparent implications.

It is perhaps first necessary—although somewhat paradoxical in view of the legends that have sprung up around her name—to ask whether Ada's intellectual legacy was actually as negligible in comparison to those of her contemporaries as it appeared to her own disappointed expectations. The scientific thought and artistic achievements of a century and a half ago cannot be judged simply from the perspective of what is known and admired today. Many names that were highly acclaimed at the time—and some that are still remembered today—were associated with beliefs no longer held, theories falsified, contributions now rejected, failures now remedied, works forgotten, books unread. Quêtelet, the still renowned economist and statistician, who was Lord Lovelace's dinner guest as well as the author of works reviewed by him, asserted that nations evolved through a well-defined life cycle patterned after that of individuals, and that the upper classes were tragically failing to reproduce themselves. Sir John Herschel, another acquaintance of the Lovelaces whose name still appears at or near the head of any list of eminent Victorian scientists, knew no way of deciding mathematically whether a given set of observations indicated a significant trend or not. Faraday, the greatest experimental physicist of his age, had no knowledge of mathematics and could not understand how its use might advance his discoveries. Many prominent doctors and scientists were practitioners of bleeding or phrenology—both now in eclipse—or believers in the more occult versions of mesmerism. Even Babbage himself was occasionally careless about such matters as statistics and mathematical convergence, was obfuscatory about just what his machines could or could not do, and rejected the now accepted concept of limits as a basis for the infinitesimal calculus. Critic as he was of scientific fraud and humbug, he was not himself above misleading the public as to his own achievements.

In such a context Ada's repeated attempts to inquire into the truth of mesmeric phenomena, her plans to experiment with the effect of electrical currents on nervous tissue, and even her notions about developing a mathematical analysis of molecular biology were not as fanciful or as naïve as they seem now, when the subjects of physics and biology are better understood. Moreover, the relatively crude and undeveloped state of scientific understanding in many fields, the informality and lack of credentialism in the arts and sciences, meant that it was fairly easy for the educated gentry to pass into a quasi-expert status. This is demonstrated by Lord Lovelace's ventures into architecture, engineering, and even the social sciences, as well as by the careers of Mary Somerville and Harriet Martineau. The measure of the failure of Ada's intellectual aspirations is not in her lack of realism so much as in her almost total failure to leave behind any tangible evidence of achievement or even progress along the lines she had mapped out for herself.

For one with serious intellectual and literary interests, she left a remarkably small body of work, published or unpublished; furthermore, most of what she left was incomplete. If we discount her juvenile essays and put aside the two works that actually reached print—both already discussed in detail—nothing resembling a completed essay on a scientific subject survives, except for one review. There are no completed works of literature except for a single review and a single sonnet. It is barely possible that the "collection of papers" mentioned by Babbage, whether or not it included the "book," contained other writing of substance on scientific or literary subjects, especially the former; indeed, there is a suggestion in Cornelia Crosse's memoirs that this was the case. If so, it is surprising that they were not preserved along with Babbage's other correspondence and papers. It is also barely possible that some of Ada's work remained in the possession of John Crosse. Yet, if this is so, it is surprising that Greig, who from reading their correspondence would undoubtedly have learned of it, made no attempt to recover it along with her letters.

Whether or not a significant portion of Ada's unpublished writings (and no attempt can be made to assess her musical accomplishments) has failed to come to light, if the surviving examples are taken as typical of her productions, the conclusion is inevitable that the Menabrea Notes did in fact represent the peak of her achievement, from both stylistic and intellectual standpoints. Yet the Notes were produced under the close supervision of Babbage, who must at least share responsibility for both the success and the failure of their intellectual content. As she herself was well aware, Ada required the collaboration and support of an intellectual partner of impressive gifts; so indeed did other notable contemporaries—Harriet Taylor, for example, who with John Stuart Mill was able to construct a fulfilling, if cryptic, career. Without such a partner Ada failed to live up to her promise on her own or any other terms.

At once an explanation on the basis of her social environment suggests itself. As a woman, she was outside the male intellectual network formed and reinforced by attendance at universities and membership in learned societies. The few men from impoverished backgrounds, such as Faraday and Boole, who managed to break into the network from other directions were exceptions that proved the rule, since they had access to such resources as the Mechanics Institutes and to employment, effectively closed to women. During her student days Ada was deprived of the peer stimulation that Babbage, for example, enjoyed with his friends in the Analytical Society and that, perhaps, she was reaching out for in her correspondence with Annabella

Acheson. Nevertheless, her early isolation cannot be compared with that experienced and surmounted by Mary Somerville and Harriet Martineau, to name but two.

Then, too, Ada had advantages and opportunities such as fall to few in any age. In her own eyes, her social position and, even more, her hereditary genius as a Byron exempted her from both the disabilities of her sex and many of the restrictions on it. And none among her friends and acquaintances were blessed with teachers of the stature and gifts of Augustus De Morgan and Babbage (even if the uncertain role of Mary Somerville is neglected). If she lacked contact with like-minded peers in early youth, she cannot be said to have been educationally disadvantaged, nor restricted or unstimulated, in the highly intellectual circle in which she moved as an adult. Moreover, she was supported and encouraged in her pursuits—at least in her scientific ones—by friends such as Wheatstone and by both mother and husband, who, with only occasional protest, consented to remove every domestic and social impediment from her path. In this she was more fortunate than any of her intellectual female contemporaries.

There may be some temptation, in view of these circumstances, to reverse the hypothesis and ask whether the strong support and encouragement she received from her guardians and governors (for mother and husband were both) did not actually have an inhibiting effect on a spirit at once as rebellious and as fearful as Ada's. This is an explanation often advanced in the case of bright children whose school performances are below parental expectations: they have been pushed too hard. In Ada's case, at least, the evidence is largely against such an explanation. Ada pursued her studies, investigations, and writing, when she pursued them, with great relish from a very early age, even hesitating to accept an invitation for a trip, on one occasion during her bedridden period, lest her Latin lessons be disrupted. However, when she lost interest in any or all of her pursuits, she seems not to have lacked pretexts to change or desist from her activities.

Another hypothesis, advanced not only by Ada herself from time to time but also by her doctors and even by Augustus De Morgan, was that excessive study or concentration by one whose constitution was frail, but particularly by women, was conducive to mental or physical breakdown. The treatment was a complete cessation of all mental effort, with perhaps a gradual and moderate resumption at some distant future date. This view was extremely common in the nineteenth century and has a popular version in the twentieth, in the form of "psychosomatic" theories. It has been suggested, for example, that Ada's adolescent paralysis was a result of scholastic overwork.

All that can be said of this is that her lessons continued and even, in some respects, intensified, for the duration of her lengthy recovery, which may or may not have been slowed by them. And the incapacity of her final illness was at times welcomed by her as an opportunity at last to devote herself to sustained work.

Whether or not brought on by her mental exertions, can Ada's numerous illnesses account for the fact that career and fame, or at least tangible achievement, eluded her? Once more, it is possible to point to a number of her contemporaries, including Darwin, Faraday, and Harriet Martineau, who also experienced lengthy periods of continuous or intermittent bad health, which for many years they accommodated to productive intellectual lives. Ada's physical illnesses were severe and at times incapacitating, yet it is notable that during the weeks and months in which she labored over the Menabrea Notes she made frequent mention of nagging and debilitating illnesses that were not permitted to interfere with publication deadlines. Even more significant, perhaps, was her anticipation, as the symptoms of her final illness developed, of using her enforced immobility at last to achieve some work of substance. (As it turned out, though, this final determination to create and to make coherent her heritage, her questing, and her suffering was as unproductive as most of her previous efforts.)

It is difficult to determine the effect of Ada's bouts of mental instability on her intellectual career. Periods of exaltation, of hubris, which she herself labeled "mania," accompanied her most ambitious plans; but were they reflections of her daring imagination or the causes of her undertaking projects she was ill-equipped to carry out? The periods of depression that followed may have been responses to the realization of her inability to effect these projects, or may themselves have prevented that persistent application, the lack of which both she and her guardians recognized as characteristic of all her undertakings. Possibly her lack of persistence in the end prevented her from acquiring sufficient knowledge to turn her philosophical speculations into fruitful ideas for investigation; the ideas and suggestions she expressed remained always too vague and mystical to be successfully pursued.

Yet she did work hard enough for long enough to make clear that only extremely dogged application would have sufficed to compensate for a deficiency in native quickness and talent, especially for assimilating the mathematical techniques of symbolic manipulation. It is unusual to find an interest in mathematics and a taste for philosophical speculation accompanied by such difficulty in acquiring the basic concepts of science as she clearly displayed. We can only be touched and awed by the questing spirit that induced her to launch so slight a craft upon such deep waters.

Appendix
Unnatural Feelings Mental & Bodily

Ada's letters, like those of many of her contemporaries, were freighted with discussions of diseases and symptoms, of treatments and cures, both mental and physical; but of all the questions that arise from an examination of her life, that of the assessment of her chronic ill health seems more difficult to address decisively than any other. Nonetheless, her illnesses, her speculations over their causes, and the extent to which illness and disablement contributed to her inability to realize her early intellectual promise were of such significance to her that an attempt to account for these afflictions should at least be made.

It is not at all surprising that even the politest society in the first half of the nineteenth century was sick a good deal of the time. Queen Victoria had typhoid fever as a young girl, and her consort Prince Albert died of it in 1861. The germ theory was a development of the latter half of the century, and so too were effective means of controlling and preventing infectious disease. One book of household advice of the period, for example, warned against the boiling of drinking water, which, it said, concentrated in it impurities such as lead. If you did not know about bacteria, that advice was reasonable enough.

Illness and attempts to control its ravages had important effects on public as well as private events. Asiatic cholera reached England in 1831. Ada was stricken with what may well have been cholera in 1837 and considered herself marked by it; later she used its prevalence in the neighborhood as an excuse to stay away from her husband's country seat. But the advent of cholera in England was also indirectly the cause of many popular disturbances. In an attempt to isolate the disease, the authorities ordered that its victims be treated in special hospitals and buried in special cemeteries, public health measures that aggravated fears on the part of the underprivileged that their stricken relatives, dead or dying, might be commandeered for the dissecting tables so necessary to the training of medical students. (The Anatomy Act of 1832 enabled medical schools to take the bodies of unclaimed paupers and felons.) The result was a number of riots and attacks on cholera hospitals and medical schools.[1] Middle-class people, even the kindliest, whose relations were not in danger of dissection, tended to be insensitive to the depth of feeling of the poor in this matter. Augustus De Morgan, for example, overhearing a conversation about the epidemic of grave robbing, is reported to have burst into a parody of "Comin' Thro' the Rye":

Should a body want a body
Anatomy to teach,
Should a body snatch a body
Need a body peach?[2]

Despite this bad beginning, a later cholera outbreak led to a celebrated public health success. In 1854, when an epidemic raged in the Soho area of London, Dr. John Snow persuaded the Soho Board of guardians to remove the handle of the pump at Broad and Cambridge Streets, which was the source of drinking water at the center of the affected area, and the epidemic died down. Yet the cause of the disease and the way in which the infection had entered the water were still unknown. It was not until thirty years later that Robert Koch identified the organism that actually caused the illness and was carried in the drinking water.

Antibiotics were of course unknown. When one was ill, all one could hope for was the relief of some of the symptoms and support in enduring the others until a spontaneous recovery occurred. The lack of understanding of the nature and mechanism of infectious diseases, let alone others against which even present medical science has made less spectacular inroads, had a number of consequences. Even in the absence of knowledge, theories abounded. The well-to-do went from doctor to doctor, from regime to regime, in search of a cure—as indeed they still do when faced with the more mysterious and baffling complaints. The result was both a proliferation and a conflation of the symptoms and the causes to which they were attributed. For example, Ada's early childhood illness was mysteriously identified as, or attributed to, "a determination of blood to the head." When bleeding is the treatment of choice, almost any condition can be referred to as an excess of blood, either in one part of the anatomy or suffused over the whole body.

What is the historian to make of such a welter of unfamiliar terms, conditions, and causes? Even the most up-to-date medical science can easily be perplexed when faced with only a few vague symptoms, often couched in strange terms, with neither the possibility of questioning the patients nor recourse to the now indispensable tools of laboratory testing. Furthermore, medical historians have been only too thoroughly converted to the diagnosis of psychosomatic illness, just as the pendulum of medical science itself is slowly swinging toward the identification of biochemical mechanisms of "mental" as well as physical illness. Yet, with all these obstacles, it does seem clear that Ada and her mother were chronically and unusually ill compared with most of their acquaintance, a surprising number of whom lived to very advanced ages. Contrasting what Ada and her contemporaries made of her recurrent and varying complaints with the interpretations to which they might now be subject, we can perhaps shed light on the intervening changes in the meaning of illness, in the nature of medical science, and in the relationship of both to their social context.

To begin with, it will be convenient to summarize Ada's medical history. Her most common and earliest-noted characteristic was a delicate stomach, which may or may not have been similar to her mother's bouts of "biliousness." Certainly by the time she was in her late twenties, and referring to her condition as "my but too common *Gastritis*," her mother felt in command of her own symptoms; for Lady Byron wrote sagely to Augustus De Morgan, in answer to his worries over the strain on Ada's health caused by her studies, "if she would but attend to her stomach, her brain would be capable even of more than she has ever imposed on it." Ada's "gastritis" was severely painful and accompanied by vomiting and other unpleasant effects that caused her to seclude herself during her attacks.

Her first serious illness occurred when she was seven and a half. It involved headaches and apparently affected her eyesight and prevented her from reading for some time. Her second serious illness began in May, 1829, at the age of thirteen and a half. Her mother wrote at the time that "the loss of all power to walk or stand . . . followed other effects of the measles, and too rapid growth." A year after the illness she was still permitted to sit up for only half an hour a day; in the course of the summer that period was increased to one hour. In 1831 she was able to walk with crutches, and in March 1832 her mother noted that she could walk unassisted if she carried weights in her hands. By September 1832 she had discarded the crutches, but she was still frequently weak and giddy over a year afterward.

After Ada's acquaintance with Mary Somerville began, Lady Byron wrote confidingly to the latter of her daughter's delicacy: "The complaints to which she has been liable were connected with the spine, and therefore particularly affect the nervous system."[3] This letter refers specifically to Ada's paralysis of the legs, from which she had by then recovered. But, if written in 1835, as seems likely, it also refers to the nervous breakdown, or "weak state," described in Ada's own letters to Mrs. Somerville on 20 February and 4 April of that year. We will consider later whether the two conditions could have been connected elsewhere than in Lady Byron's medical opinion.

Ada's next really serious illness seems to have occurred after the birth of her second child. It lasted some time, perhaps months, and from this period she was wont to date her decline in health. Her illness was possibly cholera; she called it cholera in a later letter to her mother and speculated that her subsequent troubles resulted from her having overeaten during her recovery, developed asthma, and then been unable to eat for some time.[4] Indeed, after a somewhat stout adolescence she seems to have become progressively more liable to periods of anorexia, ending in almost skeletal thinness.

There are many almost unmistakable allusions in her correspondence with both mother and husband to menstrual disorders. Her periods were not only very heavy but accompanied by the symptoms of water retention:

I have been a stupid, heavy idle LOUT these 2 days. Now I think it is blowing over again, & I am returning to *activity*. It is not that I have been *ill* at all. But I have been alternately FLIGHTY, (hating to settle to anything), & *stupid*; & have grown rather *puffy & large* also.

The fact is it has been the critical time; & as I seem to be coming out of my stupidity now, *without* RESULTS, I conclude that as before it has very probably *missed* for this time.[5]

In November 1840, about six months into her study of mathematics with De Morgan, she remarked that while mathematics was a "good ballast," sometimes she had "whims."[6] These soon became manifest in some very grandiose letters, first to her mother, later to others as well. Around the same time she mentioned "a sensation like a current in the head which accompanied some of my worst derangements of circulation," and also that she had often had a horror of going mad.[7]

Singing lessons, urged by her to her husband as beneficial to her health, specifically as counteracting her asthma and hysteria, were begun at about the same period. Nevertheless, she continued to be subject to fits of severe depression and also to periods referred to as "mania." On 21 December 1841 she wrote Sophia De Morgan of one such episode: "There has been no end to the manias & whims I have been subject to." These severe mood swings continued for a number of years and gave rise to many efforts on her part to connect, explain, and control her various symptoms.

In February 1844 she wrote Woronzow Greig to announce yet another strange sort of attack:

I think this attack will prove of no consequence, & is merely an accidental *parenthesis*, if you can understand the phrase. It looked a little awkward at first, but I consider it as quite proved to have been nothing but a passing state of the nervous system & Lady B– & my Doctor are *both* of them *fully satisfied*. I had been very queer of late at times in some of the sensations I have had about my *head*, (which during the latter months has plagued me *instead of my chest*) & yesterday morning the whole throat & face suddenly swelled up (in one instant) to enormous size, & I *felt* as if threatened with instant annihilation (tho' I am told that sensation is deceptive). This strange seizure lasted 3 minutes; but left behind it (of course) most strange feelings in the brain & eyes. I fancy it is of a cataleptic nature; but Dr. Locock says there is no giving a name to it, for it is completely *à moi* & not at all according to any precedents. He uses the phrase that a "sudden *tension takes place in the nervous cords* & lasts a few seconds or minutes." It is evident that he completely understands the case & how to treat it.[8]

Despite Dr. Locock's "understanding" of the case, the swellings did recur, and by the end of the same year he had diagnosed a "watery deposit," which he treated at first with bleeding and then with laudanum or morphine, not only against the "head sensations & swelling" but to relieve her "mad look" as well: "He told me that is was *really* peculiar, & horrible to the spectator, altho' he did not believe *I* ever could be *mad*, or scarcely even delirious."[9] Delirium was a common alternative hypothesis to "mania" when patients seemed to be raving; its advantage was that it was attributable to physical causes, such as fever. She found that her hands swelled if she drank more than a pint of liquid a day; "also this fact: that I have known myself gain 8 lbs weight under a week, mainly from yielding to a thirst for fluids."[10]

The last serious condition (before the symptoms of the cancer that killed her) to appear in her correspondence was the "heart attacks"; but on mentioning these she commented that she had been subject to such seizures for some twenty years, hence from about the time of the onset of her adolescent paralytic illness. In addition, there are numerous references to her shivering fits, treated with warm baths, and to frequent and severe colds; and finally her mother noted that "some of the senses are also below par—Smell & Taste—& always were."[11]

If we take Ada's illnesses as simply an unfortunate conjunction of painful, inconvenient, and debilitating conditions, it is impossible not to be struck by their number and variety. That is, she seems to have had frequent and severe gastric upsets, asthma and hay fever, paroxysmal tachycardia, a kidney complaint of some sort, and, possibly most important for our purposes, a manic-depressive condition, to cite only the most distressing and persistent.

Her three-year paralysis of the legs is as mysterious as any of her afflictions, precisely because of the rarity of this combination of lengthy immobility followed by total recovery. (She was later an enthusiastic waltzer and mountain climber.) Complete recovery from poliomyelitis, for example, if it occurs, usually does so within a few months or a year. In Ada's case the paralysis was said to have followed an attack of measles and "too rapid growth." There are several possible complications of measles, among them encephalomyelitis and Guillain-Barré syndrome, that result in severe paralysis from which total recovery is possible. The latter syndrome is even accompanied by such neurological symptoms as tachycardia (heart palpitations) and difficult breathing, from which Ada also suffered. Yet recovery usually occurs within four to six months. One can only surmise the weakening and prolonging effects that Lady Byron's "lie still" therapy might have had on Ada's lengthy convalescence; but the observation that Ada could walk unassisted if she held weights in both hands—a good six months before her crutches were discarded—combined with Greig's recollection that she was still prone to dizziness and a recumbent position over a year later, suggests that some disturbance of balance, perhaps an inner-ear disorder such as Ménière's syndrome, was as much responsible for her laggard return to mobility as the weakness or paralysis of her leg muscles.

For Ada herself, however great her physical suffering, it was the effect on her intellect of each successive affliction that concerned her most; on the other hand, she sometimes almost seemed to welcome the prospect of an invalidism that would leave her confined to her room yet free to be intellectually productive. After the strange psychotic episode of 1841, reported to Sophia De Morgan, there appeared numerous cautionary references, both on Ada's own part and on those of her mother and husband, to the need to avoid "mania." Her deep depressions seemed scarcely to need explanation, or treatment other than "stimulants" (by which she seemed to mean interesting or exciting activity, rather than the tincture of opium and alcoholic draughts that were prescribed for so many conditions). The term "mania," of course, was sometimes loosely used, as it still is, to mean an enthusiasm judged excessive, rather than clinical insanity; but in a number of these letters the word seems quite clearly to be used in the latter sense.

It has been suggested by Ada's biographer, Doris Langley Moore, that Ada's "mania," or at least the symptoms that appear in her letters—consisting of intimations of extraordinary powers and abilities, her presumptuous arrogance toward Babbage, and a fitful religiosity that at times bordered upon frenzy—were themselves the side effects of the opium preparations with which her doctors so liberally dosed her for her digestive, respiratory, and nervous complaints. This, however, is to misunderstand the social climate in which opium was taken in the nineteenth century, its medicinal purposes, and its effects on users.[12]

Laudanum, a solution of opium in wine spirits, had been created by Dr. Thomas Sydenham in the 1660s. It was very freely consumed in the eighteenth and nineteenth centuries, with women users outnumbering men, among all classes of society. The "Black Drop" that Lady Byron found while rifling her husband's trunk, for example, was a laudanum preparation. Ada's contemporary, Elizabeth Barrett Browning, was a devotee of the drug, while Harriet Martineau did not favor it during her first lengthy siege of disabling pain because of its soporific effects. Later, as she neared death, she, like Ada, abandoned mesmerism for the surer anodyne of opium.

Morphine, a stronger derivative of opium, did not begin to be used on an appreciable scale until the 1820s or 1830s. Heroin was introduced only at the end of the century. Although opium's addictive effects were well known, they were of little concern, and it was not these that eventually led to the drug being outlawed, but rather the occasionally lethal effects of overdose, and the increasing control of drugs by the medical profession. In other words, opium use was not considered a social problem, at least among the genteel. With the drug freely and cheaply available, addiction was not so much a social threat as a personal weakness. And the effects of addiction, such as the withdrawal symptoms, were weighed against the perceived benefits of the drug.

What were these benefits? Opium has a remarkable capacity to control pain and to induce a pleasurable mood. This made it, as many doctors testified, an invaluable medicine for the many conditions for which cures were unknown. Its tendency to produce constipation made it a specific for diarrheas, and infants and children of all classes were treated with preparations whose titles, such as "Mothers' Quietness," often suggested that the benefits were as much to the mothers and nurses as to the babies.

The rural poor, who seldom saw doctors, had been medicating themselves with preparations of native-grown poppies for centuries. In the unhealthy Lincolnshire fens, which Lord Lovelace considered fit only for the Irish, the prevalence and amount of opium ingestion, for agues and pains or just for the misery of life, was especially high. A change of attitude began to take place only after 1830, when its use, in combination with or as an alternative to alcohol, began to be considered a problem among the urban working class, along with the extensive drugging of the babies of working mothers. It was perhaps this new disquiet, as well as her general disapproval of the evasion of pain, that lay behind Lady Byron's occasional demurrers to Ada's use of opium.

Many people—Thomas De Quincey for one—began taking opium on professional advice, especially for its soothing effects on a variety of gastrointestinal symptoms and for painful, smarting eyes, from which both Ada and De Quincey suffered. Probably the eye pain was the reason for Ada's first prescriptions. Later it was prescribed for the swellings due to water retention and the "head sensations" that accompanied them. But most usually, opium was the nineteenth-century tranquilizer, used—as in Ada's case—as an alternative to bleeding for "taking down" agitated patients. In fact, according to still-current expert opinion, opium, unlike the psychedelic drugs, does not usually produce "spectacular or uncanny states of mind. . . . There are no hallucinations, waking dreams, illusions, or other psychotic-like effects. . . . The most striking thing about morphine . . . is that it dulls general sensibility, allays or suppresses pain or discomfort, physical or mental . . . and that disagreeable sensations of any kind, including unpleasurable states of mind, are done away with."[13]

These do seem to be the reasons for which Ada was prescribed laudanum and later morphine, and to be the effects she reported. "I am indebted tonight to Laudanum for such sense and tranquility as is really creeping over me this evening," she wrote at one point;[14] and at another, "The Opium has a remarkable effect on my eyes, seeming to *free* them, & to make them *open & cool.* Then it makes me so philosophical, & so takes off all *fretting* eagerness & anxieties. It appears to harmonize the whole constitution, to make each function act in a *just proportion*; (with *judgment, discretion, moderation*)."[15] Her opium-induced sense of well-being was quite different from, and indeed employed to counteract, the excited, grandiose elation and "mad look" that her mother exclaimed and her doctor prescribed against.

Even among addicts, opium is taken not for extraordinary effects like those produced by hallucinogens, but often simply to "feel normal," although that "normal" state may be idealized as one more cheerful and optimistic than is really the case for run-of-the-mill normality. But was Ada actually addicted? Certainly she used opium over extended periods—amounting to years—and reported, if she did not recognize, withdrawal symptoms, some of which—runny nose, restlessness, cramps, vomiting, breathing difficulties—were similar to the reasons for which she began to take the drug. Yet it is also true that she discontinued its use for years, returning to it only in the agonies of the last stages of cancer. It seems that, not recognizing herself as addicted, she was able to lay aside the "Opium system" as she had so many other regimes, in favor of newer and seemingly more promising treatments.

But if opium was a treatment for, rather than an explanation of, Ada's "mania," it is necesary to examine that too in the context of nineteenth-century theories of madness and its cure. The early part of the century was dominated by the notion that insanity was "morally" caused—that is to say, environmentally or experientially caused—and indeed constituted a kind of moral derangement. In health the mind was under the control of the will, but in insanity the grosser, more animal-like elements of the brain and body emerged to make the sufferer more irresponsible. This belief was complicated by debates over the relative efficacy of "management" or "moral" treatments and physical treatments such as drugs, emetics, and bleeding.[16] Nevertheless, it was the assumed weakness of will and moral inferiority of the mentally ill compared to the physically ill that underlay Ada's repeated pledges to live a moderate and regular life, "avoiding *mania*," and the satisfaction of her announcement that her doctor had decided that "all my *vagaries* are purely *physical*."[17]

For a while, in the 1830s and 1840s, phrenology seemed to provide the link between physical and moral approaches to psychiatry. Its ideas eventually permeated attitudes and treatments, just as psychoanalytic theories do today, even when its influence was unacknowledged, and even after the persistent absence of physical abnormalities in the brains of the insane contributed to the discrediting of the "craniology" on which phrenological assumptions were based. At present a similar sort of reconciliation between theories of the mental (or psychogenic) and the physical bases of mental illness, the psychoses in particular, is coming about through attention to the biochemical influences on neural functioning.

The symptoms of Ada's "mania" closely resemble those of a manic-depressive condition, for which there is now an impressive body of evidence of a biochemical basis created through some kind of hereditary mechanism.[18] The hallmark of this condition is the presence of manic episodes, since depression is a much vaguer, more variable element, considerably more difficult to identify. But the symptoms of mania too can differ in number, kind and magnitude; they include rapid, excitable speech, "hypersexuality," restlessness, euphoria, "flight of ideas," grandiosity and/or religiosity, sleep problems, and delusions, though hallucinations are less common. The condition seldom appears in childhood; the first attack usually occurs in the twenties but may appear considerably later. The attacks are self-limited to periods of a few weeks or months. Not only do patients recover both intellectually and socially from each episode, but there is even a tendency for complete remission in later years. Psychological and sociological approaches to treatment ("talk therapy" or environmental changes— that is, "moral management") have not been shown to be particularly effective, but treatment with lithium salts has been successful enough to be considered almost a specific, for the manic component of the illness at least.

As already stated, there is rather strong evidence of a hereditary basis for the tendency to this condition. Its occurrence in the general population is roughly 1 to 2 percent of men and 2 to 3 percent of women, but about 23 percent of the mothers and 14 percent of the fathers of those affected have also been found to be affected. The hereditary theory seems to involve two dominant genes at separate loci, at least one linked to the x chromosome, which mothers pass to their children of both sexes and fathers pass to their daughters but not to their sons. (It is a paternally donated y chromosome that determines the sex of male offspring.) However, the inheritance of a tendency to manic-depressive illness must be more complicated than the inheritance of x-linked color blindness or hemophilia, because while some studies have found no father-son pairs of manic-depressives, others have. Of Ada's two children who survived youth, and her forebears of whom much is recorded, only her grandfather, "Mad Jack" Byron, seems to have displayed the sort of behavior that might link him with this condition, although Byron himself might have had a "cyclothymic" personality, characterized by swings of mood between high optimism and deep disgust.

The alternative to considering Ada's mental and physical ill health as an unlucky chance conjunction of a variety of unconnected ailments is to look for some single condition or syndrome that could unite all, or at least many, of her symptoms under a single rubric. Interestingly enough, there are actually several disorders that include not only all of Ada's reported symptoms (except those related to her fatal cancer) but a good many others as well. Here we will consider two: one of physical, or at least biochemical, origin, the other psychogenic.

The first is one of a family of hereditary (in a few cases acquired) ailments known as "porphyrias"; the name comes from the Greek word for purple and refers to the dark red urine that can suddenly appear and constitutes the most startling, though not necessarily the most characteristic, symptom. All the symptoms of the porphyrias proceed from a metabolic disorder in which excessive amounts of certain chemicals called porphyrins are produced and affect the functioning of both the voluntary and the involuntary nervous systems. [19] By far the most common symptom of the disorder is acute abdominal pain, followed, in descending order of prevalence, by vomiting, muscle pain, muscle weakness, paralysis, delirium or psychotic-like behavior, tachycardia, water retention, breathing difficulties, sensory loss, double vision, and convulsions—all of which Ada suffered from at one time or another. In addition there are such symptoms as constipation, diarrhea, high blood pressure, and the dark urine, which were not reported of her, but which she might very well have had. In short, almost anything and everything, and in almost any combination, so that the absence of certain symptoms, such as the dark red urine, does not necessarily rule out the presence of the disease.

While it seems to be possible to acquire one or another form of porphyria from the damage caused by ingesting certain poisons or even by alcohol abuse, the most studied forms are those due to the inheritance, from either parent's side, of a particular defective but dominant gene. Only about half of those who inherit the defective gene, however, actually develop the disease, and there is a tendency for the condition to show up in two or three successive generations and then go underground for several more. The first attack, which almost never occurs before adolescence, can be triggered off by infections, barbiturates, alcohol, or the increased levels of estrogen that accompanies one phase of the menstrual cycle. Pregnancy exacerbates the symptoms in almost all the women victims of the most common form of hereditary porphyria, known as the acute intermittent variety. Another notable feature of this disease is that although an attack can be very severe and can even cause death, it can also clear up very suddenly—as Ada's sieges of illness often did.

Because of the number and variability of the symptoms, the porphyrias were identified with certainty as a related set of maladies only in the twentieth century, when chemical tests were developed to detect the high levels of excreted porphyrins in the urine and feces of the victims and carriers, whether or not the urine was discolored. In most populations the disorder is very rare, affecting only one person out of about a hundred thousand. This frequency, however, can be very much enhanced among groups of people where cousin marriage is common. Such a group, par excellence, is the English royalty and aristocracy. By tracing the royal family tree and subjecting a number of his living descendants to the appropriate tests, the medical historians Ida Macalpine and Richard Hunter in the 1960s produced convincing evidence that King George III was a porphyria sufferer and that his intermittent "madness" proceeded from this cause.[20] The form of porphyria they found carried by the king's descendants is called "variegate"; it is quite common among the inbred European settlers of South Africa and is very similar to the acute intermittent form, except that, in many of its sufferers, the abnormal biochemical processes also give rise to sensitivity and eruptions of the skin on exposure to the sun.

From reports of the illnesses of his ancestors, descendants, and collaterals, Macalpine and Hunter were also able to identify a number of other possible or probable sufferers among the king's relations. King James I of England, for example, like George III, reportedly had the occasional telltale dark urine, and so did James's cousin, Lady Arabella Stuart. James I's mother, Mary Queen of Scots, also had illnesses that fitted the pattern of clinical signs of the disease, though in her case (and in a number of others) the historians knew of no reports of discolored urine. Both the unfortunate queen of Scots and her descendant Princess Charlotte of Wales, experienced illnesses surrounding childbirth that, it has been suggested, were manifestations of this hereditary disease; the latter, indeed, did not survive the birth of her child. Ada's severe illness after the birth of her second child may have been not cholera but an episode of porphyria.)

Because King James and Lady Arabella were both descended from Margaret Tudor, sister of Henry VIII, Macalpine and Hunter concluded that Queen Margaret was the source of the defective royal gene. A later writer even traced possible clinical signs back to Queen Margaret's Bourbon ancestors in the fourteenth century.[21] To go backward (or forward) over so many generations, however, is to encounter a royal family tree impossibly entangled, and even Macalpine and Hunter neglected to notice that King James and Lady Arabella were not uniquely descended from Margaret Tudor, but also claimed James II of Scotland (who was neither forebear nor descendant of Margaret) as common ancestor.

To attempt to trace the ancestry of a member of the aristocracy is more difficult than in the case of royalty, both because the genealogies are not so well publicized or well authenticated and because one runs into the inevitable problem of by-blows—particularly, and of special interest in cases like the present one, of royal by-blows. The illegitimate offspring of kings, sometimes very numerous—as, for example, those of James V of Scotland—were created or married into the highest nobility to such an extent that many "good" families continued to claim them among their forebears, with varying degrees of tradition and evidence. The reliability of standard works of genealogical reference is no better than their sources, usually the heads of families with reputations to sustain, who in several instances, such as those of Lord Lovelace's father and John Crosse, supplied dubious, incomplete, or simply false information.

Surprisingly, Byron's own descent from a daughter of one of the kings of Scotland, which has been made so much of, is on shaky ground, to say the least. Thus, Ada cannot definitely be associated with the royal porphyriac gene on her father's side. Nor on her mother's side, which seems more likely from the evidence of the numerous reports of Lady Byron's illnesses. In the end, the elusive possibility places Ada's case in exactly the same position as that of her contemporary Charles Darwin, whose mysterious chronic illness has been the subject of a number of fascinating, thorough, and inconclusive historical investigations.[22]

Darwin's symptoms and the course of his illness were in many ways very similar to Ada's, being particularly marked by severe abdominal pain, nausea, and other gastrointestinal disorders, but also including heart palpitations, headaches, "excitement, violent shivering and vomiting attacks," numbness of the extremities, trembling and muscle twitching, eczema attacks that mysteriously energized him, and sudden astonishing recoveries. In short, anything and everything but the characteristic (though not essential) wine-colored urine. Several of his children too displayed symptoms and syndromes compatible with a diagnosis of porphyria.

Ada's family and Darwin's were of a similar social position, and even distantly connected by marriage into the Pole family. Both families included ancestors who had been very close to sovereigns in the sixteenth and seventeenth centuries. His forebears, as traced by his cousin Francis Galton, included members of many of the royal houses of Europe. So, although in neither case can a direct link with an apparently afflicted member of the royal family be demonstrated, in both cases porphyria remains an intriguing hypothesis—one, of course, that neither Ada nor her contemporaries ever entertained.

Another hypothesis concerning the nature and relatedness of her various symptoms, and one that Ada herself adhered to for a time at least, was that they were not physical but psychosomatic in origin: that she was hysterical. Hysteria is an extremely ancient category of disease, the belief in which has passed in and out of fashion a number of times. It has, moreover, attracted the attention of a large number of the most celebrated doctors in history, including Hippocrates—who named it from the Greek word for "womb"—Pinel, Charcot, Janet, and, of course, Freud. Like phrenology, it was originally based on a physical hypothesis that has since been discredited. But, unlike the case of phrenology, and more like that of Pavlovian behaviorism, the falsification of the physiological cause of the phenomena involved has not led to the abandoning of belief in the phenomena.[23]

The ancient Egyptians and Greeks believed that the womb was like an animal with a life of its own. Disorders, or "starvation" proceeding from denying it its childbearing function, caused it to wander about the body, producing a variety of unpleasant symptoms (much as excessive porphyrins wander about the nervous system). Attempts were made to drive it back into position by placing sweet-smelling substances at the opening of the vagina and by breathing or ingesting foul ones at the opposite end of the body. Movement of the womb to the lungs was supposed to cause breathing difficulties; other peregrinations could produce headaches, eye troubles, stomach upsets, palpitations, abnormal perspiration, swellings of the throat and neck, numbness and paralysis of the limbs, fits and convulsions.

The belief in the wandering womb, despite anatomical evidence to the contrary, persisted more or less unchallenged into the seventeenth century, and many remedies and practices originating in this belief persisted into the twentieth century, including the prescription of pungent herbs, such as valerian, as "antihysterics." Even in the seventeenth century, Dr. Sydenham, renowned for his practical, no-nonsense approach, rejected the wandering-womb hypothesis, but he continued to recommend marriage and horseback riding to "hysterical" young women. The latter prescription, originally intended to jolt the wayward uterus back into place, was considered by him as beneficial exercise for toning and improving the blood. The benefits of horseback riding for young women subject to "the vapors" (originally, vapors thought to arise from the disordered womb) appear in the novels of Jane Austen and are echoed in Ada's youthful letter to Mary Somerville recommending it "as a nervous medicine for weak patients."[24]

Sydenham also resorted to such classic techniques for purifying and fortifying the blood as bleeding, purging, and the ingestion of iron filings. Opium too was used for its calming effects, though he cautioned against giving it to those who were already "low spirited" and might be further depressed as a result.

Because of the wandering-womb theory, the issue of whether hysteria could afflict men was a vexed one. Some physicians held that it could not, and tended to designate similar afflictions of the male as "hypochondriasis." Sydenham and his followers, who considered hysteria to be more a reflection of a disordered brain, thought that certain types of men, especially those who were pale and sedentary, might be subject to it too. (Hysteria was one of the mid-twentieth-century hypotheses concerning Darwin's affliction.) There was also controversy over whether hysteria could affect the poor as well as the genteel, and only in the nineteenth century did it become well established as a truly democratic disease. Some modern studies have even found evidence that there is a tendency for hysteria to run in families, much like the manic-depressive disorder or, at one time, tuberculosis. If true, this finding could indicate a physical, possibly genetic factor — or could indicate something about the bias still built into epidemiological medical research.

Whether or not hysteria was womb-connected, that its victims were predominantly female was always a truism. Almost equally unchallenged was the idea that it was in some way connected with unsatisfied sexual longings. Indeed, one reason its discomforts have so often been denied to the poor was the common belief that only the refined could know what sexual frustration was. In the eighteenth century, Pinel, like Sydenham, after restoring the roses to his female patients' cheeks with the equestrian cure, insisted on marriage to forestall a relapse.

Some nineteenth-century physicians, attempting to reconcile the womb theory with the notion of nervous involvement, suggested that local diseases of the uterus could induce nervous susceptibility, and recommended vaginal examination of hysterical patients. Others exclaimed in horror over this course, predicting that a few examinations could enslave and addict the excitable patient to this form of sexual stimulation, which would be followed inevitably by masturbation and eventual prostitution. In the latter half of the century, S. Weir Mitchell, who became famous for his masterful ways with women patients — though he denied any overtly sexual element in his relationships with them — returned to the belief in strengthening the blood. This he' claimed to achieve by means of a rest cure, described in his book with the nourishing title *Fat and Blood*. And while Freud was formulating his sexually-based theories, his contemporary Janet used his own findings to challenge the notion that the hysteric was abnormally preoccupied with sexual thoughts and feelings.

In the twentieth century the term "hysteria" is used to refer to an almost limitless variety of bodily and nervous symptoms that have no detectable bodily origin. It is thus a kind of residual diagnosis, indicating that the physician was unable to find positive evidence of any other disease entity; and some studies do show that a large proportion of those classified as suffering hysterical symptoms are later found to have physical afflictions, sometimes ignored with fatal results. A related criticism is the failure of medical theorists to specify any precise mechanism by which conscious or unconscious thoughts, fears, and longings are converted into physical, neurological, or psychological symptoms as diverse as paralysis, menstrual disorders, pains, palpitations, blindness, convulsions, and memory loss, so that the hypothesis can be confirmed or refuted. It remains somewhat puzzling, too, that if hysterical symptoms are thought to be "conversions" of anxieties, patients should *also* be anxious.

Sometimes, in answer to these accusations, it is claimed that the person likely to display hysterical symptoms may be identified by an associated hysterical personality: seductive, dependent, manipulative, egocentric, self-dramatizing, with emotions both exaggerated and labile. This hysterical personality is said to particularly characterize victims of "Briquet's syndrome," or "St. Louis hysteria," who suffer a multiplicity of somatic symptoms, the list of which certainly covers all of Ada's ailments. Curiously, Briquet's syndrome is found among women only, and it has been pointed out that the hysterical personality is something of a caricature of the feminine stereotype.[25]

From what has been outlined here and from the discussion of mesmerism in chapter 4, it is not surprising that in the nineteenth century, hysteria and "animal magnetism" were thought to be closely connected. In particular, hysterics were thought to be especially susceptible to hypnotic trances and the suggestions of the mesmerizer. And this, of course, was the connection that drew Ada into her belief that she had been the victim of mesmeric experiments and into her interest in the effects of electricity and magnetism on the nervous system.

Oddly enough, the possibility of harm to human beings from oscillating electric, and particularly magnetic, fields has recently been raised again by investigations of reports of increased frequencies of mental and physical illness among people living under or near power lines and among those exposed to microwave irradiation of the American embassy in Moscow. The symptoms reported range from headaches and depression to convulsions, heart disease, cancer, and suicides.

The electromagnetic fields associated with power lines are at far higher voltages and far more rapid alternation than anything achievable by the crude circuitry of the nineteenth century, let alone by the mesmerizer's "animal magnetism." Ironically too, headaches, eyestrain, and even fetal damage have been reported among the women who now work all day at the video display units housing computers whose predecessors Ada abandoned in favor of her interest in electrical experiments.

Along with this renewed interest in the effects of artificially produced electric fields on biological systems, and the complementary attempts now in progress to produce computer microchips from biological materials, it is not surprising to find continued attention to the effects of the natural magnetic field of the earth. It is as if you cannot keep an old idea down. In discussing the building of her Lake Country home in her autobiography, written in 1855, Harriet Martineau commented with satisfaction, "I did not then know the importance of placing beds north and south, in case of illness, when that position may be of last consequence to the patient: but it so happens that all my beds stand or may stand so."[26]

The 1984 meeting of the British Association for the Advancement of Science was served with a very similar idea, but—a sign of current scientific thinking—now furnished with speculations concerning the mechanism of the effect. A reader in zoology at Manchester University claimed to have found that people who sleep with their bodies aligned on a north-south axis have a better sense of direction than those who sleep pointing east and west. He suggested that the effect may be due to deposits of magnetite, a magnetizable oxide of iron of which lodestone is one variety, in the bones between the eyes and the ears.[27] Often it seems no easier to sort fact from hypothesis in the present than in the past, despite the increasingly sophisticated wrappings in which the findings are presented.

Notes

Works frequently cited have been identified by the following abbreviations:

Add. MSS   British Library Additional Manuscripts
LP   Lovelace Papers, Bodleian Library, Oxford University
Memoir   "Sketch of the Analytical Engine Invented by Charles Bab-
           bage, Esq., by L. F. Menabrea, of Turin, Officer of the
           Military Engineers," Article 29, *Taylor's Scientific Memoirs*,
           vol. 3 (1843): 666–731 (translated and with notes by "A.
           A. L." [Augusta Ada Lovelace])
Moore   D. L. Moore, *Ada, Countess of Lovelace: Byron's Legitimate
           Daughter* (London: Murray, 1977)
Passages C. Babbage, *Passages from the Life of a Philosopher* (London:
           Longman, Green, and Co., 1864)
SP   Somerville Papers, Bodleian Library, Oxford University

Preface

1. Moore.

2. *Passages*, 136.

3. J. Rosenberg, *The Computer Prophets* (New York: Macmillan, 1969), 71.

4. C. Evans, *The Making of the Micro* (London: Gollancz, 1981), 10.

5. P. and E. Morrison, eds., *Charles Babbage and His Calculating Engines* New York: Dover, 1961).

6. M. Moseley, *Irascible Genius: A Life of Charles Babbage, Inventor* (London: Hutchinson, 1964).

7. Moore, 163; see A. Hyman, *Charles Babbage, Pioneer of the Computer* (Oxford: Oxford University Press, 1982), 196.

8. Moore, 99.

9. C. Babbage, *Reflections on the Decline of Science in England* (London: B. Fellowes, 1830), xiii.

Chapter 1

1. LP 2, 23 January 1792.

2. LP 29, 2 May 1813.

3. LP 71, 25 November 1806.

4. LP 71, fol. 1.

5. LP 131, fol. 186, "autodescription."

6. Quoted in M. Elwin, *Lord Byron's Wife* (London: John Murray, 1962), 98.

7. Quoted in F. M. L. Thompson, *English Landed Society in the Nineteenth Century* (London: Routledge, 1963), 18.

8. Thompson, *English Landed Society*, 19.

9. Thompson, *English Landed Society*, 100.

10. British Library, Egerton Papers 2611, to John Hanson, 30 January 1809.

11. Q. Bell, *Virginia Woolf* (New York: Harcourt Brace Jovanovich, 1972), vol. 1: 15.

12. J. C. Hobhouse (Lord Broughton), *Contemporary Account of the Separation of Lord and Lady Byron* (London: privately printed, 1870), 23.

13. LP 30, 23 February 1816.

14. LP 30, 14 March 1816.

15. LP 126.

16. LP 118, fol. 54, 14 December 1816.

17. LP 31, 2 August 1816.

18. LP 118, fol. 57, 1818.

19. LP 118, fol. 96, n.d.

20. LP 69, to Emily Fitzhugh, 27 September 1852.

21. *Childe Harold's Pilgrimage*, canto 3, stanza 116.

22. *Don Juan*, canto 1, stanzas 12–30.

23. LP 118, fol. 84, 1832.

24. LP 71, 23 August 1818.

25. LP 71, 20 October 1818.

26. In E. C. Mayne, *The Life and Letters of Anne Isabella, Lady Noel Byron* (London: Constable, 1929), appendix IV, 489.

27. Mayne, *Lady Noel Byron*, appendix IV, 492.

28. LP 118, governess's journal, 4.

29. LP 118, "Ada's" journal.

30. LP 118, governess's journal.

31. LP 78, June 1821.

32. Quoted in M. Elwin, *Lord Byron's Family* (London: John Murray, 1975), 220.

33. SP Dep. b. 206, Greig memoir, n.d.

34. LP 29, to her mother, 30 August 1815.

35. LP 54, fol. 76, n.d.

36. LP 34, 6 October 1819.

37. LP 55, fol. 150, n.d.

38. L. Marchand, ed., *Byron's Letters and Journals*, vol. 11 (London: John Murray, 1981), to Augusta Leigh, 12 October 1823.

39. LP 172, fols. 212–214.

40. LP 71.

41. LP 71.

42. LP 41, fol. 91, n.d.

43. *Memories of Old Friends: The Journal of Caroline Fox* (London: Smith and Elder, 1883), 207.

44. LP 337, May 1838.

45. LP 52, 14 February 1841.

46. LP 45, 17 March 1841.

47. LP 41, 19 May 1833.

48. SP Dep. b. 206, Greig memoir.

49. LP 182, S. De Morgan memoir, 7 April 1875.

Chapter 2

1. *Passages*, 29.

2. J. M. Dubbey, *The Mathematical Work of Charles Babbage* (Cambridge: Cambridge University Press, 1978).

3. S. E. De Morgan, *Memoir of Augustus De Morgan* (London: Longmans, Green, 1882), 89.

4. LP 336, Lady Byron to Dr. King, 21 June 1833.

5. LP 172, 9 March 1834.

6. LP 172, 15 March 1834.

7. LP 172, 24 March 1834.

8. LP 172, 13 April 1834.

9. LP 172, 24 April 1834.

10. LP 172, 1 September 1834.

11. LP 168, 10 November 1834.

12. LP 168, 26 November 1834.

13. SP Dep. c. 355, 1st draft, 23.

14. SP Dep. c. 355, 110.

15. SP Dep. c. 355, 17.

16. SP Dep. c. 355, 1st draft, 17.

17. SP Dep. c. 355, 2nd draft, 51.

18. M. Somerville, *On the Connexion of the Physical Sciences* (London: John Murray, 1834).

19. SP Dep. c. 355, 2nd draft, 51.

20. SP Dep. c. 355, 2nd draft, 136–137.

21. SP Dep. c. 355, 2nd draft, 124.

22. SP Dep. c. 367, Friday, 20 February [1835].

23. SP Dep. c. 367, 4 April [1835].

24. SP Dep. b. 206, Greig memoir.

25. SP Dep. b. 206, Greig memoir.

26. Primary sources consulted on William King, Lord Lovelace, include the Lovelace Papers; the Somerville Papers; the Babbage Papers and the Wentworth Bequest, both in the British Library; and his published articles. The only secondary source is S. Turner, "William, Earl of Lovelace, 1805–1893," *Surrey Archaeological Collections*, vol. 70, 1974.

27. LP 165, 8 June 1835.

28. SP Dep. b. 206, Greig memoir.

29. LP 182, fol. 72, 7 April 1875.

30. LP 176, analysis of the King-Byron settlement.

31. LP 47, 23 July 1835.

32. LP 41, 29 October 1835.

33. SP Dep. c. 367, 1 November 1835.

34. SP Dep. c. 367, 18 November 1835.

35. LP 174, 28 November 1835.

36. SP Dep. c. 367, 25 March 1836.

37. LP 174, fol. 37, n.d.

38. SP Dep. c. 367, 10 April 1836.

39. SP Dep. c. 367, 22 June 1837.

40. SP Dep. c. 367, 20 February 1835.

41. LP 174, fol. 19, n.d.

42. LP 174, fol. 69.

43. LP 165, fol. 64; n.d.

44. LP 41, fol. 99, probably 1838.

45. LP 337, 11 February 1838.

46. H. Martineau, *Autobiography* (London: Smith, Elder, 1877), vol. 1: 11, 22.

47. Martineau, *Autobiography*, vol. 1: 142.

48. Martineau, *Autobiography*, vol. 1: 138.

49. Martineau, *Autobiography*, vol. 3: 469.

50. See Turner, "William, Earl of Lovelace," 106.

51. LP 372, Friday, 9 August [1839 or 1844].

52. LP 165, fol. 104.

53. LP 41, fol. 194, Saturday, 12 December [1840].

54. LP 171, Tuesday, 21 December [1841].

55. LP 166, fol. 137, n.d. (probably 1851 or 1852).

56. LP 338, 22 January 1841.

57. Add. MSS 37191, fol. 87.

58. LP 168.

59. Add. MSS 37191, fol. 331.

60. LP 168, fol. 103, to Mrs. Barwell, 13 December 1840.

61. S. E. De Morgan, *Memoir of Augustus De Morgan* (London: Longmans, Green, 1882), 89.

62. De Morgan, *Memoir*, 81.

63. A. M. W. Stirling, *William De Morgan and His Wife* (New York: Henry Holt, 1922), 32.

64. Senate House, London University, MS 775/328/1.

65. LP 170, fol. 1, n.d.

66. LP 170, 17 August 1840.

67. LP 170, 15 October 1840.

68. LP 170, fol. 91, n.d.

69. LP 170, 4 September 1840.

70. LP 170, Sunday, 13 September [1840].

71. See Dubbey, *Mathematical Work of Charles Babbage*, chap. 5.

72. LP 170, 15 September 1840.

73. LP 170, 10 January 1841.

74. LP 170, 3 February [1841].

75. LP 170, 10 January [1841].

76. LP 170, fol. 34, n.d.

77. LP 170, 10 November, no year.

78. LP 42, fol. 8, Monday, 11 January 1841.

79. LP 170, 15 September 1840.

80. LP 170, 10 January 1841.

81. LP 170, 6 February [1841], 9 September [1841], and 19 September [1841].

82. LP 171, Tuesday, 21 December [1841].

83. LP 57, fol. 3, 17 January [1844].

84. LP 67, fol. 127, 20 January [1844].

85. LP 344, 21 January 1844.

86. See M. I. Kline, *Mathematics: The Loss of Certainty* (Oxford: Oxford University Press, 1980), 47–58.

Chapter 3

1. Add. MSS 37191, fol. 343, 14 March 1840.

2. For the full story of Medora Leigh see Moore. She argues that Medora was not in fact sired by Byron; but if one accepts that an incestuous relationship between him and Augusta existed, the possibility cannot be precluded.

3. LP 42, fol. 12, Saturday, 6 February[1841].

4. LP 42, 27 February [1841], 3 March [1841].

5. Add. MSS 37191, fol. 543, Tuesday, 12 January [1841].

6. Add. MSS 37191, fol. 645, 3 October 1841.

7. *Memories of Old Friends: The Journal of Caroline Fox* (London: Smith, Elder and Co., 1883), 39.

8. LP 175, fol. 8, 25 August 1843. Here Wheatstone says, "I made no stipulation whatever with Mr. Taylor when he sent Menabrea's paper for translation."

9. *Passages*, 136.

10. LP 170, fol. 149.

11. *Memoir*, 687.

12. H. P. Babbage, ed., *Babbage's Calculating Engines* (London: Spon, 1889); P. and E. Morrison, eds., *Charles Babbage and His Calculating Engines* (New York: Dover, 1961). H. P. Babbage for some reason changed the infinity symbol to 1/0, and the Morrisons copied this form. Only B. V. Bowden, in *Faster Than Thought* (London: Pitman, 1953), caught the error and corrected it, along with a number of printer's errors in the Notes. He too, however, failed to note its significance.

13. LP 168.

14. "Notions sur la Machine Analytique de M. Charles Babbage, par Mr. [*sic*] L.-F. Menabrea, capitaine du génie militaire," *Bibliothèque Universelle de Genève*, vol. 41 (1842): 376. The translation here is my own.

15. *Memoir*, 690.

16. For this information I am indebted to Dr. Allan Bromley, Basser Department of Computer Science, University of Sydney, Australia.

17. Add. MSS 37192, fol. 357, n.d.

18. *Memoir*, 698.

19. LP 171, Monday, 1 March 1841.

20. C. Babbage, *On the Economy of Machinery and Manufactures* (London: C. Knight, 1832), 132.

21. Babbage, *Economy*, 157.

22. *Memoir*, 722.

23. Add. MSS 37192, fol. 370.

24. For descriptions of the plans for the Analytical Engine at this early period, see M. V. Wilkes, "Babbage as a Computer Pioneer," *Historia Mathematica* 4 (1977): 415; C. Babbage, "On the Mathematical Powers of the Calculating Engine," from an 1837 manuscript, in B. Randell, ed., *The Origins of Digital Computers* (New York: Springer-Verlag, 1973); A. Bromley, "Charles Babbage's Analytical Engine, 1838," *Annals of the History of Computing* 4 (1982): 196–217.

25. *Memoir*, 692.

26. See E. Koppelman, "The Calculus of Operations and the Rise of Abstract Algebra," *Archive for History of Exact Sciences* 8 (1971): 155–242.

27. See J. M. Dubbey, *The Mathematical Work of Charles Babbage* (Cambridge: Cambridge University Press, 1978).

28. *Memoir*, 694.

29. *Memoir*, 696.

30. SP Dep. c. 367, 22 June 1837.

31. LP 168, 30 June 1843.

32. LP 168, 2 July 1843.

33. SP Dep. c. 369.

34. C. Babbage, "Scribbling Books," Science Museum Library, London, vol. 2, on microfiche, D3, 3/5.

35. Babbage, "Scribbling Books," vol. 3, D4, 3/9, 13 December 1837.

36. *Memoir*, 694.

37. *Memoir*, 713.

38. *Memoir*, 702.

39. Add. MSS 37192, fol. 422, 14 August 1843.

40. Add. MSS 37192, fol. 362.

41. Add. MSS 37192, fol. 349.

42. Add. MSS 37192, fol. 382, Friday, 4 o'clock (21 July in another hand).

43. Add. MSS 37192, fol. 379, Wednesday night (19 July in another hand).

44. Add. MSS 37192, fol. 401, Saturday, 6 o'clock.

45. Add. MSS 37192, fol. 337, Sunday, 6 o'clock.

46. Add. MSS 37192, fol. 360, Saturday, 3 o'clock.

47. Add. MSS 37192, fol. 414, 1 August 1843.

48. Add. MSS 37192, fol. 407, Sunday, 30 July [1843].

49. Add. MSS 37192, fol. 349, Wednesday, 5 July [1843].

50. Add. MSS 37192, fol. 339, Monday afternoon; fol. 393, Thursday, 27 July [1843].

51. Add MSS 37192, fol. 407, Sunday, 30 July [1843].

52. Add. MSS 37192, fol. 342, Tuesday morning.

53. Add. MSS 37192, fol. 348, n.d.

54. C. Babbage, *Reflections on the Decline of Science in England* (London: B. Fellowes, 1830), chap. 5, section 3.

55. *Memoir*, 699.

56. LP 42, fol. 73, 8 August [1843].

57. Add. MSS 37192, fol. 386, Tuesday morning; fol. 364, Tuesday, 5 o'clock.

58. Add. MSS 37192, fol. 412, 1 August 1843.

59. LP 168, fol. 78, Saturday, 5 August [1843].

60. LP 168, fol. 42, Tuesday, 8 August 1843.

61. LP 42, fol. 73, 8 August 1843; fol. 86, 15 August 1843.

62. C. Crosse, *Red Letter Days of My Life* (London: Bentley, 1892), vol. 2: 279ff.

63. Crosse, *Red Letter Days*, 1: 170.

64. LP 54, fol. 123, 24 November [1843].

65. Add. MSS 37192, fol. 422.

Chapter 4

1. Nottingham Public Libraries, M5590, fol. 22, Sunday [10 September 1843].

2. LP 168, 12 September 1843.

3. Add. MSS 37193, fol. 132, Ada to Babbage, October 1844.

4. Add. MSS 37193, fol. 46, 27 March 1844.

5. Add. MSS 37192, fol. 342, Monday afternoon.

6. LP 172, fol. 85, n.d.

7. LP 54, fol. 156, n.d.

8. LP 42, fol. 101, Ada to Lady Byron, 22 August 1843.

9. LP 339, 22 October 1843.

10. LP 174, fol. 95, 5 February 1844.

11. LP 173, fol. 155, to Robert Noel, 9 August 1843.

12. LP 166, fol. 74, 15 September 1843.

13. Add. MSS 37192, fol. 339, Monday afternoon.

14. LP 166, fol. 159, 29 November 1844.

15. LP 175, fol. 199, Tuesday, 5 January 1841.

16. Add. MSS 37192, fol. 335, 2 July 1843.

17. LP 168, fol. 59, 4 December 1848.

18. Add. MSS 37192, fol. 445, 1 September 1843.

19. L. Pearce Williams, *Michael Faraday* (London: Chapman and Hall, 1965), 19.

20. LP 171, 1 March 1841.

21. SP Dep. c. 369, to Agnes Greig, 5 February 1841.

22. LP 42, fol. 28, 3 March 1841.

23. *Memoir*, 722.

24. LP 42, fol. 142, 10 October [1844].

25. H. Bence-Jones, *The Life and Letters of Faraday* (London: Longmans, 1870), vol. 2: 168. The first part of the original letter is among the Lovelace Papers (171, fol. 142, 10 October 1844), but the latter part has vanished.

26. LP 42, fol. 152, 11 November 1844.

27. LP 56, fol. 20, 16 January, no year; LP 55, fol. 172, 21 November 1844.

28. Bence-Jones, *Life and Letters of Faraday*, 2: 168.

29. LP 43, fol. 200, n.d.

30. LP 67, fol. 143, 3 July 1853.

31. SP Dep. c. 367, 15 November 1844.

32. SP Dep. c. 367, Thursday, 30 January 1845.

33. *The Argosy*, London, 1 November 1869, pp. 359–360.

34. LP 166, fol. 126, Saturday [23 November 1844].

35. LP 166, fol. 157, Sunday, [24] November [1844].

36. LP 166, fol. 130, Monday, 25 November [1844].

37. *The Argosy*, 1 November 1869, p. 361.

38. LP 372, 27 November [1844].

39. LP 166, fol. 159, Friday, 29 November [1844].

40. SP Dep. c. 367, 5 December 1844.

41. LP 175, fol. 223, n.d.

42. LP 175, fol. 211, n.d.

43. Add. MSS 37193, fol. 239.

44. LP 41, fol. 138, n.d.

45. Add. MSS 37194, fol. 172, 9 August [1848].

46. Add. MSS 37194, fol. 174.

47. Add. MSS 37194, fol. 176, 27 August [1848].

48. Add. MSS 37194, fol. 181.

49. Add. MSS 37194, fol. 184, Saturday, 30 September [1848].

50. Add. MSS 37194, fol. 187, 9 October 1848.

51. LP 172

52. Add. MSS 37190, fol. 167.

53. Add. MSS 37194, fol. 174, 20 August [1848].

54. Earl of Lovelace, "On Climate in Connection with Husbandry, with reference to a work entitled 'Cours d'Agriculture, par le Comte de Gasparin. . . ,' " *Journal of the Royal Agricultural Society* 9 (December 1848): 311–340.

55. Lovelace, "On Climate," 339.

56. LP 43, fol. 50, n.d.

57. LP 59, fol. 252, n.d.

58. LP 166, fol. 200, Ada to Lord Lovelace, n.d.

Chapter 5

1. SP Dep. c. 367, 22 June 1837.

2. LP 377, to Lady Byron, 22 November 1844.

3. LP 166, fol. 62 [August 1842].

4. LP 377, 19 October [1843].

5. LP 165, fol. 176, n.d.

6. LP 165, fol. 179, n.d.

7. LP 165, fol. 198, n.d.

8. LP 165, fol. 198, n.d.

9. LP 166, fol. 1.

10. LP 166, fol. 194, n.d.

11. LP 166, fol. 1, n.d.

12. LP 165, fol. 209, n.d.

13. LP 165, fol. 202, n.d.

14. SP Dep. b. 206.

15. SP Dep. b. 206.

16. LP 171, fol. 153, 15 January 1841.

17. LP 171, fol. 157, 21 January 1841.

18. SP Dep. c. 367, Friday, 16 December [1842].

19. SP Dep. c. 367, Friday, 31 December (Greig's date) [1842].

20. SP Dep. c. 367, 10 February 1843.

21. LP 172, Thursday, 14 October 1841.

22. LP 175, fol. 3, 2 April 1850.

23. Add. MSS 43748, 3 June 1846.

24. Add. MSS 43748, 10 May 1849.

25. LP 172, n.d.

26. LP 175, fol. 205.

27. LP 175, fol. 190.

28. SP Dep. c. 367, 5 December 1844.

29. LP 54, fol. 164, n.d. [1844?].

30. Add. MSS 43748, 15 May 1847.

31. SP Dep. c. 367, 4 February 1845.

32. SP Dep. c. 367, Friday, 7 [February 1845].

33. LP 42, fol. 34 [14 November 1841].

34. LP 42, fol. 47; fol. 142, Monday, 18 July [1842].

35. LP 41, fol. 183, Tuesday, 20 October [1840].

36. LP 42, fol. 138, n.d. [1844?].

37. LP 42, fol. 195, n.d.

38. LP 42, fol. 203, Wednesday, 7 October [1846].

39. LP 372, 10 August 1845.

40. LP 43, fol. 56, to Lady Byron, 1 October [1848].

41. LP 42, fol. 119, to Lady Byron, n.d.

42. LP 372, to Lady Byron, 25 August 1848.

43. LP 59, fol. 212, n.d.

44. LP 54, fol. 96, n.d.

45. SP Dep. c. 368, to Agnes Greig, 19 June 1867.

46. LP 177, fol. 62, to Annabella, October 1868.

47. LP 173, fol. 126, to Charles Noel, n.d.

48. LP 372, to Lady Byron, 7 June 1850.

49. LP 372, August 1845.

50. LP 42, fol. 124.

51. LP 42, fol. 214, to Lady Byron, Thursday, 19 November [1846].

52. LP 57, fol. 49, 22 June [1850].

53. LP 372, 25 June 1850.

54. LP 43, fol. 107, Friday, 21 June [1850].

55. LP 57, fol. 29, to Ada, n.d.

56. LP 372, 22 February 1850.

57. LP 43, fol. 180, to Lady Byron, n.d. [1852].

58. *Westminster Review* 47 (April 1847): 110.

59. LP 371, 7 January 1838.

60. LP 372, to Lady Byron, 10 October 1844.

61. LP 372, 26 April 1848.

62. Add. MSS 37193, fol. 551, 31 May 1847.

63. Add. MSS 37194, fol. 1, 15 June 1847.

64. LP 372, August 1847.

65. LP 372, 1 November 1847.

66. Add. MSS 37194, fol. 30, 22 September [1847].

67. *Westminster Review* 48 (January 1848): 415.

68. LP 372, February 1849.

69. LP 372, 3 September 1847.

70. Add. MSS 37194, fol. 133, 28 February [1848].

71. Add. MSS 37194, fol. 172, 9 August 1848.

72. L. A. J. Quêtelet, *Du Système Social et des lois qui le régissent*, 1848.

73. *Prospective Review* 5 (February 1849): 16–35.

74. "Settlement and Poor Removal," *North British Review* 8 (February 1848): 336–377.

75. "On the Subdivision of Real Property . . . ," *Quarterly Journal of the Statistical Society of London* 11 (November 1848): 305–322.

76. LP 372, to Lady Byron, 17 November 1844.

77. LP 372, to Lady Byron, October 1849.

78. Add. MSS 37194, fol. 72, n.d.

79. Add. MSS 37194, fol. 189, 18 October [1848].

80. LP 170, fol. 157, n.d. [1848].

81. See S. Turner, "William, Earl of Lovelace," *Surrey Archaeological Collections,* vol. 70 (1974): 101–129.

82. SP Dep. c. 367, 1 May 1848.

83. SP Dep. c. 367, 5 January 1849.

84. SP Dep. c. 367, Tuesday, 13 March [1849].

Chapter 6

1. Add. MSS 37192, fol. 232.

2. LP 372, 17 November 1844.

3. Add. MSS 43747, fol. 117, 23 May 1845.

4. Add. MSS 37192, fol. 234, n.d.

5. *Westminster Review* 44 (1845), no. 5: 152.

6. *Westminster Review* 44, no. 5: 202.

7. *Transactions of the Royal Society of Edinburgh* 9 (1820): 153–177.

8. LP 43, fol. 93, 17 May 1850.

9. LP 43, fol. 96, 30 May [1850].

10. LP 43, fol. 111, Wednesday, 28 August [1850].

11. LP 372, 15 September 1850.

12. LP 43, fol. 118, 8 September [1850].

13. LP 43, fol. 122, Sunday, 15 September [1850].

14. LP 58, fol. 23, 17 September 1850.

15. LP 372, 24 August [1844].

16. LP 43, fol. 126, Monday, 23 September [1850].

17. LP 372, 22 September 1850.

18. LP 171, fol. 99.

19. LP 171, fol. 101.

20. LP 171, fol. 111, 25 January 1851.

21. LP 171, fol. 109, 27 January 1851.

22. LP 43, fol. 184, n.d.

23. LP 43, fol. 192, 17 May 1851.

24. LP 43, fol. 180, n.d.; LP 43, fol. 192, 17 May 1851.

25. LP 90, fol. 41, 14 April 1853.

26. LP 58, fol. 169, 19 June 1851.

27. LP 43, fol. 195, to Lady Byron, 19 June 1851.

28. LP 166, fol. 169, to Lord Lovelace, Thursday, 24 July [1851].

29. Murray Collection, 50 Albemarle Street, London.

30. Add. MSS 31037, fol. 140, to Augusta Leigh, 22 February 1830.

31. LP 89, fol. 161, 1 July 1851, at Ockham.

32. LP 58, fol. 181, 3 July 1851.

33. LP 43, fol. 206, Friday, 4 July [1851].

34. LP 43, fol. 216, to Lady Byron, 16 August [1851].

35. LP 173, fol. 9, 25 March 1852.

36. LP 43, fol. 219 [16 August 1851].

37. LP 43, fol. 230, to Lady Byron, Monday, 1 September [1851].

38. LP 372, 10 August 1851.

39. LP 372, November 1851.

40. LP 44, fol. 1, to Lady Byron, Tuesday, 6 January [1852].

41. LP 44, fol. 25, to Lady Byron, 29 February 1852.

42. LP 44, fol. 50, to Lady Byron, 11 March [1852].

43. LP 44, fol. 60, to Lady Byron, n.d.

44. LP 44, fol. 69, to Lady Byron, Saturday, 3 April [1852].

45. LP 44, fol. 90, to Lady Byron, n.d.

46. LP 44, fol. 80, Thursday, 29 April [1852].

47. LP 44, fol. 37, n.d.

48. LP 44, fol. 95, n.d.

49. LP 58, fol. 208, n.d.

50. LP 44, fol. 87, Wednesday, 26 May [1852].

51. LP 44, fol. 99, Tuesday, 1 June [1852].

52. LP 69, fol. 132, Wednesday, 2 June [1852].

53. LP 69, fol. 134, 3 June [1852].

54. LP 69, 9 June [1852].

55. LP 69, 9 June [1852].

56. LP 69, 11 June [1852].

57. LP 69, 17 June [1852].

58. LP 69, fol. 152, 17 July [1852].

59. LP 69, 14 August [1852].

60. LP 44, fol. 131, to Lady Byron, n.d.

61. SP Dep. c. 367, 30 July [1852].

62. LP 372, 23 August 1847.

63. LP 372, Greig to Lord Lovelace, 6 August 1852.
64. LP 372, Greig to Lord Lovelace, 6 August 1852.
65. LP 372, Lord Lovelace to Ada, n.d.
66. LP 138, fol. 3, 12 August 1852.
67. SP Dep. c. 367.
68. LP 175, fol. 161, n.d.
69. LP 174, fol. 133, 27 September [1851].
70. LP 174, fol. 133, 27 September [1851].
71. LP 175, fol. 230.
72. Add. MSS 54089, n.d.
73. LP 174, fol. 135, 14 November [1851].
74. LP 174, fol. 137, 25 November [1851].
75. LP 44, fol. 122, n.d.
76. LP 44, fol. 146, 21 August [1852].
77. Add. MSS 54089, Sunday, 15 August 1852.
78. LP 69, fol. 171, 18 August [1852].
79. Add. MSS 54089, Friday, 20 August [1852].
80. Add. MSS 54089, Saturday, 21 August [1852].
81. LP 44, fol. 144.
82. LP 69, fol. 168, n.d.
83. Add. MSS 54089, Sunday, 22 August [1852].
84. Add. MSS 54093, 3 August [1852].
85. Add. MSS 54089, Saturday, 21 August [1852].
86. Add. MSS 54089, Tuesday, 24 August [1852].
87. Add. MSS 54089, Tuesday, 31 August [1852].
88. LP 69, fol. 188, n.d.
89. LP 69, fol. 148, 14 July [1852].
90. LP 69, fol. 148, 14 July [1852].
91. LP 89, fol. 168, 21 September 1852.
92. LP 372, Lord Lovelace to Lady Byron, 6 January 1853.
93. LP 69, fol. 235, 18 October [1852]; LP 69, fol. 156, 24 July [1852].
94. LP 372, 20 January 1852.
95. LP 43, fol. 72, to Lady Byron, n.d.
96. LP 69, fol. 168, 17 August [1852].
97. LP 175, fol. 137.
98. LP 69, fol. 204, n.d.
99. LP 69, fol. 207, n.d.
100. LP 69, fol. 185, n.d.
101. LP 175, fol. 127, n.d.
102. LP 69, fol. 237, to Miss Fitzhugh, 20 October 1852.
103. LP 69, fol. 254, 27 November 1852.
104. LP 69, fol. 263, n.d.
105. LP 175, fol. 152.

Chapter 7

1. LP 372, 9 February 1852.
2. SP Dep. c. 367, 2 December 1852.
3. LP 46, fol. 243, 17 December 1852.
4. LP 372, 14 December 1852.
5. LP 46, fol. 238, 16 December 1852.
6. LP 46, fol. 243, 17 December 1852.
7. LP 372, 11 January 1853.
8. LP 46, fol. 249, 1 January 1853.
9. LP 46, fol. 249, 1 January 1853.
10. LP 171, fol. 189, 5 January 1853.
11. LP 46, fol. 253, 6 January 1853.
12. Fitzwilliam Museum, Cambridge, MS 317, manuscript journal of Wilfrid Scawen Blunt, "Alms to Oblivion," chap. 7: 30. The period when Lovelace actually learned of Ada's other "grave crimes" is incorrectly given.
13. LP 46, fol. 257, 9 January [1853].
14. LP 46, fol. 257, 9 January [1853].
15. LP 372, to Eliza Follen, 5 February 1853.
16. LP 46, fol. 264, 16 January [1853].
17. LP 59, fol. 173, n.d.
18. LP 372, 17 April 1853.
19. LP 59, fol. 37, n.d.
20. LP 90, fol. 48.
21. LP 138, fol. 79, 30 December 1852.
22. LP 118, fol. 110; SP Dep. c. 368; copy dated March 1853.
23. Add. MSS 37194, fol. 415, Tuesday, 23 July [1850]. Several other letters from Ada to Babbage have been kindly brought to my attention in various connections. They are like those in the British Library.
24. Add. MSS 37194, fol. 430.
25. Moore, 278.
26. LP 168, fol. 72.
27. LP 168, fol. 73.
28. Add. MSS 37193, fol. 232, Ada to Babbage.
29. Add. MSS 37194, fol. 14, Saturday, 19 June [1847].
30. Add. MSS 37193, fol. 228, Friday, 22 August [1851 or 1845].
31. Add. MSS 37199, fol. 610, Greek letter, 7 December 1831; Ada's comment, n.d. This document is accompanied by a description written by Babbage's son and dated 20 April 1899.
32. SP Dep. c. 367, Lady Byron to Mrs. Somerville, 9 March 1853.
33. LP 138, fol. 111, 12 January 1853.
34. LP 372, 3 April [1848?].
35. LP 372, 24 August [1844?].

36. See T. G. Ehrsam, *Major Byron: The Incredible Career of a Literary Forger* (London: Murray, 1951).

37. LP 171, fol. 179.

38. LP 90, fol. 38, 2 April 1853.

39. LP 171, fol. 195, 8 January 1853.

40. LP 171, fol. 203, 11 January 1853.

41. LP 171, fol. 203, 11 January 1853.

42. LP 171, fol. 204, to Lord Lovelace, 22 January 1853.

43. LP 90, fol. 38, 2 April 1853.

44. LP 90, fol. 41, 14 April 1853.

45. LP 90, fol. 37, Greig to Lushington, 9 February 1853.

46. LP 89, fol. 208, Lady Byron to Lushington, 19 December [1852].

47. LP 90, fol. 45, Greig to Lushington, 30 April 1853.

48. LP 138, fol. 148, 8 April 1853.

49. Add. MSS 54092, 25 June 1853.

50. M. Lovelace, *Ralph, Earl of Lovelace: A Memoir* (London: Christophers, 1920), 7.

51. Add. MSS 54092 [1854].

52. Add. MSS 37197, fol. 96, 17 September 1856.

53. Add. MSS 54093, 8 December 1852.

54. Add. MSS 54093, to Lady Byron, 4 December 1853.

55. Add. MSS 54093, to Lady Byron, 11 April 1854.

56. SP Dep. c. 368, 19 June 1867.

57. Fitzwilliam Museum, Cambridge, MS 317, manuscript journal of Wilfrid Scawen Blunt, "Alms to Oblivion," chap. 7: 30.

Appendix

1. J. Stevenson, *Popular Disturbances in England, 1700–1870* (London: Longman, 1979), chap. 12.

2. A. M. W. Stirling, *William De Morgan and His Wife* (New York: Henry Holt, 1922), 33.

3. SP Dep. c. 367, 4 March, no year.

4. LP 41, fol. 150, n.d.

5. LP 166, fol. 218, to Lord Lovelace, n.d.

6. LP 41, fol. 187, to Lady Byron, 26 November 1840.

7. LP 42, fol. 8, to Lady Byron, Monday, 11 January 1841.

8. SP Dep. c. 367, 10 February 1844.

9. LP 42, fol. 160, to Lady Byron, n.d.

10. LP 42, fol. 160, to Lady Byron, n.d.

11. LP 69, Lady Byron to Emily Fitzhugh, 11 June 1852.

12. See V. Berridge and G. Edwards, *Opium and the People* (London: Allen Lane, 1981).

13. A. R. Lindesmith, *Addiction and Opiates* (Chicago: Aldine, 1968), 26.

14. LP 42, fol. 136, Ada to Lady Byron, n.d.

15. LP 42, fol. 160, to Lady Byron, n.d.

16. A. Scull, ed., *Madhouses, Mad Doctors and Madmen* (Philadelphia: University of Pennsylvania Press, 1981); V. Skeltans, *Madness and Morals* (London: Routledge, 1975).

17. LP 164, fol. 207, Ada to Lord Lovelace, n.d.

18. See G. Winakur, P. J. Clayton, and T. Reich, *Manic Depressive Illness* (St. Louis: Mosby, 1969).

19. J. Lyndal York, *The Porphyrias* (Springfield, Illinois: Charles Thomas, 1972).

20. I. Macalpine and R. Hunter, *George III and the Mad-Business* (London: Allen Lane, 1969).

21. L. C. Hurst, "Porphyria Revisited," *Medical History* 26 (1982): 1983–1990.

22. See R. Colp, *To Be an Invalid* (Chicago: University of Chicago Press, 1977).

23. See I. Veith, *Hysteria* (Chicago: University of Chicago Press, 1965); A. Roy, ed., *Hysteria* (Chichester: John Wiley, 1982).

24. SP Dep. c. 367, 4 April 1835.

25. E. Slater, "What Is Hysteria?" in A. Roy, *Hysteria.*

26. H. Martineau, *Autobiography* (London: Smith, Elder, 1877), vol. 2: 228.

27. "Homing In on Power of Sleep," *The Guardian* (London), 12 September 1984.

Index